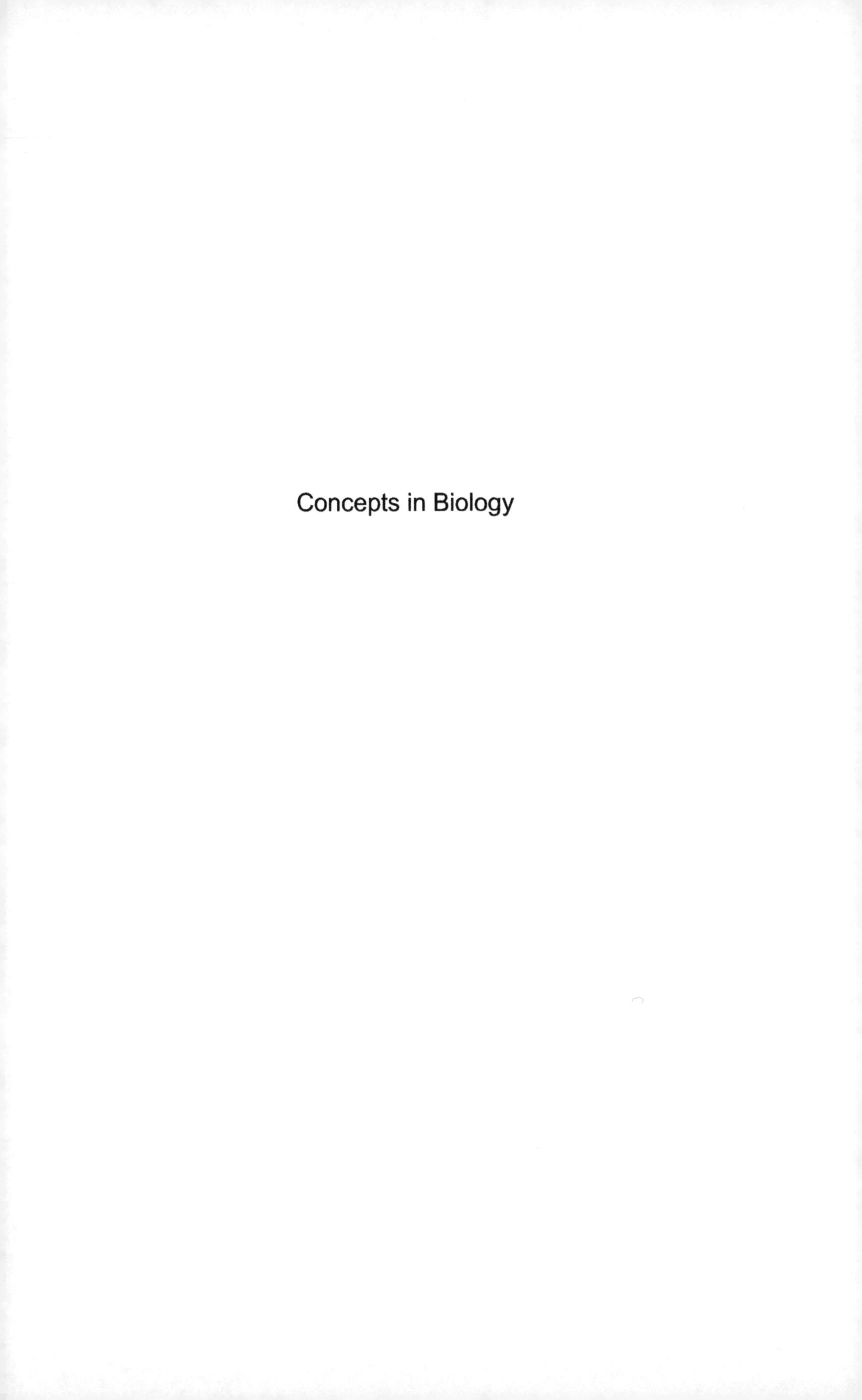

Concepts in Biology

Concepts in Biology

A Historical Perspective

Marc Gilbert
Sergej Pirkmajer

WILEY

First published 2023 in Great Britain and the United States by ISTE Ltd and John Wiley & Sons, Inc.

ISTE Ltd
27-37 St George's Road
London SW19 4EU
UK

www.iste.co.uk

John Wiley & Sons, Inc.
111 River Street
Hoboken, NJ 07030
USA

www.wiley.com

Library of Congress Control Number: 2023942286

British Library Cataloguing-in-Publication Data
A CIP record for this book is available from the British Library
ISBN 978-1-78630-940-2

Contents

Preface

This book has the ambitious aim to provide a comprehensive view of the history of concepts in biological sciences that organized our scientific knowledge and shaped our understanding of the endocrine, nervous and immune systems throughout the last century. Biological concepts take a central place in science and their meanings should be clearly defined because they influence the direction of basic research. They are vectors of scientific knowledge communication among non-experts, and consequently enable them to understand the findings of scientific research. It should be kept in mind that some concepts have undergone continuous changes. One of the best examples is the gene concept, which was initially defined in Mendelian genetics as a factor that modifies the development of a trait (classical gene concept). Thereafter, in the 1950s, a new concept emerged that claimed the gene was made of DNA, which contains information about a specific molecular product (molecular gene concept) (Kampourakis and Stern 2018).

It is noteworthy that the three fields, mentioned above, contributed to the emergence of major concepts in the first decade of the 20th century. Leading biologists considered that there were no links between these disciplines. Consequently, they functioned independently, which therefore erected disciplinary barriers, for practical formulations. Over the last 60–70 years, advances in genetics and cellular and molecular biology techniques have led to major discoveries that have highlighted how these three systems shared several molecular entities such as cytokines, hormones and neurotransmitters with their cognate receptors, broadening the communication between the three systems. This network of signals coordinates mechanisms that aim to regulate diverse biological responses including cell growth, differentiation, inflammation, metabolism, etc. It gave rise to a multiplicity of concepts so that biologists realized that these systems are indeed related and should come together in a common discipline. It became evident that these networks add another level of complexity, which is amplified at the cellular level and often reveals the universality or singularity of most cellular responses from a molecular perspective.

In the first part of this book, we present a chronological panorama of the history of these three disciplines: endocrinology, neurology and immunology. It should be emphasized that it is quite common in the history of science that an important question arises long before the development of experimental strategies or techniques needed to resolve the issue. Thus, an important but often overlooked approach to discovery is the revisiting of old questions with new techniques. This principle is often illustrated in these pages. We will emphasize how new tools, techniques and instrumentation associated with conceptual changes have helped to elucidate the complexity of the different fields. All new discoveries have consolidated our understanding in the homeostatic control mechanisms that should now be viewed as a complex interacting system. It is noteworthy that most fundamental discoveries have been initiated by researchers over many years, building intuition, clear reasoning, observation, tenacity, coherence and creativity coupled with technical excellence. To keep this spirit of creativity alive, most eminent researchers have also been the scientific progenitors of numerous careers in biology, and one day these exceptional curiosity-driven traits may be recognized by the scientific community.

Writing this book has been an immense learning experience, namely on how the emergence of new concepts has led to major breakthroughs in physiology and cell biology. Emphasis is often put on technology, which is an important tool and can be a limiting factor or a driver in biological discoveries. This book will serve as a general reference for those interested in embracing the field, as well as for experts. Readers will likely be convinced that it is difficult to predict the future trends in biology, but it is obvious that unsolved problems can potentially generate new concepts. K. Popper, a philosopher of science, stated in the mid-1930s that knowledge should never be static and suggested that "humans had no ability to accept a theory as true for all time" (Popper 2014). One of the best examples is the discovery of iPS cells, which was a paradigm shift. Old ideas may also re-emerge in a new context of research. This is exemplified by the theory of evolution that continues to evolve. In 1802, J.B. Lamarck proposed the concept that the environment can directly alter phenotype in a heritable manner. His simplistic idea that acquired characteristics are transmitted from the parents to offspring was wrong, and Darwin's theory advanced. However, we now believe that epigenetics also impacts phenotypic variation and therefore facilitates natural selection. Both theories can therefore be reconciled, as seen in a chapter of this book.

In every domain, theories and proposed mechanisms to support them continue to shift over time. A. Lwoff, winner of the Nobel Prize in Physiology or Medicine in 1965, liked to remind us that science is a "permanent revolution" and that changing ideas are a sign of good scientific health (Lwoff 1962).

June 2023

1

Historical Overview of Endocrinology, Neurology and Immunology

1.1. The history of endocrinology

The study of endocrine function was born in the second half of the 19th century and the first decades of the 20th century. However, it should be emphasized that the humoral concept derived from the speculation of pre-Socratic philosophers. Later, the concept of internal secretions was put forward in the anatomical studies of the Renaissance, when organs without ducts were described. After that, the discovery of circulation by W. Harvey (1578–1657) fueled the development of the concept of internal secretions in 1775 by T. de Bordeu (1722–1776), physician of Louis XV: "Each organ in the body released emanations which were useful to the whole body" (Eknoyan 2004). In 1855, C. Bernard (1813–1878) used the term "internal secretion" when he determined that the liver secretes glucose into the hepatic veins and bile into the intestinal tract. Even though glucose is certainly not a hormone, his concept of internal secretion was the first step in defining the endocrine system. In 1902, W. Bayliss (1860–1924) and E. Starling (1866–1927) discovered secretin (Bayliss and Starling 1902) and introduced the concept of blood-borne chemical messages. In 1905, the word "hormone" (from the Greek *hormao*, meaning "I activate") was coined by Starling. He defined it as "any substance normally produced in the cells of some part of the body and carried by the blood stream to distant parts, which it affects for the good of the organism as a whole". It was originally suggested by the linguist Mr Hardy of Cambridge, UK, to distinguish glandular extracts from other internal secretions. The term "endocrinology" was proposed shortly thereafter by N. Pende (1880–1970) in 1909. According to H. Dale (1875–1968), clinical endocrinology achieved respectability only with the discovery of insulin (Banting et al. 1923). By 1922, the discipline was in the forefront of biomedical science, since it was possible – by using extracts of endocrine glands – to treat three human endocrine disorders successfully: hypothyroidism, diabetes insipidus and diabetes mellitus (Dale 1935). The importance of the anterior pituitary gland as "conductor of the endocrine orchestra" was not understood until the early 1930s, when P.E. Smith published his parapharyngeal approach for removing the gland ("hypophysectomy") (Smith 1932). By the 1950s, the major hormones had been identified, and the history of their discovery started with the surgical ablation of endocrine glands (testes, ovaries, adrenal glands, thyroid, etc.), followed by studies of the physiological changes in the subject. Then, transplantation of the gland or gland extracts were given in replacement therapy to see whether they reversed the effect of the gland removal. Such an experiment was first carried out by A. Berthold (1803–1861) in 1849, which subsequently enabled the characterization of the pathophysiology of most hormone deficiencies. However, a straightforward interpretation of the outcomes was often misleading because it was originally thought that endocrine glands or cells released only a single hormone, but it is now known that endocrine glands or even cells may produce two or more hormones. For instance, ablation of the adrenal gland removes the corticosteroids (i.e. glucocorticoids and mineralocorticoids), as well as the catecholamines (primarily adrenaline (epinephrine), since noradrenaline is also secreted from the sympathetic nerve endings), and the outcome is a combination of deficiency of both hormone types. Although there was a growing number of newly identified hormones, investigators could not measure their blood levels. The lack of quantitative data describing hormonal changes led to the development of the radioimmunoassay technique in the laboratory of Yalow and Berson (1959), and in 1960 it enabled an accurate insulin measurement. Once this had been achieved, it was possible to prove that insulin action may be deficient despite hyperinsulinemia, meaning that diabetes is not always caused just by insufficient secretion of insulin. This technique was also used to measure numerous hormones or other substances in body fluids and to diagnose hormone-related diseases (Lepage and Albert 2006). This period was followed by the molecular biology revolution, and endocrinology became one of the most dynamic disciplines, which was mainly attributed to the applications of advances in other fields. Fundamental discoveries have been made over the last decades and shed light on the molecular processes that mediate the conversion of the hormonal signal from outside the cell to a functional change in the cell. It started in the 1960s when E.W. Sutherland discovered that epinephrine induces the formation of cAMP in liver cells and that the nucleotide converts the inactive glycogen phosphorylase to the active enzyme, which leads to the formation of glucose. It raised the question of how the hormone regulates the conversion of ATP to cAMP. He proposed a model in which the enzyme that catalyzes this breakdown was located in the plasma membrane and composed of a regulatory subunit (R) and a catalytic unit (C). This enzyme was called adenyl cyclase, and he hypothesized an interaction of the hormone with R, which in turn may influence C. It led to the description of a "possible model of adenyl cyclase as related to adrenergic receptors" (Robison et al. 1967). Later, Sutherland suggested that the effects of many other hormones could also be explained through their binding to plasma membrane receptors, causing the formation of cAMP, which then stimulates or inhibits different metabolic processes. This hypothesis was initially met with strong resistance by scientists because it seemed impossible that a single agent could lead to the diverse biological effects in response to different hormones. Subsequently, the concept was shown to be correct and Sutherland was awarded the 1971 Nobel Prize in Physiology or Medicine. The focus in the field shifted from the hormones themselves to the receptors, second messengers and signaling molecules acting within complex networks. It broadened our understanding of endocrine physiology, but it also challenged many long-held beliefs in endocrinology, for example, that endocrine cells released only a single hormone. Likewise, there is evidence that hormones can bind to different receptors and receptors can bind to a wide number of different hormones. If we consider the way that most hormones were discovered, it was tedious and time-consuming work. Thus, the discovery of the TSH-releasing factor (TRF) required 80,000 sheep hypothalamic fragments (Guillemin et al. 1965). Thanks to the reverse pharmacology strategy, this process has been markedly improved, thus leading to the discovery of many endogenous compounds, including hormones. It revealed the chemical diversity of cell-to-cell signaling molecules, and over the last few decades endocrinology has undergone major paradigm shifts with the discovery of endocrine functions by organs not hitherto considered as endocrine organs, such as adipose tissue, liver, muscles, lungs and heart. It also highlighted that these organs are important regulators of biological responses through organ "crosstalk". These discoveries increase the complexity of the endocrine system, and recently, new methods and techniques of investigation have demonstrated that the endocrinology world is expanding as a multifaceted discipline. Thus, using a comprehensive bioinformatics approach, the concept of "molecular mimicry", coined by R.T. Damian (1964) and which gained ground in type 1 diabetes (T1D) research (Atkinson 1997), was reinvestigated. Recently, it was discovered that the mechanism of molecular mimicry applies to viruses, as they create factors that mimic host features that may give rise to hormone ligands (Huang et al. 2019). These ligands do not have sequence similarity with the host proteins, but instead have structural or functional similarities. Thus, the viral insulin/IGF1-like peptides have been identified, which can bind the human IGF1 receptors and stimulate classic post-receptor signaling pathways. This system of viral hormones can be viewed as a paradigm shift for host–virus interactions. These observations also give new insights into the evolution of peptide hormones. However, today, there are only a few studies investigating this concept. Similarly, it took a long time to recognize the hypothalamic–pituitary regulation of peripheral endocrine organ function.

1.2. The history of neurology

The term "brain" was initially used by the ancient Egyptians in about 1700 BCE (Breasted 1905). The distinction between the "cerebrum" (*enkephalon* in Greek) and "cerebellum" (*parenkephalis* in Greek) was first made by Aristotle around 300 BCE. Next, R. Descartes proposed his view of the brain anatomy in "Traité de l'homme" (1648), which had a major influence on the conception of human physiology in the last half of the 17th century. The internal surface of the brain is its most important part, and is riddled with pores, which are simply the gaps between fine nervous threads that form a kind of mesh or network. Descartes also separated the notion of mind, which holds abstract thoughts, from that of the physical body.

The terms "neuron", "synapse" and "neurotransmitter" all have an interesting history (López-Muñoz and Alamo 2009). The introduction of the "anatomo-clinical method" in the first half of the 19th century resulted in the extensive development of the histological disciplines, which led to the formulation of the neuron theory in the 1830s. A group of four researchers contributed to this theory: J.E. Purkinje (1787–1869), F.G.J. Henle (1809–1885), M.J. Schleiden (1804–1881) and A.V. Kölliker (1817–1905), and published their findings in 1852. It was also stated that cells "should be conceived as the essential formal units of the body".

The concept of the synapse emerged in the late 19th century, and the assimilation of the neurotransmission phenomenon was difficult. In the 1870s, F. Kühne mentioned that nerve endings terminate in the formation of the muscular membrane (subsequently named the motor end-plate/the neuromuscular junction). R. y Cajal (1852–1934), in his anatomical drawings (1889), anticipated the concept of the synapse, but it was C.S. Sherrington (1858–1952) who provided a functional explanation of the structural postulates of Cajal (1894). He combined the anatomical and physiological concepts into a single unit, which was then called a "synapsis" (from the Greek, "junction, connection"). Likewise, he provided the first data on the existence of excitatory and inhibitory synapses. This duality of stimulating or inhibiting inputs regulating a cellular response is a major concept in biology.

The discovery of neurotransmitters was a great scientific advancement in the 20th century. In 1877, E.H. du Bois-Reymond (1818–1896) suggested that nerves could stimulate muscles by means of chemical substances, whereas the prevailing theory was that of electrical transmission. It took another three decades before the chemical hypothesis of neurotransmission was consolidated and firmly confirmed. In 1904, synaptic transmission was postulated by T.R. Elliot (1905) and J.N. Langley (1905), and their experiments on sympathetic stimulation strongly suggested that the action potential could cross the synapse through chemical substances, and they were named chemical mediators. However, the concept of chemical neurotransmission, as well as the existence of neurotransmitters, was only confirmed decades later. In

1905, Langley suggested the existence of the receptor. He called it the "receptive substance of target cells" (Langley 1905, 1906).

Throughout the following decades, the theory of neurotransmission was consolidated between 1921 and 1926 when O. Loewi carried out experiments on the vagal stimulation of the isolated heart of amphibians (Loewi 1921; Loewi and Navratil 1926). He postulated that the vagal nerve releases a substance ("Vagusstoff"), and in 1926 he successfully identified it as acetylcholine (Loewi and Navratil 1926). Then, in 1936, H. Dale confirmed the neurotransmitter role of acetylcholine in the peripheral nervous system of mammals (Dale et al. 1936). However, the idea of a direct communication via electrical synapses, especially in the central nervous system, was persistent, and it took a further decade for chemical transmission in the central nervous system to be accepted. It is marvelously recounted in E.S. Valenstein's book "The War of the Soups and the Sparks" (Valenstein 2007).

In the mid-1950s, the electron microscopy technique allowed us to describe the individuality of the neurons and the discontinuity at the synaptic level. Another technical advancement in the early 1960s was the fluorescence histochemical method, which enabled the visualization of catecholamine neurons and their pathways in the brain. Using this technology, A. Carlsson provided evidence for the existence of dopamine as a neurotransmitter in the central nervous system (Carlsson et al. 1958), which was an essential step towards rational pharmacotherapy of disorders, such as Parkinson's disease and schizophrenia

The history of interaction between drugs and their targets started with C. Bernard and O. Schmiedeberg (1838–1921), who put forward the idea of a specific receptor structure on which drugs would act. As mentioned above, in 1905, Langley suggested that curare and nicotine produced their effects through the stimulation of a "receptive substance" (Langley 1906). A year later he proposed the concept of the transmitter receptor (neuroreceptor), but it was not well accepted by the scientific community. The idea was developed further by P. Ehrlich, who postulated that microorganisms and cancer cells possess surface chemoreceptors for dyes, which led to the concept of chemotherapy of infectious diseases and cancer. Between 1970 and 1974, S. Langer and K. Starke suggested the presence of α-adrenergic autoreceptors at synaptic terminals of the sympathetic system, which would act as a regulator mechanism of neurotransmitter release (Langer and Lehmann 1988). This autoreceptor concept was then extended to other transmitter systems. Since the postulate of the "receptive substance" theory, proposed by Langley in the early 1900s, diverse disciplines such as electrophysiology, pharmacology and biochemistry joined together with a common goal of successfully identifying the first neurotransmitter receptor, the nicotinic acetylcholine receptor (Changeux

2020). This receptor has become "the founding father of the pentameric ligand – gated ion channel superfamily", and the sequence of 20 amino acids comprising the N-terminal of the α-subunit of the *Torpedo marmorata* was obtained by J.P. Changeux's group (Changeux 2012).

Over the last half of the 20th century, advances in the discipline of molecular biology have allowed us to identify, sequence and clone numerous types and subtypes of receptors. Likewise, it became possible to identify the chain of molecular events that is initiated by receptor activation, and which triggers the biological response. These sequential events, which are tightly regulated, are called signaling pathways.

Given the diversity of extracellular signals that may potentially regulate our sensations, emotions and creativity, it has been suggested that the transmission of information could be mediated by neurotransmitters (neuropeptides, amino acids or their derivatives and monoamines), neuromodulators (cytokines, neurotrophins, purines, adenosine) that would modify the neurotransmitter response, and neuroregulators (NO, prostaglandins) that would influence the excitability of neurons.

The history of neuroendocrinology is quite fascinating. It emerged in the mid-20th century after discovering that the brain produces neurohormones (hormone production by neurons) that control hormonal secretion, and in turn how hormones affect brain function. However, it should be kept in mind that the concept of "neurosecretion", which was first proposed by C. Speidel in 1917, was vigorously rejected over the next two decades, as it was said that a neuron could not have a glandular function (Sarnat 1983). Then, the field of neuroendocrinology took a major step forward when H.B. Friedgood in 1936, and J.C. Hinsey in 1937, postulated that the anterior pituitary gland was controlled by substances liberated into the hypophysial portal vessels from nerve terminals in the median eminence (for review, see Fink (1976)). This hypothesis was not accepted until work from G.W. Harris in 1955 clearly demonstrated the flow of blood from the hypothalamus at the median eminence to the anterior pituitary gland (Harris 1955). This supported the concept that the hypothalamus controlled anterior pituitary gland function. Subsequently, Schally's and Guillemin's groups identified several peptide hormones released by the hypothalamus, which control the endocrine function of the gland (Guillemin 1975, 2005). This scientific breakthrough was then followed by general descriptions of neuroendocrine systems, hypothalamic–pituitary–end-organ axes and feedback mechanisms. One of them, which regulates the cortex of the adrenal gland, is a multisystem axis that used feed-forward and feedback loops to regulate glucocorticoid hormone levels. Likewise, the crosstalk between axes has emerged as potential regulators of key biological processes such as growth hormones and sexual hormones, which play a major role in the termination of linear growth. Similarly, gut and adipose hormones interact with the reproductive axis to regulate both energy homeostasis and reproductive function.

1.3. The history of immunology

Immunology is a relatively new discipline and its origin is attributed to E. Jenner (1749–1823), who reported in 1796 that cowpox induced protection against human smallpox, which is often a fatal disease (Jenner 1801). However, in ancient China and Persia, there were previous observations showing that inoculation of vesicle fluid from cases of smallpox into people protect them from this disease, which was referred to as variolation (*variola* – smallpox). Jenner called his procedure vaccination (from *vacca* – Latin for cow), and this term is still used to describe the inoculation of healthy individuals with attenuated disease-causing agents to protect them from an infectious disease. However, Jenner did not identify the nature of the infectious agent.

It was not until late in the 19th century that R. Koch (1843–1910) proved that infectious diseases are caused by microorganisms/pathogens, each one suspected of being responsible for a particular disease (for a historical review, see Kaufmann and Schaible (2005)). This discovery stimulated the strategy of vaccination, and in the 1880s, L. Pasteur successfully developed a vaccine for cholera and rabies, while E. von Behring (1854–1917) introduced serum (i.e. antibody) therapy of diphtheria, a breakthrough for which he received the first Nobel Prize in Physiology or Medicine in 1901. Since this time, this discipline has made major advances in deciphering the immune system. It is currently considered as a multilayered system comprising three major defense mechanisms: 1) physical and chemical barriers, 2) innate and 3) adaptive responses.

This scientific discipline was kicked-off by two seminal discoveries. E. Metchnikoff (1845–1916) discovered that many pathogens are immediately destroyed by phagocytic cells, which he called macrophages, and thereby introduced the concept of phagocytosis (Tauber 1992). They are rapidly recruited at the site of infection, thus providing the first line of defense that initiates the inflammatory response. It is noteworthy that this complex response is stereotyped in nature, since a subsequent infection will cause the same cascade of events with similar kinetics and intensity.

The second discovery was a neutralization of bacterial toxins by antibodies. These two discoveries led to the concept of antigen-unspecific innate immunity mediated by cells, and antigen-specific acquired immunity mediated by humoral factors. In parallel, immunobiology identified lymphocytes, which were segregated into antibody-producing B-cells (also known as plasma cells), as well as T-cells, which are central regulators of immunity.

To elicit an adequate immune response, a close cooperation between different immune cell types is required. One of the main issues which has been addressed and investigated was the identification of the cellular and molecular mechanisms triggering the activation of the system and regulating its intensity and duration. A key process in initiating an adaptive response is in the delivery of antigens to the lymphocytes (Moser and Leo 2010). It was suggested that macrophages could initiate this crucial step due to the fact that they are likely among the earliest immune cells to encounter antigens. It was logical to assume that they could present the antigen, or a fragment of the latter, to the lymphocyte. Numerous lines of evidence confirmed these hypotheses, which gave rise to the concept of "antigen-presenting cells (APC)" for any cells involved in this process (Burgdorf and Kurts 2008). Their role is therefore to capture and transfer information from the outside world to the cells of the adaptive immune system. Subsequently, it was noted that the antigen had to be associated with a special class of molecules to transfer the information to the lymphocytes. It gave rise to the concepts of "major histocompatibility complex (MHC)" and that of the "T-cell receptor" (Bonilla and Oettgen 2010). Regarding the mechanisms of T-cell activation and function, they have recently been clarified, and the concepts of co-stimulatory and co-inhibitory receptors have revealed that the immune system is a tightly regulated network that is able to maintain a balance of immune homeostasis (Chen and Flies 2013). However, there are circumstances in which the balance is not maintained and the immune system either overreacts against innocuous antigens (allergy), mistakenly attacks our own antigens (autoimmunity), or is inefficient (cancer, immunodeficiency).

The topic of autoimmunity began in 1901, first with Ehrlich's doctrine of "horror autotoxicus", then interpreted as "autoimmunity cannot happen", after he noted that goats immunized with their own red blood cells do not develop antibodies (Ehrlich 1900). This implied that autoimmunity was "dysteleological" such that "contrivances" must exist to prevent immune reactions from harming the body. However, by 1904, the antibody nature of the autohemolysin responsible for hemoglobinuria was reported, but without generating a concept of autoimmunization as the cause of the disease. About 30 years (1915–1945) thereafter, was the era of "eclipse" for autoimmunity, followed by a series of studies that put "horror autotoxicus" into a realistic frame. The year 1965 was the beginning of the acceptance of autoimmunity, when a wide range of autoimmune diseases were reported. Since the original formulation, much has been learned and investigators focused mainly on the factors responsible for the initiation and progression of the autoimmune process. A large amount of data supports the notion that autoimmune diseases arise from a combination of genetics and environmental factors. The latter are probably the most intriguing players and the largest group consists of bacterial, viral and parasitic infections. The mechanisms by which an infectious agent results in autoimmune disease have not been fully elucidated. However, an attractive explanation has been introduced by G. Snell and gave rise to the concept of molecular mimicry (Oldstone 1987). Note that this concept was also introduced in endocrinology (see above). G. Snell suggested that there are structural antigen similarities between the infectious agent and the host cell, resulting in the generation of autoantibodies. Regardless of the mechanisms involved, genetics and environmental factors disrupt the fragile balance between self-recognition and protection from non-self that depends on a tight control of the activity of numerous immune cell populations. A detailed description of cell types and their respective roles highlighted the diversity of mechanisms required to establish and maintain immunological tolerance. These studies revealed some key mechanisms or events that have been conceptualized, and specific vocabulary was used, such as central (thymus) and peripheral (lymphoid tissue) tolerance, autoreactive T lymphocytes and immune tolerance checkpoint inhibitors.

Immune deficiency is exemplified in cancer patients. In the early 1990s, it was known that the immune system can identify and eliminate tumor cells, and this process was termed as cancer immune surveillance. Unfortunately, some tumor cells can evade immune surveillance, and this immune deficiency is related to a decreased function of components of the immune system (immunosuppression concept). In 2001, an important study provided evidence that the immune system both protects against tumor development and promotes their outgrowth. This dual role led to the formulation of the immunoediting concept, which is developed in detail in section 4.3.5.

For decades it was assumed that the CNS was ignored by the immune system due to the presence of the blood–brain barrier that was supposed to protect the brain, but it turned out that the view was overly simplistic. It is becoming increasingly evident that there is an intricate relationship between the immune and nervous systems (Ransohoff and Brown 2012). A range of mechanisms exist to limit the immune response in the CNS. This type of interaction is explicated through the production of molecules (cytokines, hormones and peptides) from the CNS and through the activation of afferent and efferent neurological pathways, with both immunosuppressive and immunostimulating effects. On the contrary, some peripheral signals also communicate with the CNS, and some of them, like leptin, are key players linking the immune system, metabolism and brain function. Finally, cytokines, the prototypical mediators of intercellular communication in the immune system, affect brain functions, modulate activity of the hypothalamo–pituitary axis and have hormone-like metabolic effects in various peripheral tissues, which again highlights the intricate links between the endocrine, immune and nervous systems. These interactions among these systems determine the maintenance of health or the susceptibility to diseases.

It should be emphasized that many scientific concepts arose from hypotheses, which often turned out to be incomplete, combined with false paths, erroneous suppositions, inconclusive experiments and mistakes. Nevertheless, they provided a rationale for predictions that generated innovative experimental approaches, which then gave rise to new conceptual development. Last, but not least, it should be kept in mind that all the issues that scientists address bring an additional layer of complexity. The picture that emerges is that there is a wide variety of regulatory networks involving multiple mechanisms with different temporal and spatial properties.

This book has been written with the aim of discussing some of the emerging and evolving concepts that have shaped these three disciplines over the last century.

2

Regulatory Systems Integrating External and Internal Changes

2.1. Regulatory systems: endocrine, nervous and immune

The purpose of this section is to review how the experimental physiology introduced by C. Bernard has enabled us to progressively evaluate the respective roles of endocrinology, neurology and immunology over time, when investigators brought up the first fundamental issues to be addressed. At the beginning, each discipline used very simple experimental approaches, but they brought about tremendous changes in our knowledge of biology, some of which are reported here. Independently, they made discoveries that highlighted the importance of their respective systems in the regulation of homeostasis in physiology and physiopathology. As the general understanding of major regulatory mechanisms progresses, it has emerged that immune and nervous systems share several molecular entities. The challenge was then to integrate knowledge of the roles of key players into a coherent sequence of mechanisms, such as a network of communication pathways, which are recruited during physiological and pathophysiological processes. Similarities were extended to the endocrine system, and gave rise to a multiplicity of concepts which are presented in different chapters.

To avoid extensive citation, several review articles containing many of the original references are included.

First, a key observation should be mentioned as it was likely a starting point for scientific thinking. It was the notion of equilibrium among bodily humors that was already mentioned by the Hippocratic school and transmitted to several schools of medicine, and it was resumed in purely physical terms by J.B. Lamarck. Then, the experimental physiology from the 1850s, conducted by C. Bernard, introduced the

For a color version of all the figures in this chapter, see www.iste.co.uk/gilbert/concepts.zip.

concept of "milieu intérieur", and in the 1920s, W. Cannon introduced the concept of "homeostasis" of the "milieu intérieur" (Cannon 1929; Cooper 2008). After the first hormone discoveries, researchers have been interested in their respective role in regulating the stability of internal environment. They sought to understand how multicellular organisms sense and respond to external environmental changes/variations to maintain the whole-body homeostasis. Investigators first noted that there was a basal stability of the homeostasis, and it was constantly oscillating around a set-point and always ready to reset itself when the organism faced changes in the environment.

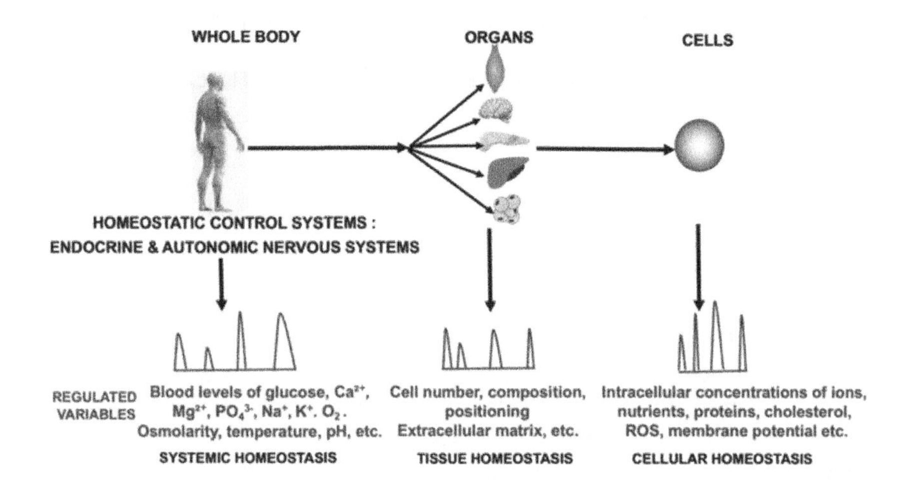

Figure 2.1. *Schematic illustration of systemic, tissue and cellular homeostasis through the combination of regulated variables*

As mentioned above, any living organism faces a great diversity of physical or chemical changes that can threaten this stability, and these systems have the role of safeguarding, i.e. to integrate these signals and elicit appropriate responses, qualitatively and quantitatively, so that they reestablish and maintain the homeostasis. It should be emphasized that homeostasis operates at three levels: entire organism, tissues and cells. In each of these compartments, regulated variables like blood levels of glucose, ions and osmolarity in the whole organism are maintained within an acceptable dynamic range by the endocrine and autonomic nervous systems. It therefore enables cells of different organs to function under optimal conditions despite widely fluctuating factors outside of the body. For instance, all humans can experience a microbial invasion, or a physical injury, and this initiates a protective response, called inflammation. It must be fine-tuned and tightly regulated because a deficiency or an excess of inflammatory responses result in immunodeficiency or morbidity such as systemic inflammatory response syndrome, respectively. Homeostasis is restored when inflammation is resolved, and it is achieved through a rapid, localized and coordinated communication between immune, nervous and endocrine systems.

To summarize, these three systems have developed an intricate network of cellular specializations and interactions, which allow them to sense specific changes and produce an appropriate response through regulated variables. When a perturbation occurs, the functional integration of organs, tissues, cells and molecules enables a dynamic stability of homeostasis (Figure 2.1).

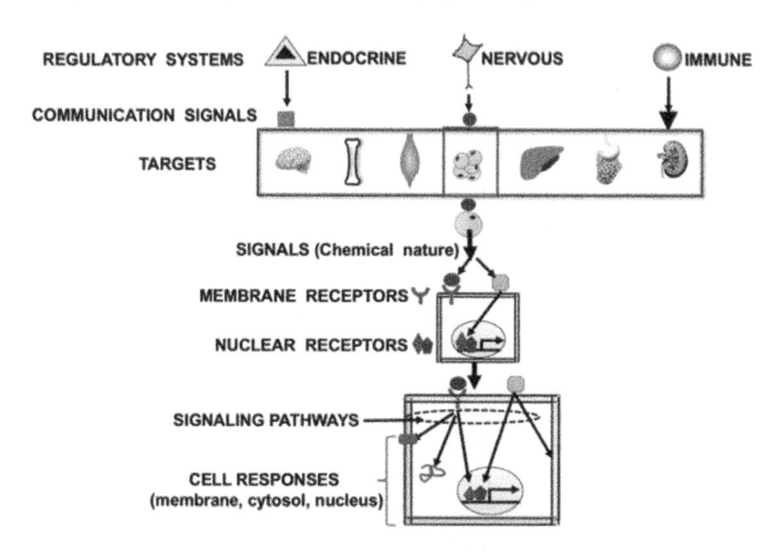

Figure 2.2. *Schematic illustration of the different components involved in the regulation of cellular responses to maintain homeostasis. The three regulatory systems (endocrine, nervous and immune systems) release chemical signals to communicate with their targets (organs/tissues). At the cellular level, they bind to plasma membrane receptors or nuclear receptors which then activate signaling pathways and elicit an appropriate biological response*

Over the few last decades, we have learned a great deal about the molecular mechanisms underlying diverse cellular processes in these three systems. It has also been revealed to scientists that the level of complexity in the regulation of these systems and interactions between them are growing. Although they steadily bring a brick to the puzzle of our basic knowledge in biology, it seems to be more and more difficult to construct an overall picture of the network of an endless, diverse list of signals that elicit cell type-specific responses. From a philosophical standpoint, there is a master word: "integration", which is often used when we elucidate the mechanism(s) that govern different biological responses at the body, cellular and molecular levels. For instance, integrative physiology describes the major functions and then detailed studies address one of the basic, but complex questions: "how do cells integrate the diversity of signals?" In a broad sense, a cell should translate them in terms of a cascade of molecular events for eliciting an appropriate biological response. These aspects are summarized in Figure 2.2, and we put a special emphasis on the regulatory systems, communication modes and molecular processes that mediate the conversion of the extracellular signal to a functional change in the cell, through signaling pathways.

2.1.1. *Endocrine system*

2.1.1.1. *Role of glands*

To demonstrate the potential role of a gland in human physiology, investigators used two complementary experimental approaches that were carried out on rodents (rats and mice) as well as dogs and sheep. At first, an ablation of the gland was performed to create a deficiency of a potential signal(s) emanating from the gland. This was followed by studies of the impact of gland deprivation on a limited number of biological processes (growth, organ development, blood substrate levels, etc.) because of a lack of biochemical assays and proper measurement equipment. The second step was to administer a crude extract of the gland to correct the deficiency-induced responses. The readout of these first experiments was mostly descriptive, but shortly thereafter, sometimes simultaneously, the chemical identification of the active principle, of the crude extract of the gland was undertaken. In parallel, investigators set up chemical assays, radio-immunoassays which allowed a quantitative measurement of proteins, steroids, etc. (Figure 2.3).

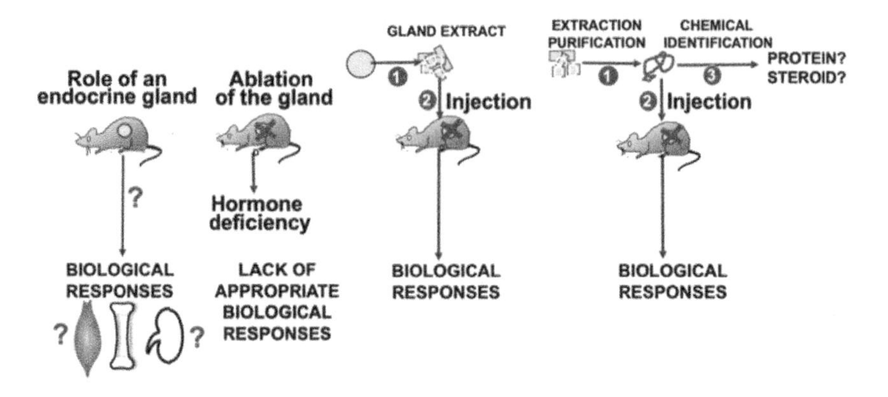

Figure 2.3. *Experimental approach to demonstrate the potential role of an endocrine gland in the regulation of biological processes*

This technique of gland ablation was performed on the thyroid, adrenal, gonads, pancreas and pituitary glands. A second experimental approach involves pharmacological inhibition of hormone synthesis and/or secretion, which mimics the gland ablation used to define the role of the thyroid gland. It consists of an administration of propylthiouracil (PTU), which specifically blocks the hormone synthesis (thyroxine).

To interpret the role of these glands, through the administration of a crude extract of the gland, caution should be taken. For instance, when a pituitary gland extract was administrated (Houssay 1943), it had a diabetogenic effect, but it was not possible to say whether the adenohypophysis contained one or more anti-insulin, diabetogenic principles. We now know that the growth hormone is the critical hormone involved, and the carbohydrate metabolism does not depend merely on the action of insulin. These experiments laid out the general outline of the endocrine system that regulates the major biological responses. Key metabolic hormones, secreted by different glands, have been successfully identified, but over the last two decades, it was discovered that metabolic organs also produce hormones and their identification increased exponentially in recent years. They have been named adipokines (Fasshauer and Blüher 2015), myokines (Giudice and Taylor 2017) and hepatokines (Meex and Watt 2017) to describe that they are produced by white adipose tissue and brown adipose tissue, skeletal muscle and liver, respectively. They are secreted in response to changes in the metabolic status of the body, such as exercise, cold exposure, feeding and fasting periods, which therefore participate in metabolic adaptation/flexibility. Dysregulation of their secretion is currently known to contribute to a wide spectrum of endocrine diseases.

In this chapter, we summarize well-established mechanisms of communication between the endocrine, nervous and immune systems, and focus on how they gave rise to concepts such as feedback, crosstalk between organs, ratio, spatio-temporal order, critical period, etc. They are part of our scientific vocabulary, and have brought a new angle to the reading of experimental data.

2.1.1.2. *Communication between organs: crosstalk concept*

Communication between organs was first identified when C. Cori and G. Cori demonstrated that lactate produced by anaerobic glycolysis in muscles can be recycled by the liver and converted to glucose through the gluconeogenesis pathway. Then, the glucose is returned, through the blood, to muscles where it is metabolized into lactate (Cori and Cori 1946). It was named the "Cori cycle", and was one of the first examples of a communication system between organs, and its main role was to facilitate metabolic adaptation to energy availability and demand (Figure 2.4). Note that under anaerobic conditions, the Cori cycle is the most important process permitting a re oxidation of NADH to NAD^+ and it is controlled by the lactate dehydrogenase, which catalyzes the conversion of pyruvate to lactate, which is then released in the blood to fuel the hepatic gluconeogenesis.

It should be underscored that the Cori cycle also involves the renal cortex, particularly the proximal tubules, another site where gluconeogenesis occurs.

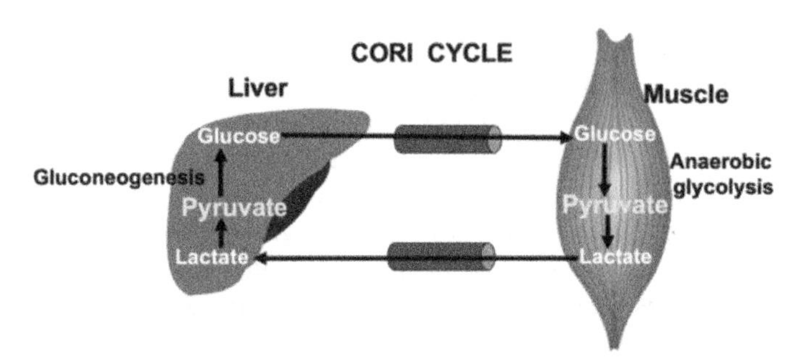

Figure 2.4. *The Cori cycle. It shuttles lactate to the liver to fuel the gluconeogenesis pathway and the glucose is transported back to the muscle, thus closing the cycle*

As noted above, peripheral organs produce a plethora of bioactive molecules that carry a signal in an endocrine, paracrine or autocrine manner (see section 2.2.3.3). Likewise, immune cells can also produce chemical signals, namely cytokines. Recently, their respective role has been explored and there is evidence that all of them play a role in the inter-organ communication to maintain homeostasis in physiological situations. For instance, interest has been focused on the communication between the white adipose tissue and the brain to understand the mechanisms controlling the energy balance, both in physiological and pathological (obesity) conditions. In section 4.3.3, we summarize well-established mechanisms of inter-organ communication and focus on how recent research has highlighted the importance of the crosstalk.

2.1.1.3. *Ratio concept*

The concept was introduced by R. Unger when considering the control of hepatic glucose production (Unger 1971). Knowing that the net flux of glucose is regulated by insulin (I) and glucagon (G) that work in tandem with each other, but in opposite directions, it was thought that the relative concentrations of these hormones may determine the hepatic glucose balance. In a series of experiments, it was clearly shown that the glucose flux was dictated by the I/G ratio and not the absolute concentration of either hormone. It also applies to the lipolytic and lipogenic fluxes across the adipose tissue. These findings also established that an increased I/G ratio favors an "anabolic" response, and the inverse promotes a "catabolic" response.

Similarly, a two-site, bi-hormonal concept was proposed for the control of ketone body production (McGarry and Foster 1977). The metabolic process of ketosis, which occurs in the fasting state, is initiated by an increased release of fatty acids from the adipose tissue, associated with an increase of their uptake by the liver, accompanied by an enhanced capacity to convert fatty acids into ketone bodies. The former event is triggered by a fall in insulin levels and the ketogenesis is under control of the glucagon rise. Hence, the investigators postulate that the I/G ratio may constitute a bi-hormonal system for the control of the ketogenic process. Since then, the molecular mechanism underlying the ketogenesis regulation has been reinvestigated and other players are now involved, but it is beyond the scope of this section to discuss this issue.

2.1.1.4. *Critical period concept*

This concept is also known as a sensitive period or time window. In 1921, C. Stockard was the first to formulate this concept after noting that almost any chemical was capable of producing malformations in fish embryos if applied at the "proper" time during development (Stockard 1921). He also reported that the most rapidly growing tissues are also the most susceptible to any change in environmental conditions. In other words, the faster rates coincide with the critical periods. To support this concept, three studies are briefly presented. Two studies demonstrate that hormone alterations during development result in central nervous system abnormalities. A third one deals with the alterations of the immune system during development, when exposed to chemicals.

The first study deals with the sexual differentiation of the rodent brain. Steroid hormones have significant effects on the developing brain and are known to program adult functions required for sex-specific reproductive behaviors. Many behavioral sex differences observed in the adult are the consequences of sex-specific exposure to gonadal steroid hormones during critical windows of development. In rodents, the absence of testosterone during a critical perinatal period results in the expression of the default program of the brain, which is female-like (McCarthy and Arnold 2011), whereas perinatal treatment of females with testosterone will defeminize the brain and prevent the expression of female reproductive behaviors in adulthood.

A second study illustrates this concept when studying the cerebellar cortex maturation of the newborn rat (Clos et al. 1974). When the surge of the thyroxine plasma level, which occurs in the newborn throughout the first three weeks of postnatal life, is prevented with the administration of propylthiouracil, which blocks the thyroxine synthesis, the differentiation of the Purkinje cells is incomplete. This abnormality can be corrected only if the thyroxine is administered during the first two weeks of postnatal life.

A third study has clearly defined two critical windows of vulnerability in fetal development of the immune system of rodents (Landreth 2002). Gestational days 7–10 encompass a period of hematopoietic stem cells formation from mesenchymal cells. Exposure of the embryo to toxic chemicals during this period results in failures of stem cell formation, abnormalities in the production of all hematopoietic lineages and immune failure. Then, between day 10 and day 16, there is a tissue migration of hematopoietic cells and an expansion of progenitor cells. This developmental window is particularly sensitive to agents that interrupt cell migration and proliferation.

The concept of critical window also applies to genomic imprinting (see more details in Chapter 5). This term was first used in biology by K.Z. Lorenz in the late 1930s. Genomic imprinting refers to the differential expression of genetic material, at either a chromosomal or allelic level, depending on whether the genetic material has come from the male or female parent. It involves modifications of the nuclear DNA, which do not involve a change in nucleotide sequence, and the term imprinting implies that something happens during a critical or "sensitive" period in development. Hence, the stage during which germ-line cells are formed may represent one critical period during which genetic information is "tagged", temporarily changing this information to permit differential expression.

Similarly, genetic studies of people conceived during the Dutch famine in 1944 revealed that transient prenatal malnutrition induced permanent epigenetic alterations. These findings highlight the critical role of the gestational timing during which environmental conditions result in lifelong phenotypic consequences (Heijmans et al. 2008).

The concept of critical periods has also been well documented for neuronal circuits across several systems and species. It was initiated in 1958 by Lorenz (1958) and the concept has profoundly influenced not only biologists, but also psychologists, such as philosophers, physicians, parents and educators. The reader is referred to the excellent review for more details of the field that delineates critical phases in brain development that carry a social impact far beyond basic neuroscience (Hensch 2004).

2.1.2. *Nervous system*

The nervous system receives a wealth of information from an individual's surroundings and body, and its main task is to ensure that the organism adapts to the environment. For a long time, it held the "top rank" or first position of all the organ systems, but the early 19th century saw a revolution in scientific thinking about its anatomy and functions. On an anatomic basis, it can be subdivided into the central nervous system (CNS: brain and spinal cord) and peripheral nervous system (PNS:

the cranial and spinal nerves). From a functional perspective, somatic (voluntary) and autonomic (involuntary or vegetative) nervous systems are recognized. The autonomic nervous system is subdivided into the sympathetic system and the parasympathetic system. With respect to its functions, it should be emphasized that it exerts a highly centralized control over other structures and physiological functions of the body, through an incredibly complex network. This latter aspect has been investigated, and numerous studies highlight the key role of the bidirectional communication between the brain and the endocrine and immune systems.

Interactions between peripheral organs and the brain are well documented, and recent advances describe how the inter-organ communication networks regulate energy homeostasis (Castillo-Armengol et al. 2019). For instance, to maintain homeostasis in fasting (Al Massadi et al. 2017) or feeding (Turton et al. 1996; Ronveaux et al. 2015) conditions, a great number of studies have described how the brain integrates chemical signals, released from the peripheral tissues, and responds through the release of neurohormones and/or an activation of the autonomic nervous system of the target tissues, in order to maintain energy homeostasis. Other studies focused on cold-induced thermogenesis highlighted the role of the brain in response to cold stress. Cold is sensed by the sensory neurons in the skin, and they convey this signal to the hypothalamic preoptic area (Zhang and Bi 2015). The transduction of the signal stimulates other hypothalamic areas, namely orexin-producing neurons located in the lateral hypothalamus area (LHA) that in turn stimulate the sympathetic firing in brown adipose tissue, which promotes thermogenesis.

Other examples of communication between peripheral organs and the brain are presented in detail in section 4.3.3.

2.1.3. *Immune system*

In all living organisms, the immune system provides a high level of protection from invading pathogens in a robust manner. The immune system consists of billions of cells dispersed in the body, and it has long been evident that this system must be subjected to internal regulatory mechanisms to maintain its homeostatic balance. It was also thought that the molecular signals they release when mounting an immune response seemed to be completely devoted to the control of the function of the immune system. For a long period of time, there was a kind of consensus among immunologists claiming that there was no reason to incorporate other more integrative host systems into immunological thinking.

Over the last three decades, numerous studies performed with new experimental approaches have revealed that extensive communication between the immune, endocrine and nervous systems operates through intricate chemical messengers. A detailed discussion of these communications is provided in section 4.3.3.

2.2. Origin and diversity of signals and communication modes

This section highlights the fact that circulating factors are generated not only from the endocrine glands, but also from certain organs or tissues. A brief overview of their origin is discussed.

2.2.1. *Origin*

2.2.1.1. *Endocrine glands*

In all textbooks on physiology or endocrinology, the reader can refer to a detailed description of the following glands: thyroid, parathyroid, adrenal, anterior pituitary, posterior pituitary and gonads. As mentioned earlier, their respective role in physiological and pathophysiological states was established in two steps (see Figure 2.3). First, does a specific gland play a role in metabolism, thermoregulation, reproduction, growth, etc.? If so, its surgical ablation should impact the regulation of one or several biological parameters compared to a control. Second, does administration of a gland extract or a gland grafting correct the observed defects? If so, the extract undergoes a series of purification, which aims to identify the active principle which will correct the ablation-induced deficiency. It should be emphasized that a partial ablation of a gland (adrenal and thyroid glands) results in a hypertrophy and hyperplasia of the remaining part, which enables the gland to maintain normal hormone blood levels. In other words, this increased capacity of the remaining gland to secrete the hormone compensates for the absence of secretory cells. Similarly, ablation of one kidney results in a hypertrophy of the second one.

The basic approach of ablation–restitution is now replaced by a more sophisticated one, but the principle is the same, only the scale has changed: from the gland to the molecular level. For instance, when addressing the role of a protein, investigators eliminate the gene encoding for the protein. They perform a “gene knockout” in mice, which represents its “molecular ablation”, by analogy with the surgical ablation. This technique is designed to completely remove one or several exons from a gene of interest, resulting in the production of truncated protein or the complete abolition of protein production. Although this technology represents a valuable research tool, some important limitations exist. A gene alteration expressed in the germ line may cause embryonic lethality, resulting in no viable mouse to study gene function, and the gene may serve a different function in adults. On the other hand, a functionally related gene or genes may compensate for the gene that was ablated. For instance, acetylcholinesterase knock-out mice are viable due to butyrylcholinesterase activity, although pharmacological inhibition of acetylcholinesterase, which is essential for degradation of acetylcholine in the neuromuscular junction, is lethal (Xie et al. 2000). Furthermore, a gene alteration may exert its effect in multiple different cell and tissue types, creating a mosaic of phenotypes in which it is difficult to distinguish direct function in a particular tissue from secondary effects resulting from altered gene function in other tissues. To overcome these drawbacks, another technique termed "Cre/LoxP" allows for gene activation or repression at different time points, and in a tissue-specific manner during the whole life (see Hall et al. (2009)). The interested reader can visit the website: https://services-web.research.uci.edu/facilities-services/tmf/online-resources.html. It is a database of general and tissue-specific and inducible Cre recombinase expressing strains of mice. Conversely, the insertion of multiple copies of a gene into the genome results in an overexpression of the encoded protein. The aim of this approach is to test whether it results in exacerbation of the physiological response, which in some instances would explain pathophysiological states.

In 2007, three scientists, M. Capecchi, M. Evans and O. Smithies, received the Nobel Prize in Physiology or Medicine for their discoveries of principles for introducing specific gene modifications in mice using embryonic stem cells.

2.2.1.2. *Endocrine organs/tissues*

Over the past 30 years, our views about the origin of circulating factors have changed. It began in 1987 when the Spiegelman and Flier groups defined white adipose tissue as a secretory organ after the identification of a secretory protein, named adipsin (Cook et al. 1987), also known as complement factor D, a protease that participates in the activation of the complement. Thereafter, Friedman's laboratory identified leptin (Zhang et al. 1994), and it was followed by the discovery of new secreted molecules such as adiponectin and resistin. With the discovery of leptin secreted by the adipose tissue, the term "adipokine" was coined. To date, hundreds of bioactive adipokines secreted by the white adipose tissue, which have a role in lipid accumulation, energy homeostasis as well as the regulation of inflammation and insulin sensitivity, have been identified (Ouchi et al. 2011). Then, muscle also acquired the status of secretory organ, with the finding of muscle-derived IL-6 in 2000, followed by the identification of hundreds of other secreted proteins (Severinsen and Pedersen 2020). Recently, a novel group of signaling molecules, such as FGF21, follistatin, angiopoietin-like, has emerged and they are released from the liver (Weigert et al. 2019; Jensen-Cody and Potthoff 2020). These liver-derived factors play a role as modulators of inflammation, insulin resistance, whole-body energy homeostasis and metabolic diseases (Iroz et al. 2015). In analogy to adipokines, these signaling molecules from muscle and liver have been termed myokines and hepatokines, respectively. Note that some of these molecules are exclusively or predominantly produced by these endocrine organs mentioned above, and they are named “organokines”. For instance, skeletal muscle is a predominant source of irisin, a myokine that is thought to be involved in the browning of adipose tissue. However, others, such as IL-6, are prominently secreted from multiple tissues, including immune cells, skeletal muscle and adipose tissue, and can be classified as typical cytokines, as well as myokines or adipokines, depending on the site of secretion. “Organokines” are therefore classified according to the site of secretion and common functional characteristics (e.g. regulation of secretion by muscle contraction in the case of myokines), and do not represent specific, non-overlapping groups of secreted signaling molecules.

It is important to emphasize that, as we have improved our technologies of detection of molecules, this has made it possible to identify a great number of factors which are expressed and released by most organs or tissues. To date, there is an endless list of molecules released by different organs or tissues and they exert their effects through an endocrine, paracrine or autocrine mode (see Figure 2.7). The next challenge ahead is to describe their biological function(s). For instance, among the hundreds of muscle-derived peptides which have been identified, only 5% have been described to have a functional role.

In this section, only organs/tissues that were not suspected to play an endocrine role are briefly discussed, and more detailed information about kidney, bone, liver, muscle, enteroendocrine cells and neurons is presented in the section dealing with the inter-organ crosstalk (see section 4.3).

2.2.1.3. *Kidney, bone, endocrine and immune cells, neurons*

Kidney

Besides its excretory and other homeostatic functions, the kidney is an endocrine organ essentially involved in the regulation of bone mineralization, blood pressure and erythropoiesis. The kidneys produce and secrete the hormones erythropoietin (EPO), klotho and calcitriol, as well as a hormone-regulating factor, renin (DeLuca 1975; Kurt and Kurtz 2015) (Figure 2.5(A)). The *klotho* gene was named after a Greek goddess of fate who spun the thread of life.

– Erythropoietin

The glycoprotein hormone erythropoietin (EPO) is the primary humoral regulator of red blood cell formation in mammals.

– Klotho

Klotho exists in a membranous and a secreted form. It regulates calcium and phosphate homeostasis and thus contributes to bone mineralization. Klotho membrane is required as a co-receptor for high affinity of FGF23 to its receptor FGFR1 (Ho and Bergwitz 2021).

– Calcitriol

Calcitriol is the bioactive metabolite of vitamin D, also called 1,25$(OH)_2$D3 or 1,25-$(OH)_2$-cholecalciferol. It contributes to bone mineralization by regulating the homeostasis of calcium and phosphate by acting on the intestinal tract, the bones and the kidneys (see section 4.3.3).

– Renin

When renin (*ren* – the kidney in Latin) was discovered in 1898 by R. Tiegerstedt at the Karolinska Institute (Tigerstedt and Bergman 1898), it was defined as a "pressure substance". It was an important breakthrough in the context of the new concept of "milieu intérieur" by Bernard. Curiously, like C. Bernard, they detailed their meticulous approaches, including the design of a flowmeter to measure blood pressure changes and documentation of effects of renin, etc. The choice of the kidney as the source of renin coincided with the long-standing recognition that high blood pressure was related to kidney diseases.

In the late 1930s, angiotensin I (Ang-I) and angiotensin II (Ang-II) were isolated, and angiotensin is a hybrid of angiotonin and hypertensin. Following this identification, the renin-angiotensin system (RAS) was established, which consists of the enzymatic cascade initiated by the cleavage of the liver-derived angiotensinogen by renin. It gives rise to Ang-I, which is further cleaved to produce Ang-II. Initially, it was thought that the mode of action of Ang-II was mediated through the classical endocrine pathway. Recently, it was proven that Ang-II exerts its effects through a local paracrine or autocrine RAS. It controls blood pressure and fluid–electrolyte balance (see Inagami (1998)).

Bone

In 1941, F. Albright recognized that bone was metabolically active in addition to its structural role. In 2000, Karsenty's group revealed that the leptin controls bone mass through a hypothalamic relay. This unexpected observation laid the ground to a new understanding of skeletal physiology. Over the last two decades, a growing number of studies have clearly established bone as an endocrine organ. Among the bone-derived factors that perform true endocrine functions are osteocalcin and FGF23 (Figure 2.5(B)).

– Osteocalcin

Osteocalcin is an osteoblast-specific secreted protein and an important constituent of the bone with multiple physiological functions. Among others, once secreted in the blood, osteocalcin stimulates insulin production and increases insulin sensitivity in target organs that leads to enhanced glucose and fatty acid uptake in muscle and white adipose tissue. Moreover, a number of epidemiological analyses support the role of osteocalcin in the regulation of glucose and energy homeostasis in humans (Mizokami et al. 2017).

– FGF23

FGF23 has recently emerged as a new class of phosphate-regulating protein, termed phosphatonin. It is secreted by osteocytes and osteoblasts and acts on the renal tubule to induce phosphaturia via decreased Na/Pi transporters expression, and PTH has a similar effect on the kidney when acting via its receptor. In contrast to PTH, which stimulates the hydroxylation of 25(OH)D3 to 1,25(OH)$_2$D3, FGF23 suppresses the hydroxylation and lowers 1,25(OH)$_2$D3. FGF23 also inhibits PTH secretion in rodent models. These observations highlight the importance of the PTH–vitamin D–FGF23 axis (Blau and Collins 2015).

– Lipocalin-2

Lipocalin 2 (neutrophil gelatinase-associated lipocalin) was previously thought to be exclusively secreted by adipose tissue and is associated with obesity. It is also expressed by osteoblasts at levels about ten times higher than in white adipose tissue or other organs. It can influence energy metabolism by suppressing appetite, after binding to the MC4R in the hypothalamus.

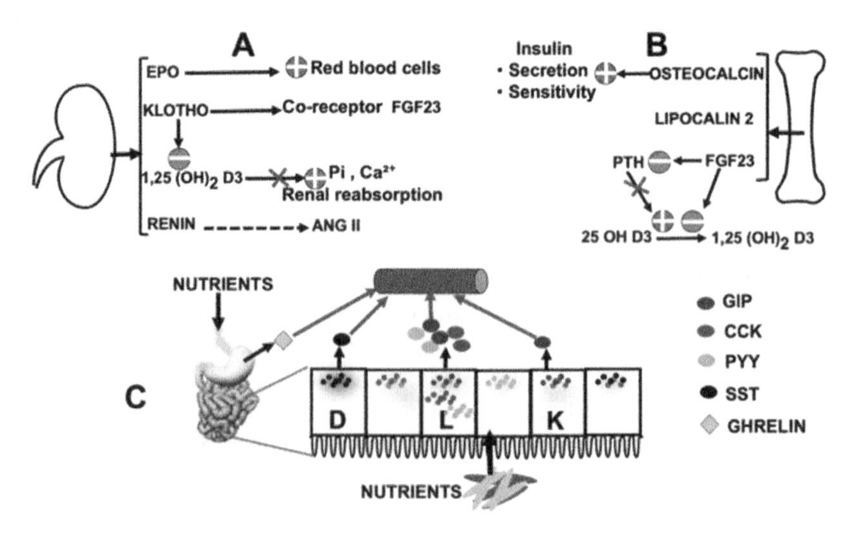

Figure 2.5. *Endocrine organs: (A) kidney, (B) bone and (C) gastrointestinal tract*

Endocrine cells

– Gastrointestinal (GI) tract

Historically, gastrointestinal hormones have never been fully accepted in endocrinology, which seems to be a paradox because this discipline was conceptualized with the discovery of secretin. Later on, the development of cell and molecular biology enabled us to identify more than 30 peptide hormone genes and discover that most individual hormones originated from one intestinal cell type. It ensued that the old concept "one cell-one hormone" regarding gastrointestinal hormones could not prevail any longer. Currently, all gastrointestinal hormones originate from the enteroendocrine cell (EEC) system, which consists of scattered hormone-producing endocrine cells within the epithelium of the GI tract, including stomach, small and large intestine. It represents the largest specialized endocrine network in terms of number of cells, integrating environmental and nutrient cues, enabling neural and hormonal regulation of metabolic homeostasis. Over the last decade, molecular cloning of the cDNAs and genes that encode EEC hormones has markedly expanded the complexity of EEC endocrinology. It also brought up new issues that challenged historical concepts in this field. Originally, gut endocrine cells were named L, K and D cells, for the predominant hormone they produced. For instance, "L cells" have long been recognized to express the proglucagon gene. This nomenclature is no longer useful because it was recently revealed that "L cells" are plurihormonal and depending on their location within the gut, they co-express cholecystokinin (CCK), glucose-dependent insulinotropic peptide (GIP), peptide YY (PYY) and secretin. This observation also applies to the other enteroendocrine cells K and D, which were initially considered to secrete glucose-dependent insulinotropic polypeptides (GIP) and somatostatin (SST), respectively (Figure 2.5(C)). There is now evidence that EEC biology should be considered as a complex of multihormonal cells with pleiotropic interactive networks (Drucker 2016). Recently, there has been considerable interest in the role of EEC in gut–brain bidirectional communication, emphasizing their function in nutrient sensing and food intake control. The reader is referred to a series of excellent review articles (Breer et al. 2012; Dockray 2014).

As mentioned above, it is now clear that a mixture of hormones can be produced by a single intestinal cell type, and this concept also applies to certain type of neurotransmitters. For instance, neuropeptide Y (NPY) and γ-aminobutyric acid (GABA) are co-expressed in AgRP neurons, and co-release to promote feeding (Wu and Palmiter 2011). Clearly, the concept of "one gland-one hormone", "one cell-one hormone" or "one neuron-one neurotransmitter" has been eclipsed by the complexity and diversity of cellular secretory activity. This complexity can be extended to different endocrine glands.

C cells

They are a minor component of the thyroid, comprising less than 0.1% of cell mass in the normal gland. In 1962, H. Copp discovered that the C cells produced a potent hypocalcemic hormone, which was called calcitonin. It acts predominantly on bone to lower its resorption, through inhibition of the osteoclast activity. There is also evidence for an action in the kidney to decrease tubular reabsorption of calcium, and in the brain and hypothalamus where several actions have been reported. Despite these well-documented effects, its physiological role is questionable. For instance, low levels of calcitonin observed after a thyroidectomy or high levels of the hormone in patients with medullary thyroid carcinoma do not cause any bone or calcium pathophysiology (Fuchs et al. 2019).

Neurons

– Hypothalamus as an endocrine gland

In different nuclei of the hypothalamus (preoptic area = POA; paraventricular nucleus = PVN; dorsomedial nucleus = DMH of the hypothalamus; arcuate nucleus = ARC), there are neurons that project to the median eminence and release neurohormones into the hypophysial portal system that stimulate or inhibit the function of the cells of the anterior pituitary gland. These neurohormones are as follows: gonadotropin-releasing hormone = GnRH; corticotropin-releasing hormone = CRH; somatotropin release inhibitory hormone = SRIF or somatostatin; TRH = thyrotropin-releasing hormone = TRH; growth hormone-releasing hormone = GHRH; dopamine = DA (Clarke 2015) (see section 4.3.1). The review of the secretion of neurohormones and their respective targets relies on data that was accumulated from large animal models, mainly sheep, and essentially in the time interval from the 1970s to the 1990s (see section 1.2).

– Posterior pituitary

In several regions of the hypothalamus, primarily in large magnocellular neurons situated in the supraoptic and paraventricular nuclei, there are neurons that project their axons to the posterior pituitary, where they release arginine vasopressin (AVP) and oxytocin (OXT) into the peripheral circulation. These hormones are secreted, for example, during dehydration and labor, respectively (Baribeau and Anagnostou 2015).

– Immune cells

A wide range of immune cells, including B-cells, T-cells, mast cells, neutrophils, basophils and eosinophils, release molecular signals, called cytokines, a term that was first coined by S. Cohen in 1974 (Cohen et al. 1974). Other names such as lymphokine (cytokine made by lymphocytes), monokine (made by monocytes), interleukin (made by one leukocyte and acting on other leukocytes) and chemokine (CHEMOtactic CytoKINE = cytokine with chemotactic activities) are also used. Lymphokines were originally thought to be produced only by lymphocytes, enabling a communication with other immune cells, but they are now known to also be secreted by non-lymphoid cells.

Additionally, it was also observed that most lymphokines display multiple properties. A new nomenclature was then proposed, and the lymphokines were named "interleukin" followed by a number. However, some lymphokines retain their original names, for example, the interferons that have been identified for their ability to "interfere" with viral replication in infected cells (Lacy 2015).

2.2.2. *Diversity*

Extracellular signals can be classified into two groups based on their nature. They are briefly listed, and they will be mentioned throughout the different chapters when addressing their specific role in biological responses.

2.2.2.1. *Chemical signals*

Chemical signals include: hormones (proteins, metabolites), neurotransmitters and neuromodulators (amino acids and their derivatives, peptides, purines, monoamines, gasotransmitters: NO, CO), steroids, cytokines (lymphokines, chemokines, interferons), lipid mediators (prostaglandins) (Zhu et al. 2006), nutrient metabolites (Husted et al. 2017), extracellular matrix (ECM). Anecdotally, in 1984, the cytolytic molecules used by cytotoxic T-cells were compared with neurotransmitters, as they were released into the "synapse", which represented the interactions between the TCR and MHC antigen complex (Norcross 1984). This interaction is therefore sometimes still referred to as an immune (also immunological) synapse.

2.2.2.2. *Physical signals*

Physical signals include: skin contact, gravity, sound waves, food texture, muscle stretch, air flow, temperature, shear stress (Xu et al. 2018).

2.2.3. *Communication modes*

2.2.3.1. *Endocrine*

In endocrine systems, the signaling factor is released from the cell of origin and transported by the blood or another fluid to a distant site, where it binds to a receptor of the responding cell and elicits a biological response. The signaling factor is called a hormone (Figure 2.7).

Concept of peptide hormone precursors

In the 1960s, D.F. Steiner studied the time course of insulin biosynthesis and demonstrated the existence of a precursor that has immunological properties similar to insulin, but it was less active compared with insulin (Steiner 2011b). Then, several laboratories established that a precursor larger than proinsulin is involved in the biosynthesis of insulin. These findings are summarized in Figure 2.6, and more detailed information is presented in Chan and Steiner (1977). This work was the starting point of a new chapter in the research field of peptide hormone biosynthesis. It led to the discovery of a whole range of peptide prohormones, particularly in relation with their processing. For instance, it led to the discovery of C-peptide, a proteolytic fragment of proinsulin, and the development of an assay for measuring the C-peptide in the blood, which importantly contributed to clinical assessment of hypoglycemia, as well as residual β-cell function in type 1 diabetes.

Concept of preprohormone/prohormone/hormone

The preprohormone is the initial product biosynthesized by the hormone-producing cell. It gives rise to a prohormone which is an intermediate product formed along the pathway of intracellular synthesis of the hormone. It is worth noting that a preprohormone sequence can be a precursor of one bioactive peptide (proinsulin) or three peptides (glucagon, GLP-1 and GLP-2) or identical copies (five copies of TRH) (Rehfeld 1998).

Here are some examples: preproenkephalin B, preproglucagon, preprooxytocin, preprogastrin, prepro-opiomelanocortin, preproinsulin (Rehfeld et al. 1989). The next step was to elucidate the prohormone-processing mechanism, which also applies to proneuropeptides and results in the release of bioactive peptide(s). These mechanisms are described in Rehfeld (1998), and led to the discovery of endoproteases: carboxypeptidase E (CPE) and proconvertases (PC1 and PC2) (Steiner 2011a).

Pro-opiomelanocortin (POMC) is one of the best examples of a polypeptide precursor of hormones and neuropeptides. In the 1930s, a series of clinical observations, far from the modern molecular world, suggested a processing of human POMC, and in 1978, the concept that POMC was a prohormone was firmly established. See the excellent review by Harno et al. (2018). The POMC gene generates the preprohormone POMC, which undergoes post-translational modifications (acetylation, amidation, phosphorylation glycosylation, disulfide linkage formation). Then, a series of proteolytic cleavages give rise to several peptides, and various modifications diversify the biological functions of POMC-derived peptides. In the corticotropic cells of the anterior pituitary, POMC is cleaved into three major fragments and is the precursor of N-POMC, ACTH and β LPH (Takahashi and Mizusawa 2013). In the melanotropic cells of the intermediate, the lobe (or layer in humans) of the pituitary the POMC is further processed, and ACTH is split into α-MSH and corticotropin-like intermediate peptide (CLIP) (Figure 2.6(B)). Given the similarity between ACTH and α-MSH, which stimulates melanocytes, the pigment-producing cells in the skin, it is not surprising that excessive ACTH secretion, such as in Addison disease, is associated with pathologically increased skin pigmentation. This example once more clearly demonstrates how molecular biology not only furthers our understanding of physiology, but illuminates the mechanistic underpinnings of various medical conditions, thus highlighting the importance of basic biomedical sciences for clinical medicine.

The two neurohypophyseal hormones vasopressin (AVP) and oxytocin (OXT) are also synthesized as preprohormones in the magnocellular neurons of the hypothalamus, and their precursors are the prepro-AVP-neurophysin II and prepro-OXT-neurophysin I, respectively. These precursors undergo multiple post-translational modifications during the axonal transport, which gives rise to the active neuropeptide, which is then released in response to a stimulus (Ivell et al. 1983) (Figure 2.6(C)).

Concept of cell-specific precursor processing

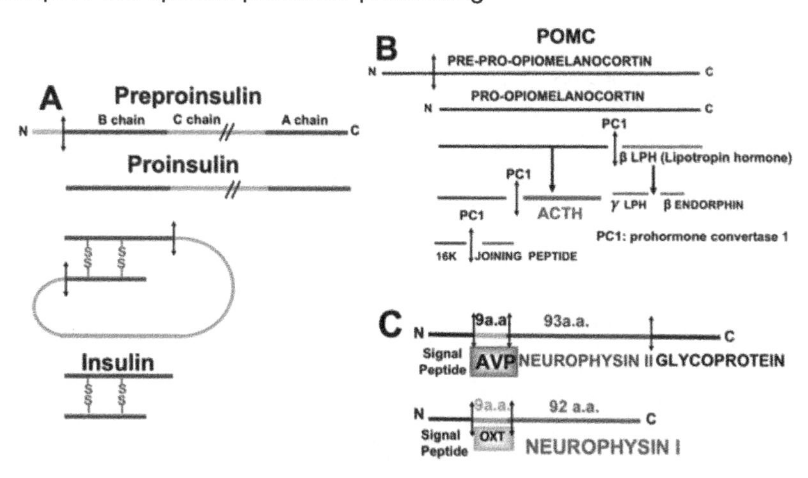

Figure 2.6. *Concept of preprohormone, prohormone and hormone*

The singular gene for a peptide hormone is expressed not only in a specific endocrine cell type, but also in other endocrine cells, as well as in entirely different cells such as neurons, adipocytes, myocytes and immune cells. It raises the question of how the body ensures that the peptide hormones in peripheral circulation do not compete with neuronal release of similar neuropeptides in the cerebral synapses. It is explained by the cell-specific processing of prohormone, which prevents the release of different fragments of the prohormones from different cells. For instance, proglucagon processing is an example of such tissue or cell-specific precursor maturation. In the pancreas, glucagon itself is the bioactive product, while in the small intestine, truncated GLP-1 is the essential proglucagon product. Similarly, the cell-specific processing of procholecystokinin takes place in the small intestine, the brain and the pituitary gland (see the review by Rehfeld and Bundgaard (2010)).

2.2.3.2. *Paracrine/autocrine*

Unlike the endocrine system, signaling molecules are released into the fluid phase acting on target cells in a local environment, which is called the paracrine mode of intercellular communication. Signaling molecules may act back on the cells of origin, and this is called the autocrine mode.

Some signaling molecules encompass the three modes of communication endocrine/paracrine/autocrine, such as the growth hormone (GH), which is synthesized in the anterior pituitary gland and involved in growth and reproduction. Similarly, prolactin secreted by the anterior pituitary (Matera 1996), cytokines and growth factors (GFs) communicate through endocrine/paracrine/autocrine modes (Figure 2.7).

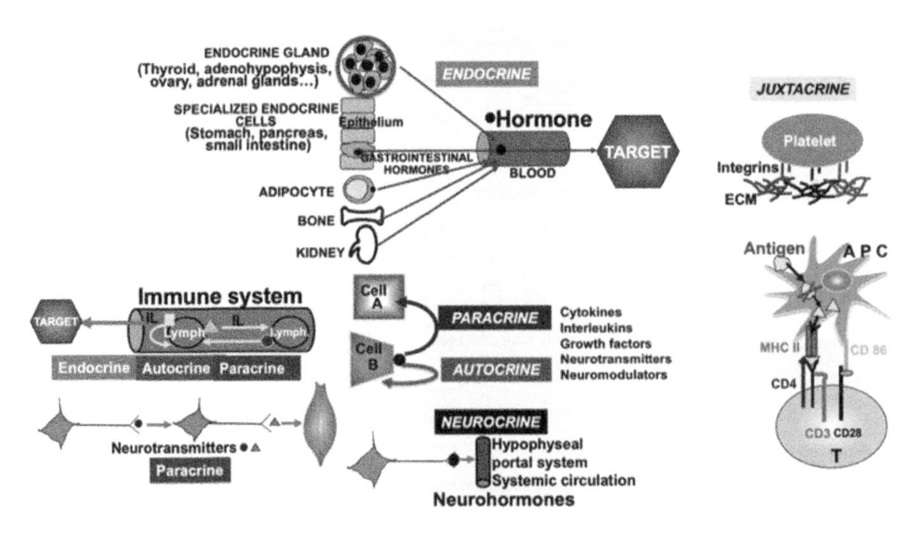

Figure 2.7. *Schematic illustration of the different communication modes, including endocrine, paracrine, autocrine, neurocrine and juxtacrine*

Cytokines

It is difficult to identify the precise starting point of studies of cytokines. The first suggestion that soluble factors modulated host reactions came from V. Menkin in 1944, who "purified" factors with fever-inducing activities from inflammatory exudates and called them "pyrexin" (Menkin 1944). Before their molecular characterization, fever-producing factors, pyrogens, were divided into exogenous pyrogens if they originated outside the body and endogenous pyrogens if they were produced in the body. While bacterial lipopolysaccharide (LPS) is now one of the best characterized exogenous pyrogens, cytokines such as IL-1, TNF-α, IL-6 and interferons were established as endogenous pyrogens.

However, in 1957, the discovery of interferons (IFNs) was a turning point for immunologists. The pioneering work by A. Isaacs and J. Lindenman should be mentioned, illustrating that a major breakthrough in biology often results from a technically simple experimental approach (Isaacs et al. 1957). Here is the conclusion of their studies:

> During a study of the interference produced by heat-inactivated influenza virus with the growth of live virus in fragments of chick chorio-allantoic membrane, it was found that following incubation of heated virus with membrane a new factor was released. This factor, recognized by its ability to induce interference in fresh pieces of chorio-allantoic membrane, was called interferon. Following a lag phase interferon [...] was released into the surrounding fluid.

Thereafter, successive discoveries between 1964 and 1967 initiated the study of cytokines by immunologists S. Kazakura and L. Lowenstein, who were the first to detect the presence of leukocyte mitogenic factors in supernatants of antigen-stimulated leukocyte cultures (Kasakura and Lowenstein 1965). The structure and function of these molecules had remained elusive because of a lack of a purification technique. But in 1978–1980, C. Weissmann and T. Taniguchi identified and cloned the complementary DNAs (cDNAs) of IFNα and IFNβ, respectively (Nagata et al. 1980; Taniguchi et al. 1980). The IFN family of cytokines is now recognized as a key component of the innate immune response and the first line of defense against viral infection.

In the late 1960s, immunologists focused on the role of T-cells in antibody production by B-cells. In 1968, it was demonstrated that besides the interactions between T- and B-cells that were essential for antibody production, other factors were released from T-cells. They were called T-cell replacing factor, but none of them had been purified and their cDNAs were not cloned yet. In the 1980s, these factors were cloned and called IL-4, IL-5 and IL-6 (see the review by Dinarello (2007)).

Many cytokines act locally in an autocrine or paracrine fashion, but some of them (IL-6) enter the bloodstream and act in a typical endocrine fashion (Figure 2.6). The latter point is illustrated by classic hormones, growth hormone (GH), prolactin (PRL), erythropoietin (EPO) and leptin (LEP), which are clearly cytokines, as evidenced by the type of receptor they bind to and their signaling pathway.

Neurotransmitters

– Synaptic and extra-synaptic transmission

In the past three decades, the concepts underlying the interneuronal communication in the central nervous system have been revised. In the conventional synapse, classical neurotransmitters, such as glutamate, GABA and acetylcholine, are considered as synaptic transmitters operating via receptors coupled to ion channels, transmitter-gated channels, within the synaptic cleft. This idea was held in general agreement until the end of the 1960s, when it was discovered that there was another form of interaction among neurons, thereby challenging the traditional view of direct synaptic communication. It was shown that classical transmitters can be released extrasynaptically (non-synaptic sites) (Vizi 1974), via a vesicular system, or leak out from the synaptic cleft and diffuse to target cells more distant than those seen in conventional synaptic transmission. Thus, glutamate, which is mainly released into the synaptic gap, can escape from the synapse and reach extrasynaptic receptors (Figure 2.8). Depending on their localization, they are named perisynaptic and extrasynaptic receptors because they are at the edges of synapses, and far from the release site, respectively (see the review by Vizi et al. (2010)).

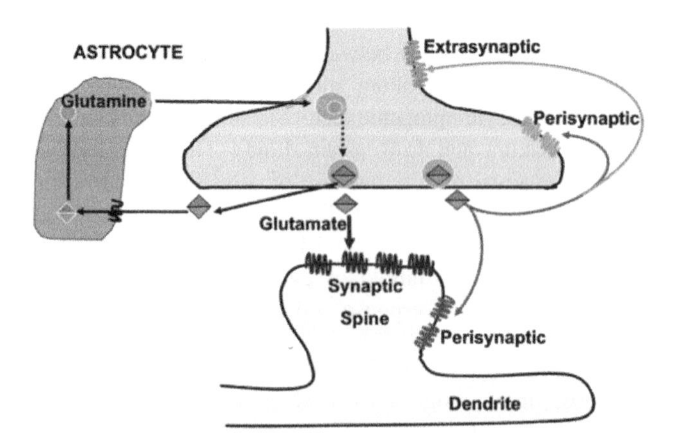

Figure 2.8. *Subcellular localization of glutamate receptors (mGlu) at synaptic and non-synaptic (perisynaptic or extrasynaptic) sites. They are concentrated at the postsynaptic membrane and extrasynaptically along axons and axon terminals. To prevent "glutamate excitotoxicity" and maintain glutamate homeostasis, astrocytes remove about 90% of its release in the synaptic cleft*

However, it should be emphasized that ionotropic receptors are also located at extrasynaptic sites. In smokers, nicotine in the brain most likely acts on high-affinity receptors (extrasynaptic nAChRs), because its concentration is too low to activate any synaptic receptor which requires a higher concentration of endogenous ligand. In light of these observations, it should be emphasized that clinical pharmacotherapy mostly targets high-affinity receptors, because drugs can reach only low concentrations in the brain during therapy.

Finally, it should be mentioned that synaptically released neurotransmitters, such as glutamate, GABA and glycine, must be rapidly cleared from the extracellular space, which accounts for 15–30% of the total brain volume (Syková and Nicholson 2008). For instance, an excess of glutamate in the synaptic and extra-synaptic space leads to neuronal hyperexcitation, known as "glutamate excitotoxicity". To prevent toxicity and maintain glutamate homeostasis, astrocytes can remove about 90% of its release in the synaptic cleft. In the astrocytes, it is converted to glutamine, which is released and used by neurons as a precursor for the synthesis of glutamate or GABA (Agnati et al. 1995; Vizi et al. 2010; Mahmoud et al. 2019).

Neuromuscular transmission

The early discoveries in the neuromuscular transmission started about 150 years ago and led to new concepts in cell biology. The reader should refer to the reviews by W. Bawman (Bowman 2006, 2013; Way 1985). In Paris, F. Magendie received samples of curare from Napoleon III and used them to immobilize animals. C. Bernard and A. Vulpian, who worked in Magendie's laboratory, tried to locate the site of action of curare between the motor nerve and the muscle, i.e. at the motor endplate, a concept clearly stated in 1875 (Vulpian 1882). Then, the mechanism of action of curare was established by Langley in 1906 when he studied the actions and interactions of nicotine and curare on frog muscles. He showed that curare blocked the stimulant action of nicotine in both innervated and denervated muscles, demonstrating that it mainly targets the muscle rather than the nerve endings. Langley concluded that the nerve impulse crossed the junction not by an electrical discharge, but by a secretion that we now call a neurotransmitter – acetylcholine. Unlike glutamate, GABA and glycine, the action of acetylcholine is not terminated by uptake into perisynaptic cells, but via degradation by acetylcholinesterase, which hydrolyses acetycholine into acetate and choline in the synaptic cleft. Since the neuromuscular junction is much more accessible for experimental observation and manipulation than most synapses, it was an elegant experimental model for the investigation of synaptogenesis, as well as for basic physiology and pathophysiology of synaptic transmission.

2.2.3.3. *Juxtacrine*

This term was coined by Anklesaria, Massagué and colleagues 1990, to describe the binding of the membrane-anchored pro-transforming growth factor alpha (TGFα) of a line of stroma cells to epidermal growth factor receptors (EGFR) on adjacent hematopoietic cells. This molecular interaction led to both adhesion and proliferation of the target hematopoietic cells (Anklesaria et al. 1990). Therefore, in juxtacrine interactions, proteins from the inducing cell are tethered to the plasma membrane and interact with receptor proteins of adjacent target cells. The juxtacrine factors are "non-diffusible", which is the essential feature that distinguishes juxtacrine stimulation from endocrine, paracrine and endocrine modes of cell–cell signaling. It should be emphasized that certain signaling molecules, such as tumor necrosis factor-α (TNF-α) appear to have the great versatility of acting in all four modes. Similarly, RANKL (see Decoy molecules – section 3.3.5), an osteoblastic factor, which activates osteoclasts, may exist in the membrane-bound or secreted form and may therefore act in juxtacrine or paracrine mode.

There are two main types of juxtacrine interactions that can be defined as follows: (1) a protein on cell A binds to its receptor of adjacent responding B-cell (example: APC/Antigen/MHC-II and I) and (2) on cell A, integrins interact with extracellular matrix proteins (Figure 2.7).

APC/Antigen/MHC-II and I

Initially, the interaction between the T-cell receptor of the $CD4^{+}$ T lymphocyte and the major histocompatibility complex (MHC-II or I)–peptide complex of the antigen-presenting cell (APC), which are in close apposition, was termed "immune (immunological) synapse" (Norcross 1984). Immunologists and neurobiologists described some similarities and differences in their respective synapses. One of the notable similarities between the immune and the neurological synapse is that both structures are activated in a "quantal" way. There is calcium flux in T-cells in response to a single "quantum" of the MHC–peptide complex. However, the T-cell synapse will not organize into a functional unit until at least 10 MHC–peptide complexes are present (Irvine et al. 2002). Photoreceptors in the eye are also maximally sensitive, with measurable responses to single photons. However, effective signaling in the retina only occurs after several photons together trigger the outer segments of the receptor (Baylor et al. 1979).

Integrins/ECM

Cells attach to the extracellular matrix (ECM) proteins through receptors, the integrins, which bind to specific sequence motifs in the major components of the matrix, such as fibronectin, laminin and collagen. In response to these matrix ligands, which are not diffusible, integrins convey signals (outside-in signaling) that control cell cycle, survival, proliferation and differentiation.

2.2.3.4. *Intracrine*

In the 1980s, it was reported that the peripheral organs were capable of synthesizing sex steroids from precursor steroids of adrenal origin. The term of intracrinology was then coined to describe the intracellular synthesis, action and inactivation of sex steroids from the inactive sex steroid precursor dehydroepiandrosterone sulfate (DHEA-S), Estrone-S (Labrie 1991). This concept of hormone action is distinct from the concept of endocrinology with a gland secreting active hormones into the blood circulation, exerting direct effects in target tissues. It is important to mention that a large proportion of androgens in men (40%), and most estrogens in women (75% before menopause), are synthesized in peripheral target tissues. Notably, DHEA-S is the most abundant product of the adrenal cortex. After cessation of estrogen secretion by the ovaries at menopause, sex steroids continue to be required and are provided exclusively by the mechanisms of intracrinology (Labrie 2010) (Figure 2.9).

Recent studies revealed an intracrine signaling for lipid mediators, such as prostaglandins (PGs), platelet-activating factors (PAF) and lysophosphatidic acid (LPA). Although their mechanism of action primarily depends on plasma membrane G protein-coupled receptors (GPCR) to trigger immediate effects, recent evidence suggests the existence of an intracellular GPCR population in the nucleus that binds these lipid mediators and mediates their long-term responses, including transcriptional regulation of major genes (Zhu et al. 2006).

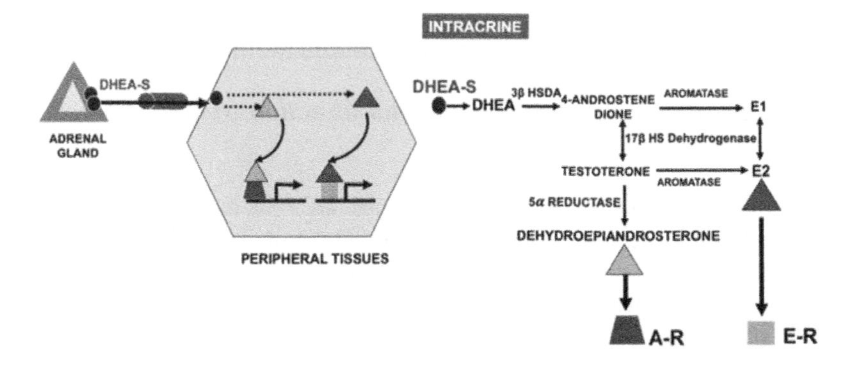

Figure 2.9. *Intracellular synthesis of sex steroids from DHEA of adrenal origin. Contrarily to sex steroids originating from gonads, they directly exert their effects in target tissues, in which they are synthesized. This mechanism gave rise to the concept of intracrinology*

2.2.3.5. *Gap junctions*

An excellent review on the history of the concept of this mode of cell–cell communication is summarized in Evans and Martin (2002). The concept of gap junctions as major mediators of coordination emerged from the outcome of the convergence of apparently unrelated studies. Pioneering work in the 1950s on fast excitatory transmission in the crayfish giant fiber system indicated that these cells communicated directly via electrical pathways. These conclusions were not apparently in keeping with the prevailing view of animal cells as independent entities surrounded by an electrically insulating membrane and communicating only by release of extracellular messengers. Controversies continued, but there was a general acceptance that both mechanisms of cell–cell communication could operate in parallel. In the late 1960s, the phenomenon of metabolic cooperation between contacting co-cultures of non-excitable cells demonstrated that it operates through direct intercellular transfer. In 1986, the biochemical identification of gap junctions was definitively established, and this mode of communication between cells is described in all cells, except in striated muscle, where the cells have fused during development.

The presence of gap junctions enables the coordination and amplification of the tissue response to a particular signal. For instance, gap junctions enable electrical coupling between myocytes in myocardium, as well as augmentation of glucose-stimulated insulin secretion from the β-cells in the pancreas. The importance of gap junctions for the β-cells is easy to demonstrate experimentally. Isolated β-cells, which have disrupted gap junctions, secrete less insulin in response to glucose than the β-cells within the islets of Langerhans, which are coupled via gap junctions (Benninger et al. 2008).

2.3. Integration of extracellular signals: plasma membrane receptors

2.3.1. *Chemical signals and mechanisms of action: receptor types and signaling modulation*

Plasma membrane receptors serve as the first layer of molecules that receive, interpret and transduce chemical signals (hydrophilic molecules) in response to dynamic changes in the extracellular milieu. There are six major receptor families: the ligand- and voltage-gated ion channels (LGICs and VGICs), G protein-coupled receptors (GPCRs), receptors with intrinsic kinase activity (tyrosine kinase, serine/threonine kinase), cytokine receptors, integrins and Toll-like receptors (Figure 2.10). As previously mentioned, the notion of the receptor is one of the most important concepts in the biological sciences introduced by J. Langley in physiology and P. Erlich in immunology, which both importantly impacted the development of pharmacology.

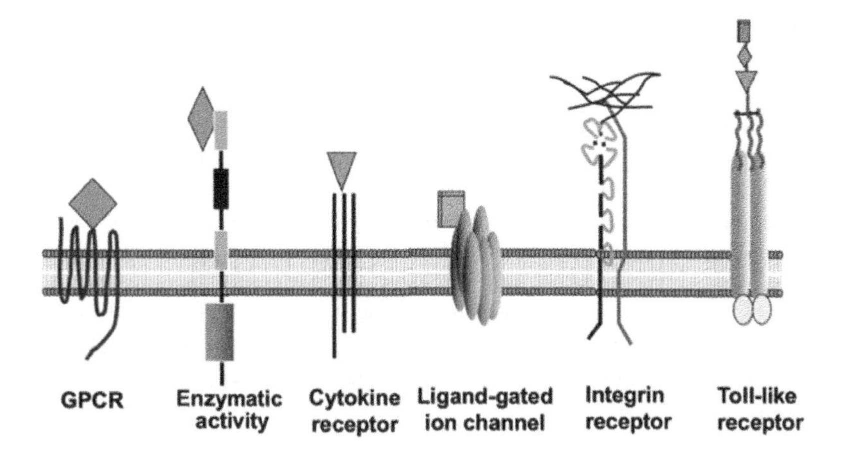

Figure 2.10. *Receptor families: GPCR, receptors intrinsic kinase activity, ligand-gated ion channels, cytokine receptors, integrins and Toll-like receptors*

2.3.1.1. *G protein-coupled receptors*

As late as the early 1970s, the physical existence of receptors remained controversial. A prominent pharmacologist, R.P. Ahlquist, wrote: "To me they are an abstract concept conceived to explain observed responses of tissues produced by chemicals of various structures" (Ahlquist 1973). Ironically, it was the 1970s that saw the unequivocal identification of the first receptor, namely the nicotinic acetylcholine receptor.

In the early 1980s, the sequencing and subsequent cloning of the bovine retinal photoreceptor, rhodopsin, revealed a novel mammalian protein structure with similarity to bacteriorhodopsin, a light-sensitive proton pump found in halophilic bacteria. With the cloning of the β2-adrenoreceptor in 1986, and other receptor sequences that soon followed, it became clear that this basic architecture, consisting of an extracellular N-terminus, seven membrane-spanning α-helices connected by intracellular and extracellular loops, and an intracellular C-terminus, was representative of a large family of membrane receptors (see Luttrell (2006).

G protein-coupled receptors (GPCRs) make up the largest and most diverse family of membrane receptors in the human genome, which comprises over 800 proteins in the human genome, relaying information about the presence of diverse extracellular stimuli to the cell interior.

Most GPCR-mediated cellular responses result from the receptor acting as a ligand-activated guanine nucleotide exchange factor for heterotrimeric guanine nucleotide-binding (G) proteins, whose dissociated subunits activate effector enzymes or ion channels. GPCR signaling is subject to extensive negative regulation through receptor desensitization, sequestration and down regulation; termination of G protein activation by GTPase-activation proteins and enzymatic degradation of second messengers. Additional protein–protein interactions positively modulate GPCR signaling by influencing ligand-binding affinity and specificity, coupling between receptors, G proteins and effectors, or targeting specific subcellular locations. These include the formation of GPCR homo- and heterodimers, the interaction of GPCRs with receptor activity-modifying proteins, and the binding of various scaffolding proteins to intracellular receptor domains. In some cases, these processes appear to generate signals.

It should be mentioned that ligand-biased signaling is a relatively new concept based on the idea that a receptor can exist in multiple active conformations, each stabilized by a specific ligand, with characteristic binding kinetics, and therefore with a particular signaling profile (Klein Herenbrink et al. 2016). For GPCRs, this would translate into different coupling behaviors towards G proteins and β-arrestins.

2.3.1.2. *Tyrosine kinase receptors*

The protein kinase family of enzymes can be classified according to the target amino acids of their protein substrates, i.e. tyrosine and serine/threonine. Both types of kinases share a high sequence similarity in their catalytic domains of 250–300 amino acid residues, which suggests that they descended from a common ancestral gene. Many protein tyrosine kinases are plasma membrane proteins or are associated with the inner surface of the cell membrane. In humans, there are 58 RTKs that are further subdivided into 20 families and function as receptors for a great diversity of ligands. Among these, growth factors are especially prominent and the discovery of RTKs is closely related to growth factor research.

The beginning of growth factor research can be traced back to 1957 when R. Levi-Montalcini discovered the nerve growth factor (NGF) (Cohen and Levi-Montalcini 1957). Five years later, S. Cohen isolated and characterized another protein that induced precocious eyelid opening when injected into newborn mice. It was called the epidermal growth factor (EGF) as it stimulated the proliferation of epithelium (Cohen 1962). They received the Nobel Prize in Physiology or Medicine in 1986.

In the 1970s, it was proposed that these two growth factors bind specifically to cell surface receptors. The first quantitative characterization of ligand binding sites was elucidated with the insulin receptor in 1973 (de Meyts et al. 1973). Then, Carpenter et al. identified the presence of specific binding sites (receptors) for EGF on the surface of target cells in 1975 (Carpenter et al. 1975), and in 1978, the EGFR was identified (Carpenter et al. 1978). Subsequently, a variety of new polypeptide growth factors that stimulate cell proliferation by binding to receptors at the cell surface were discovered. Those include PDGF, FGF and HGF.

It was then postulated that ligand/receptor complexes may play a passive role in delivering them to intracellular compartments in which they elicit their actions. It was thought that the cell surface receptor has a role of "carrier" as it delivers the ligand to different intracellular targets. It was also suggested that ligands activate their receptors leading to the production of an intracellular messenger analogous to cAMP after β adrenergic receptor stimulation. A series of experiments carried out on the insulin receptor ruled out the possibility that it functions as a passive carrier. Thus, antibodies from certain diabetic patients mimic cellular responses of insulin (Flier et al. 1977), which provides strong evidence that the receptor plays a crucial role in mediating insulin cellular responses.

Insights into the molecular mechanisms of action of growth factors and their receptors were obtained in the late 1970s and early 1980s, and they result from a convergence of two different fields. The elucidation of mechanisms involved in the transformation of cells by retroviral oncogenes (see section 2.3.2.7) was at the forefront of biomedical research, and it was discovered that a common enzymatic activity is implicated in controlling cell growth (normal and transformed cells). In transformed cells, T. Hunter and B.M. Sefton showed that the transforming protein of the Rous sarcoma tumor virus, v-Src, is a protein tyrosine kinase that displays an excessive kinase activity, strongly suggesting that a chronic tyrosine phosphorylation of cellular proteins would cause uncontrolled cell proliferation, thereby leading to the tumorigenesis process (Hunter and Sefton 1980a).

Because tyrosine phosphorylation was also observed with growth factor-stimulated surface receptors, such as EGFR, PDGFR and insulin receptors, this strongly suggested that it plays an important role in normally regulated cell proliferation, as well as in the aberrantly stimulated proliferation of cancer cells, and that these proteins all activate a common enzymatic activity (i.e. protein tyrosine kinases).

In the early 1980s, the molecular identity of EGFR was elucidated after the cloning of human EGFR cDNA. It revealed that the receptor contains a large extracellular ligand binding domain and a cytoplasmic region that contains a ligand-activated tyrosine kinase domain (Ullrich et al. 1984), which mediates autophosphorylation on multiple tyrosine residues. In the 1990s, a series of studies addressed the questions of how RTK activates multiple signaling pathways to relay information from the cell membrane to different intracellular compartments. The first clues came from the finding that PLCγ uses its Src homology 2 (SH2) domain for binding to the phosphorylated tail of activated EGFR (Margolis et al. 1990). Since then, two highly conserved and critical signaling pathways have been described, the phosphoinositide-3-kinase (PI3K/Akt) and the extracellular signal-regulated kinase (Ras/ERK), and they are activated in response to growth factors, hormones and nutrient signals (see the reviews by Hemmings and Restuccia (2012), and Morrison (2012)).

2.3.1.3. *Serine/threonine kinase receptors*

Serine/threonine kinase activity was first detected in cytoplasmic proteins. In 1987, S. Staal cloned the v-Akt oncogene from the AKT8 transforming retrovirus in spontaneous thymoma of the AKR mouse, and its product was called c-Akt. Then, in 1991, the sequences encoding the protein were named RAC (related to A and C kinases) or PKB (protein kinase B), and investigators now refer to it as Akt/PKB (Kandel and Hay 1999; Manning and Toker 2017).

Besides the cytoplasmic proteins, a myriad of genes encoding transmembrane serine/threonine kinases have been cloned. Almost all act as receptors for the transforming growth factor β (TGFβ) superfamily, which also includes structurally related proteins such as bone morphogenetic proteins (BMPs), activin/inhibin families and the glial cell line-derived neurotrophic factor (GDNF).

Transforming growth factors (TGFs) were first described in 1978 (de Larco and Todaro 1978), and it was shown that they shared properties with the v-Src gene, in that they were able to transform normal fibroblasts (Todaro et al. 1979). In contrast to v-Src, transformation did not result from cell-intrinsic, genetic changes, but from secreted factors that did not affect the genotype. The term TGF was chosen for this activity because of the induction of a transformed phenotype.

All members of the TGFβ family signal through a receptor complex formed by two distantly related types of serine/threonine kinase proteins. They are named type II and type I based on their molecular weights, and two such receptors (TβR-II and TβR-I) have now been identified. TβR-II is a constitutively active kinase, and the autophosphorylation is not clearly upregulated upon ligand binding. After binding of TGFβ to TβR-II, TβR-I is recruited into a stable heteromeric complex and is phosphorylated, by TβR-II, on its serine and threonine residues. Thus, TβR-I functions as a substrate for TβR-II; its phosphorylation and activation by TβR-II is essential and sufficient for most TGFβ-mediated signaling.

2.3.1.4. *Cytokine receptors*

Immunologists have contributed to the identification of cytokines and their receptors. Originally, the term “cytokines” meant “factors produced by cells”, and now it encompasses a wide range of molecules including interleukins (IL), colony-stimulating factors (CSF), interferons (IFN), tumor necrosis factors (TNF) and classical hormones like prolactin (PRL) and growth hormone (GH). Most cytokines are pleiotropic molecules acting in an autocrine, paracrine or endocrine manner. They also exhibit both negative and positive regulatory effects on various target cells. Cytokines are classified according to their structure and the structure of their receptors (Onishi et al. 1998). Type I cytokines are those that share a four α-helical bundle structure, and they bind to type I cytokine receptors which share similar structural characteristics. Type II cytokines are the interferons (IFNs) and related cytokines, and they bind to type II cytokine receptors which share structural similarities. Most cytokine receptors consist of multiple subunits, a ligand-binding subunit and common signal transducers.

Japanese scientists made a significant contribution to the identification of cytokine receptors and their signaling. T. Kishimoto and T. Taga discovered the IL-6R (also known as IL-6Rα) and its signal-transducing receptor component gp130. Subsequently, Takatsu discovered the IL-5Rα. The functional redundancy of various cytokines can be explained by the presence of a common signal transducer. For instance, the IL-6R family and IL-2R family share the common signal transducer gp130 and common γ (γc), respectively. Type I cytokine receptors are homodimers, heterodimers or oligomers. For instance, IL-6, IL-6R and gp130 bind to form a hexameric complex that activates the JAK/STAT3 pathway (Boulanger et al. 2003). For further information, see the review by Fujii (2007).

Receptors can have disparate functions. Two receptors for IL-1 have been identified that belong to the immunoglobulin superfamily. Both IL-1α and IL-1β bind to each of these receptors with equal affinity. The type 1 receptor, however, is solely responsible for signal transduction, whereas the non-signaling type 2 receptor sequestrates both IL-1α and IL-1β on the cell surface or, when shed in soluble form, can selectively bind IL-1α and serve as an inhibitor of IL-1β activities.

One of the most interesting observations regarding cytokine receptors has been the discovery that many of these receptors can exist either in membrane-bound or soluble forms. Such soluble versions of cytokine receptors that are capable of binding specific cytokines have led to the speculation that such molecules may represent a physiological means by which cells limit the bioavailability of cytokines to their targets. Such a hypothesis has recently been strengthened by studies in which recombinant versions of soluble cytokine receptors have been shown to inhibit immune reactivity. This has led many investigators to conclude that inhibition of cytokine function may well be of significant therapeutic use in the treatment of several disorders. For instance, during infection and inflammation, the IL-6 level is markedly increased, and it is associated with an elevation of a soluble form of IL-6R (sIL-6R), which is released by cleavage of the membrane-bound IL-6R. They form a complex IL-6/sIL-6R, but unlike with some other soluble cytokine receptors, this does not lead to a block of IL-6 signaling. Indeed, binding to sIL-6R enables IL-6 to act on the cells that express gp130, even if they lack the membrane-bound IL-6Rα. This mode of signaling, which is referred to as trans-signaling, expands the population of cells that are responsive to IL-6, which may be important for some of its pathological effects in chronic inflammation. Interestingly, once bound to the membrane IL-6Rα of a target cell, the IL-6/IL-6Rα complex may be presented to a neighboring cell that expresses gp130, again leading to activation of intracellular signaling in a cell that lacks IL-6Rα and is not responsive to IL-6 on its own. This type of signaling is called trans-presentation and may be considered as a special form of juxtracrine signaling because two cells that express the ligand-binding receptor subunit (IL-6Rα) and the signal transducing subunit (gp130), respectively, are physically linked once the ligand (IL-6) is present.

The example of IL-6 and its various membrane-bound and secreted receptors highlights the complexity of intercellular communication, which likely underpins its diverse physiological and pathophysiological actions. Importantly, the diversity of signaling modes by which IL-6 exerts its biological effects opens up the possibility for selectively targeting the pathways that are impaired in a particular condition rather than suppressing entire IL-6 signaling, which may lead to deficiency of its physiological actions. For instance, trans-signaling appears to play a major pathophysiological role in various inflammatory conditions. It would therefore be an appealing strategy to target this pathway selectively, while avoiding excessive inhibition of the canonical pathway via the membrane-bound IL-6Rα and gp130, which are thought to primarily transmit the physiological actions of IL-6 (Uciechowski and Dempke 2020).

2.3.1.5. *Ligand-gated ion channels*

The cell membrane quickly and accurately changes its permeability to various ions. This permeability is mediated by selective ion channels forming transmembrane pores. The channels are opened and closed by different stimuli such as neurotransmitters, membrane stretch, temperature and transmembrane voltage. Depending on the design of the pore, they are selective for specific ions. In this section, only ligand-gated channels are discussed, with a special emphasis on the neurotransmitter receptor.

Historically, the theoretical notion of the neurotransmitter receptor was suggested by J. Langley in 1905–1908, and it took almost 70 years before the first neurotransmitter receptor called the "nicotinic acetylcholine receptor" was isolated and purified as a protein (Cartaud et al. 1973). The discovery brought a new insight in the chemistry of brain communications. Throughout this period of time, there were unsuccessful attempts to isolate and identify this receptor, and the techniques or models applied and the reasons for the failures are described in Changeux (2020).

Meanwhile, the elementary molecular mechanisms of biological regulation were investigated, and it was assumed that the interaction between acetylcholine and the ion channel is indirect or allosteric (see mechanisms in section 3.1.3.1). This discovery was followed by the identification of many members of the pentameric receptor family in the brain. Without a doubt, the notion of allosteric modulation has created a major landmark in the strategies of drug design for ligand-gated ion channels, as well as GPCRs, resulting in the successful development of new classes of drugs used in clinics. Allosteric modulators include some of the most prescribed psychopharmaceutical drugs in the world, such as benzodiazepines, barbiturates, local and general anesthetics, and several new families of drugs. This understanding has major practical consequences in the conception of new pharmacological agents at allosteric modulatory sites in each of their conformational states. It is worth noting that the concept of allosteric modulators as pharmacological agents binding to allosteric sites differs from the classical concept of competitive drug binding, where the drug competes for the same binding site as the endogenous ligand.

It should be emphasized that the structural identification of the nAChR as the first membrane receptor for a neurotransmitter involved several laboratories around the world, and its first biochemical identification, from fish electric organ, was made in Europe. Subsequently, several receptors were isolated and identified for different neurotransmitters, again in Europe.

These ion channels were described in greater detail and it was revealed that they are composed of two functional moieties: (1) a selectivity filter (S), which determines which types of ions may pass the membrane, and (2) a gate (G), which specifies under which conditions the channel is opened.

Ion channels are subdivided into two major classes according to their gating trigger (L): voltage-gated channels (VGCs) and ligand-gated channels (LGCs). Some characteristics of the LGCs are reported.

LGCs can be classified into three superfamilies: (1) the superfamily of receptors that resemble the nicotinic acetylcholine receptors: glycine receptors (GlyR), GABAA receptors (GABAAR), nicotinic acetylcholine receptors (nAChR) and some serotonin receptors (5-HT3R), (2) the superfamily of ionotropic glutamate receptors (GluR) and (3) the ATP-gated purinoceptors (P2X).

The functional diversity of the LGCs comes from the multiplicity of receptors. Each receptor consists of an association of different subunits, leading to a huge number of combinations. Thus, 11 and 16 different subunits were found for nAChRs and GluRs, respectively. It therefore provides a tremendous functional, spatial and temporal diversity.

LGCs are not only directly regulated by their transmitter, but also by the binding of diverse additional ligands by allosteric mechanisms (see section 3.1.3.1) and post-translational modifications (e.g. phosphorylation), as well as during their biosynthesis, which depends on presynaptic and electric activity.

As new ligand-gated ion channels were discovered, investigators reported mutations (gain or loss of function) of these channels, leading to the stabilization of allosteric states. This gave rise to the concept of receptor diseases, which also applies to different categories of receptors. In 2019, the Allosteric database listed 5,983 allostery-related diseases.

A second group of ion channels, called "voltage-gated ion channels", has been mentioned above. They also regulate the transfer of ions across the cell membrane. The ions travel via a pore formed by the different subunits that constitute the channels, and the pore can be gated by changes in transmembrane voltage. They have a key role in electrical signaling by excitable cells, such as neurons. Despite advances in understanding the exact structure of these channels, the mechanisms by which the channels sense transmembrane voltage remain elusive (Börjesson and Elinder 2008).

2.3.1.6. *Integrins*

The concept that there must exist transmembrane connections between extracellular matrix proteins and the actin-based cytoskeleton developed during the 1970s. By 1976, significant circumstantial evidence had accumulated that suggested some relationship between extracellular fibronectin-containing fibrils and intracellular actin filaments, and it was suggested that an integral membrane protein or proteins may link the two (Hynes 1976). The postulated "fibronectin receptor" was cloned and sequenced in 1986 (Tamkun et al. 1986). It was named "integrin" for the transmembrane protein linking the extracellular matrix to the cytoskeleton. The integrin family emerged, but there was no consensus on whether they should be called "receptors" like the G-protein-coupled receptor or tyrosine kinase receptors. It became progressively clear that they transmit signals (Hynes 1992).

Concept of bidirectional signaling

The bidirectional nature of integrin signaling, first envisioned by R. Hynes (Hynes 1992), is now a firmly established fact. Integrins integrate the extracellular and intracellular environments. Their extracellular domains bind to ligands on the surface of other cells or in the extracellular matrix (ECM), while their cytoplasmic domains bind to cytoskeletal-associated proteins. Integrins are composed of type I transmembrane α and β subunits, each with a large extracellular domain, a single-pass transmembrane (TM) domain and a small cytoplasmic domain.

Integrin outside-in signaling

Integrin ligation triggers outside-in signals that regulate, among other responses, adhesion, cell motility, cell survival, morphology, proliferation and gene expression (Abram and Lowell 2009). Matrix binding to integrin initiates conformational changes that causes integrin microclustering within dynamic adhesion structures, including focal complexes and focal adhesions. Then, the earliest biochemical event in the outside-in signaling response is the activation of tyrosine kinases, particularly those of the Src and Syk families (Figure 2.11(A)). A number of adapter proteins that serve as scaffolds for protein–protein interactions have been implicated in this signaling (e.g. Cas/Crk, paxillin). Following these interactions, several biochemical events are induced that culminate in modulation of the actin cytoskeletal network.

A new concept emerged from the observation that intrinsic domains within ECM proteins may act as ligands for growth factor receptors. Thus, laminin (ECM protein) contains multiple EGF-like domains, as do many ECM proteins, which may bind to EGF receptors and signal (Schenk et al. 2003) (Figure 2.11(B)). It is worth noting that in the absence of an EGF ligand, a direct interaction of integrins with an EGF receptor results in stimulation of the signaling pathway (Figure 2.11(B)).

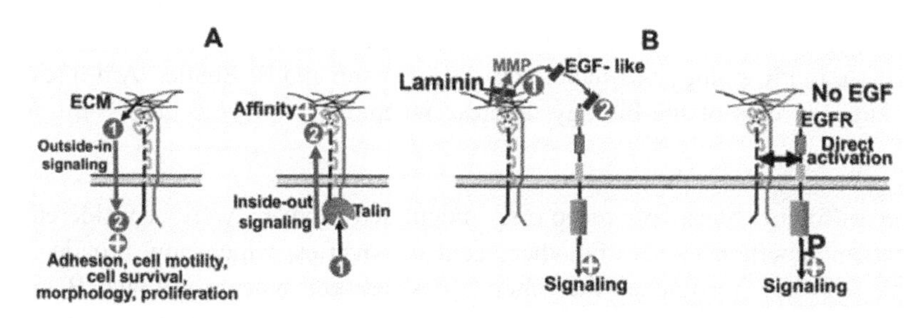

Figure 2.11. *Integrin receptors and bidirectional signaling (A). Extracellular matrix domain acts as a ligand for growth factor receptor and direct interaction of integrins with EGFR results in activation of the receptor in the absence of ligand (B)*

Integrin inside-out signaling

To bind ligands with high affinity, integrins undergo conformational changes regulated by inside-out signals and become "activated". Talin, which binds to most integrin β cytoplasmic domains, represents an indispensable mediator of this process. Knocking down its expression with small interfering RNAs suppresses β1 and β3 integrin activation. The molecular mechanisms initiating the activation are poorly understood, and there are many unanswered questions. However, structural studies have revealed how binding of talin with the integrin can transmit long-range allosteric changes in transmembrane proteins that can increase the affinity of integrins for their ligands (Ginsberg 2014, see Figure 2.11(A)).

2.3.1.7. *Toll-like receptor*

Over the past few decades, key cellular sensors responsible for triggering innate immune signaling pathways and host defense have started to be resolved. The existence of pattern-recognition receptors (PPRs) was suggested in 1989 by C. Janeway, who proposed a general concept of the ability of PPRs to recognize and bind conserved molecular structures of microorganisms known as pathogen-associated molecular patterns (PAMPs). Upon PAMP engagement, PPRs trigger intracellular signaling cascades, resulting in the expression of various pro-inflammatory molecules. These recognition molecules represent an important and efficient innate immunity tool for all organisms and include the Toll-like receptor (TLR). Toll-like receptors (TLRs) are prototype of PPRs which can detect molecules derived from viruses, fungi, bacteria and protozoa. Similar mechanisms are found to be induced by host motifs, called danger-associated molecular patterns (DAMP), which arise from insult to the host.

The discovery of Toll-like receptors (TLRs) was an important event for immunology research and was recognized as such with the awarding of the 2011 Nobel Prize in Physiology or Medicine to J. Hoffmann and B. Beutler. A full review of the key concepts of the biology of these receptors is covered in O'Neill et al. (2013).

The different types of receptors, mentioned above, were considered as independent functional units until the recent discoveries, which now broaden this concept. It is now recognized that interactions between receptors A and B in the presence of the ligand of the former can elicit a cellular response of the receptor B in the absence of its ligand. Similarly, associations of receptors with accessory proteins or co-receptors create a diversity of molecular architecture which modify their function. These different complexes enable a diversity of modulation of signaling and are classified into two categories.

2.3.1.8. *Heterocomplex and ternary complexes*

Heterocomplex (GPCR/RTK)

Since the initial discoveries of the signaling cascades induced by the two major classes of plasma membrane receptors, GPCRs and RTKs, multiple new components entered this network, thus increasing its complexity. Over the last two decades, we have moved well beyond the concept, which stated that individual receptors mediate cellular effects only through their "own" ligands. It is now well recognized that, in addition to canonical second messenger-regulated mechanisms, GPCRs exert a growth-promoting activity involving RTKs and their downstream signaling cascades. In other words, GPCRs activation can enhance RTKs signaling activity, thus coupling the wide diversity of GPCRs with the powerful signaling capacities of RTKs. It was first described by Ulrich's group (Daub et al. 1996), and these observations led to the emergence of the "transactivation" concept, which refers to the activation of RTKs by GPCR ligands and represents a pathway that links GPCRs to the classical extracellular signal-regulated kinase (ERK) signaling. It also led to the "heterocomplex" concept, including the interactions between two GPCRs, and between GPCR and other membrane receptors, such as ion channel receptors (NMDA, GABA receptors) or receptor tyrosine kinases (EGF, FGF receptors). These heterocomplexes trigger intracellular signaling and cellular response different from that induced by GPCR or other receptors (RTKs and ion channels) alone (Delcourt et al. 2007). It should be emphasized that transactivation of RTKs by GPCRs may occur via two distinct mechanisms: 1) activation of GPCRs and its signaling pathway that leads to an activation of RTK and 2) direct interaction and subsequent activation of RTKs by GPCRs. See the excellent review by Latko et al. (2019).

The first mechanism is illustrated with the activation of the EGFR by the thrombin receptor through a signaling activation. Binding of thrombin to its receptor causes an activation of its signaling pathway, resulting in increased activity of ADAM, which in turn releases the RTK ligand from its precursor, which subsequently activates EGFR (Figure 2.12(A)).

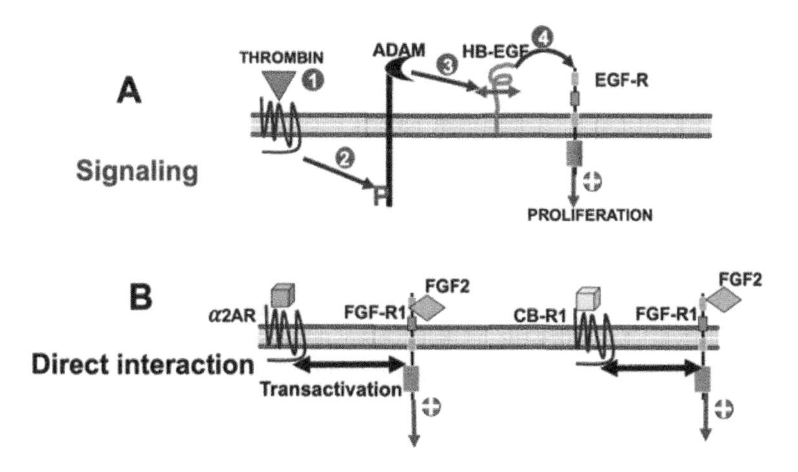

Figure 2.12. *Mechanisms of transactivation of RTKs by GPCRs. Heterocomplex of receptors GPCR/RTK and activation of RTK through two mechanisms: (1) activation of the GPCR signaling pathway, resulting in increased ADAM activity, which in turn releases the RTK ligand which subsequently activates the receptor (A), and (2) direct interaction of receptors leading to a transactivation of RTK by GPCR (B)*

The second mechanism is illustrated with the FGFRs, which are involved in the development, function and maintenance of the CNS. FGFR1 has a key role in these processes as it is a binding partner of diverse GCRs. The modulation of signaling by the FGFR1-adrenergic α_{2A} receptor (A2AR) heterocomplex was found to be important for regulation of the synaptic plasticity (Flajolet et al. 2008). The formation of cannabinoid receptor 1 (CB1R1)-FGFR1 complexes leads to activation of ERK1/2 and is important for neuronal differentiation (Figure 2.12(B)). The interaction of FGFR1 with the muscarinic acetylcholine receptor (mAChR) enhances neurite growth (Di Liberto et al. 2017).

Ternary complex (holoreceptor): co-receptors and docking receptors

– Co-receptors

The traditional activation of a signaling pathway requires the binding of a ligand to its cognate receptor. Some receptors need another binding partner which associates with its own ligand to form a complex, which then activates the signaling pathway. This transmembrane protein is named: accessory protein = co-receptor = accessory receptor; or docking receptor. Here are some examples of co-receptor/ligand/receptor signaling complexes. This ternary complex is required to activate the signaling pathway:

* *FGF23 - Klotho - FGFR1,3,4*

As previously mentioned, FGF23, the endocrine factor, released by osteoblasts and osteocytes of the bone, inhibits phosphate reabsorption in the kidney. To activate the signaling pathway, it should bind to a complex of FGFR and the co-receptor klotho. In addition, FGF23 suppresses parathormone secretion in a klotho-dependent manner (Urakawa et al. 2006; Han et al. 2018).

* *GDNF - GFRα - RET*

Physiologically, the tyrosine kinase receptor RET (REarranged during Transfection) is activated by glial cell line-derived neurotropic factor (GDNF) ligands that bind to co-receptor GDNF family receptor alphas (GFRαs), which in turn lead to a dimerization of RET and GFRα (Kawai and Takahashi 2020). RET has a key role for the development of both mammalian enteric nervous system and lymphoid organogenesis in the gut (Veiga-Fernandes et al. 2007). Given that RET has multiple ligands, and co-receptors, it may explain the differential response within the two systems. This raised the interesting possibility that multicellular, multiorgan life forms may use similar molecular tools to orchestrate the development and organization of diverse tissues within the same organ:

* *NRP-SEM/VEGF-VEGFR2/Plexins (Guo and Vander Kooi 2015)*

Neuropilins (NRP) bind to two classes of structurally unrelated classes of ligands, semaphorins (Sem) of class 3 and members of vascular endothelial growth factor (VEGF), with different biological functions. They have a role of co-receptors for these ligands, and they complex with the VEGFR2 or plexins, which convey the intracellular signaling. NRPs can also interact other types of growth factors, including PDGF, HGF, VEGF, TGFβ1 and complex with their corresponding signaling receptors, but their function has not been fully characterized (West et al. 2005; Banerjee et al. 2006; Glinka and Prud'homme 2008). NRPs also couple with multiple integrins to control cellular function and adhesion, and this coupling has been reported in multiple pathological contexts.

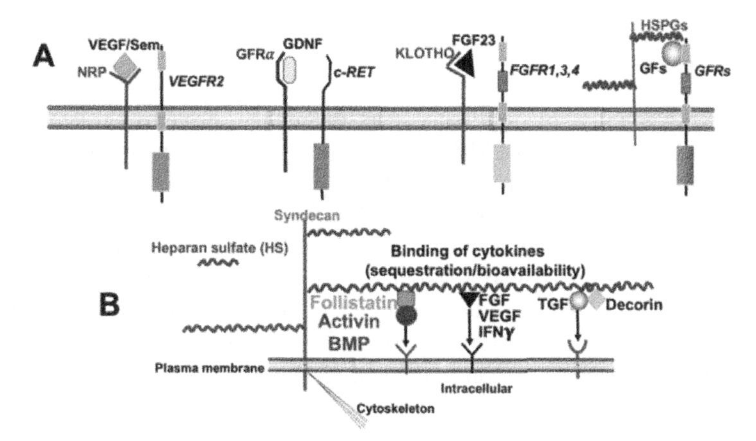

Figure 2.13. *Activation of receptors by (A) ligands, in association with co-receptors: ternary complex = co-receptor/ligand/receptor complex and (B) heterocomplex and regulation of ligand bioavailability and binding to its receptor*

– Docking receptors

They regulate the bioavailability of soluble factors and binding to their receptors:

* *GFs - HSPGs - GFRs*

Proteoglycans consist of proteins, namely syndecans, on which heparan sulfate (HS) chains are covalently attached, forming heparan sulfate proteoglycans (HSPGs). They act as plasma membrane receptors, but they do not activate a signaling pathway. However, due to their spatial structure, HS chains have two key roles: 1) as a cell adhesion receptor, because they bind to ECM, which is a unique characteristic, compared to cytokine and growth factor receptors (Kwon et al. 2012), and 2) as a docking receptor, because of the interactions of the extracellular domain with various soluble and insoluble factors in the extracellular matrix.

With respect to the second role, HS chains are responsible for recruiting soluble growth factors such as FGF, TGFβ, VEGF and IFNγ, but as previously mentioned, these interactions do not transduce intracellular signals. It allows each growth factor to bind to its cognate receptor with higher affinity than the growth factor itself, which then transduces the signal. Therefore, as a docking receptor, syndecans and HS regulate the bioavailability of soluble factors, and this binding of cytokines and growth factors to HS chains represents a reservoir of ligands, which subsequently gave rise to the concept of bioavailability/sequestration of ligands. One of the best examples is the TGFβ, whose bioavailability is under the control of the decorin. The latter binds the growth factor, which prevents its binding to its receptor (see section 3.1.5 on sequestration).

2.3.1.9. *Allosteric proteins and modulation of receptor–ligand binding*

* *GPCRs – allosteric proteins*

In 1998, a family of accessory transmembrane proteins, called receptor activity-modifying proteins (RAMPs), was identified, which can alter ligand binding of the class B GPCR (McLatchie et al. 1998). For instance, RAMP2 associated with calcitonin-like receptor (CLR) conferred high-affinity adrenomedullin (AM) binding over the hormone calcitonin gene-related peptide (CGRP). Subsequently, it was reported that RAMP1 and RAMP3 associated with calcitonin receptor (CTR) confer high-affinity amylin (AMY) binding over calcitonin (reviewed in Hay and Pioszak (2016) and Pioszak and Hay (2020)). It is worth noting that calcitonin inhibits osteoclast activity, whereas amylin is a glucoregulatory hormone.

Recent advances in our understanding of the biochemistry and pharmacology of RAMP–CLR/CTR complexes have revealed that RAMPs act as endogenous allosteric modulators of these class B GPCRs. These RAMP interactions provide new opportunities for drug development (Figure 2.14). The allostery process is discussed in section 3.1.3.1.

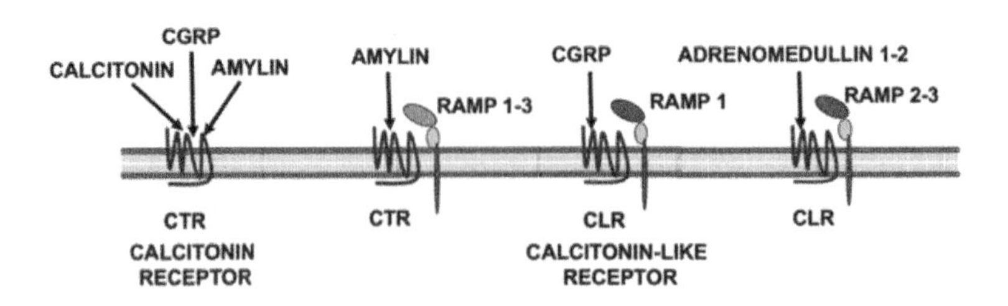

Figure 2.14. *Modulation of the binding site of GPCR by allosteric proteins (RAMPs)*

Negative allosteric modulators (NAMs) and positive allosteric modulators (PAMs) also bind to the GPCR. Their allosteric binding sites are topographically distinct from the endogenous ligand binding site, and they enhance the affinity and/or efficacy of the agonist. They are involved in the modulation of the metabotropic glutamate receptor subtype 5 (Gregory et al. 2013).

2.3.1.10. *Counter-receptors*

* *Integrins and non-matrix ligands*

The adhesion of cells is carried out through integrins that bind a wide variety of extracellular matrix (ECM) proteins through the recognition of small peptide sequences as simple as the RGD or LDV tripeptides. Once bound to its ligand, it initiates signaling pathways. Beyond their classical role as mediators of ECM attachment, integrins are also involved in cell–cell adhesion and they bind non-ECM molecules (reviewed in LaFoya et al. (2018)). These non-matrix ligands are termed counter-receptors, and the best example is the immunoglobulin superfamily cell adhesion molecules (IgCAMs). These interactions between integrins and these counter-receptors mediate leukocyte migration from the blood stream, hematopoietic stem cell homing and mobilization and metastasis of tumor cells. Binding properties of integrins continue to be uncovered, and it is no surprise that the diversity of non-ECM integrin ligands is expanding. For instance, integrins serve as cell surface receptors for growth factors, hormones and small molecules.

For instance, angiopoietin-related proteins (ARPs) display structural similarities to the growth factor angiopoietin, but do not bind to the classical angiopoietin receptor. Instead, they bind integrins through a fibronectin-like domain containing an RGD sequence (Zhang et al. 2006), which therefore mimics a classical ECM protein and implements their cellular effects.

Similarly, integrins are involved in the activation of TGFβ (reviewed in Worthington et al. (2011)), which is secreted with an RGD containing latency-associated peptide (LAP) bound to TGFβ. When LAP is bound to αV/β6 integrins, it enables a separation of LAP from TGFβ, which then activates.

Several lines of evidence indicate that integrin αV/β3 bears a receptor site for the thyroid hormones T3 and T4, and the thyroid hormone receptor site is at or near the RGD binding pocket. A corollary to the existence of these binding sites is that thyroid hormones may trigger part of their effects via the membrane in addition to the canonical pathway via their nuclear receptors.

These interactions between integrins and non-ECM ligands highlight the multifaceted biological functions of integrins and are exploited for a number of applications in the biotechnology field. Various artificial ECM are created with incorporated RGD peptides to enable cell seeding and growth (Anderson et al. 2004).

It is worth noting that the diversity of cellular responses can also result from the dimerization and oligomerization of receptors, such as IGF1-R and BDNF-R/TrkB, which is constitutive and induced, respectively.

2.3.1.11. *Pleiotropy, redundancy*

A pleiotropic effect is observed when a single molecular signal targets different cell types and elicits a wide spectrum of biological responses. Pleiotropy is a common characteristic of cytokines; for instance: IL-6 exerts its effects on multiple cells and modulates their immune, metabolic and other functions, which may result – depending on the circumstances – in divergent biological actions, including pro-inflammatory or anti-inflammatory effects, as well as insulin resistance or enhanced insulin action.

Redundancy occurs when different molecular signals elicit a similar function. This is quite common with hormones and cytokines (O'Shea et al. 2008). For instance, cytokines that belong to the same subfamily can share biological activities, which is explained by the utilization of a common receptor subunit termed the beta-subunit and a unique ligand-specific alpha-subunit. As previously mentioned (see section on cytokine receptors), the IL-6 receptor complex comprises a specific ligand-binding subunit IL-6Rα and a signal-transducing subunit gp130, which is shared by several cytokines. The concept of redundancy also applies to the functions of protein isoforms, like the glycogen synthase kinase-3α and β (GSK-3). The interpretation would be that the extra copy reflects a fail-safe in case of loss of one. However, it should be emphasized that redundancy does not necessarily depend on sharing receptor subunits or on the existence of functionally related protein isoforms. For instance, multiple counter-regulatory hormones, which act as functional antagonists of insulin, including adrenaline (epinephrine), glucagon, cortisol and growth hormone, can all increase plasma glucose concentrations, although they do not share the same receptors. Due to this functional redundancy, deficiency of only one counter-regulatory hormone does not markedly increase the risk of hypoglycemia.

Having multiple hormones that can oppose insulin and increase plasma glucose concentrations may seem wasteful from the evolutionary point of view. However, importantly, since the brain is dependent on glucose as a fuel, preventing hypoglycemia is essential for the maintenance of homeostasis and the integrity of the organism. Redundancy of counter-regulatory hormones can therefore be considered as an advantageous safety factor.

In addition to their redundancy function and pleiotropic effect, cytokines are often produced in a cascade, as one cytokine stimulates its target cells to make additional cytokines. Cytokines can also act synergistically or antagonistically. For instance, during infection, bacterial products stimulate the secretion of IL-1 and TNF-α, which increase the secretion of IL-6, which in turn induces the production of acute phase proteins in the liver, thus enhancing the systemic inflammatory response, as well as the secretion of anti-inflammatory cytokines, which help to keep inflammation in check. Importantly, IL-1, TNF-α and IL-6 act in concert to activate the HPA axis, the main result being increased secretion of glucocorticoids, such as cortisol, thereby limiting the extent of systemic inflammatory response. Clearly, a finely balanced interplay of cytokines with overlapping as well as opposing functions is required to maintain homeostasis.

Figure 2.15 gives an overview of the major issues discussed earlier in section 2.3.

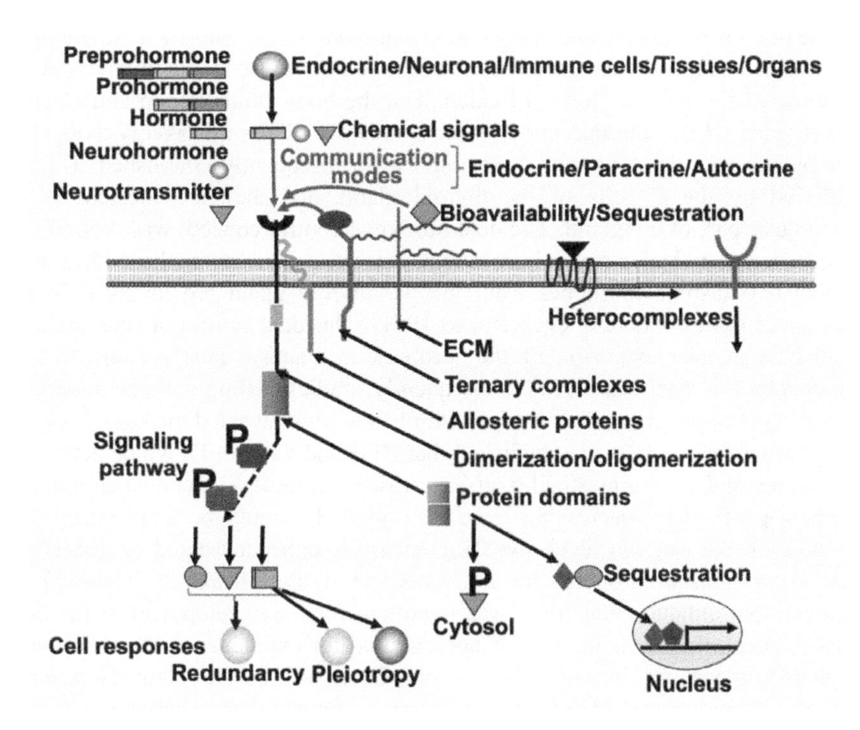

Figure 2.15. *Schematic illustration of the mechanisms enabling the diversity of cellular responses. The diversity results from (1): the type of chemical signals, their communication modes and their bioavailability; (2) the type of receptors and their arrangements as monomers, dimers, oligomers heterocomplexes and ternary complexes and (3) modulators of receptor activity (allosteric proteins). Depending on the combination of these upstream molecular events, they elicit an activation of specific signaling pathways causing (1) a cascade of protein phosphorylation and a series of protein protein interactions in the cytosol, and (2) a gene transcription*

2.3.1.12. *Dual hormonal control of biological processes*

A dual hormonal control of a biological process has been progressively established after a series of studies beginning in the 1900s and addressing the mechanisms of regulation of plasma calcium levels. It was first demonstrated that the parathyroid glands are involved in the regulation, as the ablation of the glands leads to a marked fall in blood calcium, and an extract of parathyroids restored the calcium level in dogs whose parathyroids had been removed. After purification of the extract, the active principle was named parathormone (PTH) (Copp 1963). It was thought that the release of the PTH is controlled by the level of calcium in the blood, with hypocalcemia stimulating its production, while hypercalcemia suppresses it and thus causes a fall in plasma calcium to that which follows parathyroidectomy. This interpretation did not prevail, and it was suggested that the prompt return of calcium level in response to induced hypercalcemia could not be explained based on a decrease of PTH release; it was likely due to a humoral agent. Shortly thereafter, this agent was isolated, and was named "calcitonin" (CT) because it was involved in the regulation, of the normal "tone" of calcium in the body fluids (Copp and Cheney 1962) (Figure 2.16). Note that initially, the source of CT was mistakenly thought to be the parathyroid gland, but a thyroid origin was subsequently established. The CT is secreted by the C cells of the thyroid gland, and they are located in the parafollicular part of the gland. The dual hormonal control concept was definitively established, but it had already been suggested by C. Rucart in the 1950s, who observed two distinct hormones from the parathyroid gland are involved in the regulation of plasma calcium level (Rucart 1951). This dual control of calcium level provided an elegant extension of the feedback mechanism (see section 3.1.2.1) postulated by F.C. McLean (1957). Subsequently, studies dealing with the molecular details of the hormonal regulation of calcium levels have revealed the complexity of the regulation. It is now well established that PTH and vitamin D, which act in the bone, kidney and gastro-intestinal tracts, increase calcium levels, and calcitonin has an opposing effect by decreasing bone resorption. It should be emphasized that depending on the calcium level, the PTH release is either increased or decreased, which is an unusual property for an endocrine tissue. However, it should be mentioned that although calcitonin was important for the development of the dual control concept, the hormone itself is not essential for maintenance of calcium and phosphate homeostasis. Indeed, PTH, vitamin D, as well as the relatively recently discovered phosphatonin, FGF23, are sufficient to prevent dysregulation of calcium and phosphate metabolism, indicating that calcitonin is functionally redundant.

It is noteworthy that PTH exerts two opposite effects depending on the duration of the administration of the hormone (Li et al. 2007). Continuous infusion and pulsatile treatment or daily injection cause a bone resorption and a bone anabolism, respectively. This paradoxical anabolic effect can be explained by a transient upregulation of genes classically associated with a resorptive response. In other words, the kinetic profile of PTH-responsive genes is an important factor causing the dual effects of the hormone. This observation suggests that the PTH likely regulates two signaling pathways differentially. This dual action of PTH is also relevant from the clinical perspective, because osteoanabolic actions of PTH provide the rationale for the use of PTH analogues for the treatment of osteoporosis.

Intense research efforts over the last five decades have focused on identifying the extracellular signals that positively and negatively control the diversity of biological responses. A large amount of data supports the notion that most biological processes are seldom regulated by only two opposite signals, but rather a complex balance of stimulatory and inhibitory extracellular signals. The question, which is then raised, is: how do two groups of opposite actions elicit either a stimulatory or an inhibitory effect? Several examples in different sections have addressed this issue.

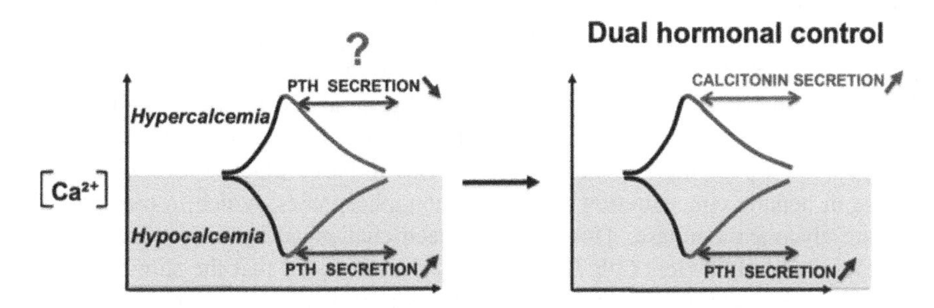

Figure 2.16. *Hormonal regulation of hyper- and hypocalcemia requires two extracellular signals: dual hormonal control concept*

2.3.1.13. *Regulation of two opposite biological processes by a common signal/duality of control*

The homeostatic system is maintained by a balance of extracellular inputs that positively and negatively influence a myriad of biological processes. Here, it is important to draw a distinction between two situations which occur when studying metabolic processes. Metabolism is a summation of anabolic and catabolic pathways. Metabolic homeostasis in an adult organism with stable body weight means that these two opposite fluxes, anabolic and catabolic, proceed simultaneously at the same rate, and consequently there is no net accumulation or degradation of biological macromolecules.

However, the whole body can temporarily face a disruption of this balance and the response should aim to restore the initial steady state. Thus, they must be effective controls of the opposite pathways. The efficiency of the readjustments is greater when a signal controls opposite biological processes. In other words, through a dual action, a signal can simultaneously stimulate a process while inhibiting the process which opposes the stimulation, thereby preventing a recycling of carbon moieties.

Here are some examples:

– Carbohydrate metabolism

The liver is an organ that plays a central role in the control of blood glucose level. Depending on the physiological situation, it can switch from net glucose output to net glucose uptake. This precise control occurs during periods of food consumption or food deprivation, and during exercise

After a meal, a fraction of the glucose entering the blood circulation is stored in the liver as glycogen. To explain the glycogen deposition, S. Soskin (1940) stated that the blood glucose level is the primary stimulus that elicits glucose uptake or glucose output by the liver. This concept was based on a series of experiments performed on hepatectomized dogs and received general acceptance (Bergman and Bucolo 1974). Until 1967, the initiation of glycogen deposition in the liver was assumed to be a push mechanism by which a rise in blood glucose level causes an increase in hepatic concentration of hexose monophosphates, which in turn should stimulate glycogen synthase. This view has been challenged by the observation of H. De Wulf and H.G. Hers (1967) that clearly demonstrated that the stimulation of glycogen synthesis by glucose is a pull mechanism, in which the stimulation of the last enzyme of the pathway causes the depletion of the intermediary metabolites. Because the elevation in glucose level elicits a secretion of insulin, its role was investigated, and it is clearly established that in addition to glucose, it regulates the activation of the glycogen synthase and the inhibition of the phosphorylase activities (Hers and Hue 1983). It should be emphasized that glycogen synthase starts to activate only when the phosphorylase activity is off. The control of these enzymatic activities by an on/off mechanism leads to a net accumulation of glycogen due to a coordinated inhibition and stimulation of the glycogenolytic flux and glycogenesis flux, respectively. This mechanism prevents a glycogen turnover or cycling, since the system is arranged in such way that glycogen synthase is active when the phosphorylase is inactive and vice versa (Petersen et al. 1998; Adeva-Andany et al. 2016). A common signal (insulin) regulates two opposite biological processes. In the fasting state, the inverse process prevails: there is a net glycogenolytic flux, due to a coordinated inversion of the fluxes mentioned above. It is attributed to a decrease and an increase of insulin and glucagon levels, respectively.

After a meal, another fraction of glucose entering the blood circulation is directed to the glycolytic pathway. Similar to the glycogenesis and glycogenolytic fluxes mentioned above, another flux (gluconeogenesis) operates in an opposite direction of the glycolytic flux. These two fluxes should also be regulated in a concerted fashion to allow a net synthesis of either pyruvate or glucose to proceed. Two hormones, insulin and glucagon, are key regulators of these pathways, through mechanisms of phosphorylation/dephosphorylation and changes in gene expression of enzymes (Pilkis et al. 1988).

In the fasting state, the elevation of glucagon levels is accompanied by a fall in insulin levels, which consequently reverses the insulin effects on these pathways, and gluconeogenesis predominates. More detailed information about the molecular mechanisms involved in this hormonal regulation is beyond the scope of this review. Readers should refer to the reviews by Pilkis et al. (1988) and Hatting et al. (2018). Despite the complexity of this regulation, it should be emphasized that the sites of hormone action occur at two substrate cycles. The first one is that between pyruvate kinase (glycolytic enzyme) and pyruvate carboxylase and phosphoenolpyruvate carboxykinase (PEPCK) (gluconeogenic enzymes). The second one involves 6-phosphofructo-1-kinase (6PF1K) (glycolytic enzyme) and fructose-1-6-bisphosphatase (F1,6Pase). In a fasting state, the stimulation of gluconeogenesis is due to both the inhibition of these two glycolytic enzymes and the stimulation of gluconeogenic enzymes mentioned above. In contrast to the regulation of glycogen metabolism, the dual control of the opposite fluxes by insulin and glucagon operates at the site of the substrate cycles. Insulin induces the transcription of genes that encode glycolytic enzymes and simultaneously represses the transcription of genes that encode gluconeogenic enzymes; glucagon has opposite effects. Finally, it needs to be mentioned that gluconeogenesis may serve two different purposes depending on the prevailing conditions. Under the fasting condition, gluconeogenesis contributes to hepatic glucose output, thus helping to maintain normoglycemia. Under the normal refeeding condition, the gluconeogenic pathway unexpectedly remains active and participates in the bulk of liver glycogen synthesis. This led to the conclusion that glycogen is formed by an indirect pathway, involving the sequence glucose-lactate-glucose-6-P-glycogen, whereas muscle glycogen is formed by the conventional pathway: glucose-glucose-6-P-glycogen (Newgard et al. 1984).

During a stress or exercise, there is an increased hepatic glucose production, which is controlled by an increase in epinephrine levels. This hormonal signal exerts a dual control on the liver. It stimulates glucagon secretion, which in turn increases glycogenolysis and simultaneously inhibits insulin secretion. The latter does not exert its opposing effect on glycogenolysis, and consequently epinephrine and glucagon become more effective.

In summary, this duality of the control of two opposite fluxes (glycogenogenesis/ glycogenolysis and glycolysis/gluconeogenesis) by an appropriate hormonal signal is a mechanism that allows us to optimize the net flux of these metabolic pathways.

– Lipid and protein metabolism

The mechanisms of regulation mentioned above also apply to lipid metabolism. In the 1960s, it was proposed that two separate sets of insulin receptors may be involved in mediating the inhibition of lipolysis and the stimulation of lipogenesis (Thomas et al. 1978). These two responses to insulin were investigated on adipocytes, and it was shown that they were mediated by the same receptor (Thomas et al. 1979). Thereafter, cellular mechanisms have been studied *in vivo*, in postprandial state, and it was shown that the elevation of circulating insulin promotes triglycerides synthesis through *de novo* lipogenesis and fatty acids esterification, and it simultaneously restrains lipolysis. Consequently, there is a net accumulation of fat in the white adipose tissue, through a regulation of metabolic fluxes that work in opposite directions, thereby confirming the dual actions of the hormone. Of note, this type of mechanism often operates in metabolism regulation. For instance, the activation of mTOR signaling stimulates anabolism and simultaneously represses catabolism to prevent a futile cycle in which newly synthesized blocks are immediately broken down.

– Regulation of food intake

Food intake is mainly regulated through a balance between orexigenic and anorexigenic signals, which are under the control of peripheral hormones. In the late 1970s, it was demonstrated that insulin regulates food intake and energy balance (Woods et al. 1979). Thereafter, subpopulations of neurons were identified in the arcuate nucleus (ARC) of the hypothalamus: 1) neurons that produce orexigenic neuropeptides, agouti-related peptide (AgRP)/neuropeptide Y (NPY) and 2) neurons that release anorexigenic neuropeptides, pro-opiomelanocortin (POMC)/ cocaine- and amphetamine-regulated transcript (CART). Intracerebroventricular administration of insulin decreases food intake in a dual manner: by downregulating the expression of AgRP/NPY to reduce appetite and by upregulating expression of POMC/CART to promote satiety (Dodd and Tiganis 2017).

Unlike peripheral administration of insulin, which often causes weight gain, central administration of insulin leads to a reduction in food intake and body weight.

2.3.2. *Integration of multiple signal inputs: ratio of stimulatory vs inhibitory signals*

Most biological processes are tightly regulated by an integration of positive and negative signals. The influence of these two opposite signals, regulating a cell response can tip the balance on one side or the other, depending on their ratio, in terms of concentration or expression. Disruption of the balance allows one group of signals to override the influence of the opposite one. Here are some examples of ratios which should be tightly regulated to prevent excessive cell response.

2.3.2.1. *Insulin/glucagon ratio*

Glucose

The insulin/glucagon ratio (I/G) concept was introduced by R.H. Unger when studying the regulation of carbohydrate and metabolism (Unger 1971). These studies established that an increased I/G ratio favors an "anabolic" response, and the inverse promotes a "catabolic" response. Furthermore, it points out that the biological response is dictated by the I/G ratio and not the absolute concentration of either hormone.

Their role on glucose metabolism has been investigated, focusing on the regulation of the net hepatic glucose balance, which is crucial in the control of glycemia (see section 2.3.1.13). This organ has the capacity to adjust its release or uptake of glucose, depending on nutrient availability. In the fasting state, the I/G ratio is low and implies that the glucagon prevails, and this results in a net release of glucose, which is caused by a stimulation of the gluconeogenesis and the glycogenolysis. Conversely, in the feeding state, the I/G ratio is increased and the effect of insulin prevails; it blocks both glycogenolysis and gluconeogenesis and stimulates both glycolysis and glycogen synthesis.

FGF21 and follistatin

During exercise, the hepatic release of FGF21 and follistatin is increased by a low I/G ratio, and the phenomenon is impaired in patients with type 2 diabetes (Hansen et al. 2016).

2.3.2.2. *Pro-apoptotic/anti-apoptotic proteins*

To maintain normal physiology and tissue function, damaged and dysfunctional cells are constantly cleared via regulated cell death and replaced by new, healthy cells. Many of the diseases that constitute the leading causes of death and disability, including neurodegenerative, cardiovascular, autoimmune and infectious diseases, involve either excessive or insufficient cell removal. Apoptosis is important for proper development, maintenance of tissue homeostasis and cancer prevention.

Apoptosis is initiated by either external stimuli or internal stimuli, and mediated via two distinct pathways: the extrinsic pathway (death receptor-mediated) and the intrinsic pathway (mitochondrial-mediated). The key regulation on the execution of intrinsic apoptosis is controlled by the Bcl-2 family of proteins, which contains both pro-apoptotic and anti-apoptotic (pro-survival) members that balance the decision between cellular life and death. The namesake of this family, Bcl-2, was discovered because of its frequent translocation and overexpression in B-cell lymphomas, where Bcl-2 is frequently grossly overexpressed.

The anti-apoptotic proteins of the Bcl-2 family are B-cell lymphoma extra-large (Bcl-XL), B-cell lymphoma W (Bcl-W), Bcl-2-related isolated from fetal liver 1 (BFL1) and myeloid cell leukemia 1 (MCL1). They contain one or more BH (Bcl-2 homology) domains (BH1–BH4). The pro-apoptotic proteins are BAX and BAK.

Anti-apoptotic proteins can block apoptosis by binding and sequestering monomeric-activated BAX and BAK proteins and or BH3-only activator or sensitizer proteins. The balance between pro-apoptotic and pro-survival proteins determines whether a cell will live or die (Deng et al. 2007). To maintain its survival, a cell expresses high levels of anti-apoptotic proteins to almost neutralize pro-apoptotic proteins. For apoptosis to occur, anti-apoptotic proteins within the cell must be overwhelmed and BAX and/or BAK must be activated. This is where the balance between anti-apoptotic and pro-apoptotic proteins comes into play. However, the expression level of anti-apoptotic proteins is not itself a strong predictor of apoptosis, and low levels of expression of BAX or BAK render cells refractory to apoptosis. For a more detailed overview of the topic, see Singh et al. (2019).

2.3.2.3. *Pro-angiogenic/anti-angiogenic proteins*

The process of angiogenesis encompasses the formation and regression of capillaries from existing blood vessels, which occurs throughout life in normal tissues. Over the past four decades, fundamental concepts of the process have been introduced and discussed elsewhere in detail (Folkman 2003). Angiogenesis is also a key process in tumor growth progression because it is strongly correlated with vasculature formation (Kazerounian and Lawler 2018). In light of this observation, a series of studies have been focused on the mechanism underlying the angiogenic process in normal tissues and in disease. It was concluded that the process is finely regulated, and the response of endothelial cells is determined by the net input of concurrent signals from pro-and anti-angiogenic factors. The endothelial cells are quiescent in healthy tissues, and this is explained by the fact that anti-angiogenic factors antagonize pro-angiogenic signals, and in tumoral tissues, the inverse situation prevails to promote the tumor growth. Among the pro-angiogenic factors, the VEGFA plays a major role, and it binds to the co-receptor neuropilin-1 and the VEGFR-2 (see section 2.3.1.2). Regarding the anti-angiogenic molecules, the thrombospondin (TSP-1) is the major player in the process and in healthy tissue the presence of basal levels of TSP-1 in the vessel wall, likely prevent angiogenesis. It is noteworthy that TSP not only activates anti-angiogenic signal transduction, but also antagonizes pro-angiogenic function of VEGFA. This duality of function of an extracellular signal is a concept that is highlighted in section 2.3.1.2. Given its multidomain structure, it can bind to multiple membrane receptors, which in turn can activate distinct signaling pathways (see section 3.3.3: functional domains).

As mentioned in this section, the protein level ratio is indicative of a net increase or decrease of biological response.

2.3.2.4. *Pro-inflammatory/anti-inflammatory cytokines*

The clinical symptoms of inflammation have been described by the Roman Aulus Cornelius Celsus in the first century CE. He mentions in his treatise De medicina that "*rubor et tumor calore et dolore*" (redness and swelling with heat and pain) are the major symptoms of inflammation. In 1794, Scottish surgeon J. Hunter wrote that "Inflammation in itself is not to be considered as a disease but as a salutary operation consequent to some violence or some disease" (Majno 1991). The immune system is essential for the host response to pathogens (bacteria, viruses), injury, tumor development and extracellular organisms. Of all of the body's systems, the immune system is likely the most challenging to coordinate, due to its organization. Despite a great variety of non-self or antigens, there are two types of immune responses, one is regulated by the components of the innate immunity: antigen-presenting cells (APC)-macrophages/monocytes, dendritic cells (DCs) and natural killer (NK), and the second one is regulated by the components of the adaptive immunity relayed by $CD4^+$ T lymphocytes (naive Th0), which are now recognized as being key regulators of immune responses. Here are the reasons.

* *Th1/Th2*

In the late 1980s, R. Coffman and T. Mossmann discovered two T-cell subsets, Th1 and Th2, and demonstrated that they arise from naive $CD4^+$. These investigators suggested that they can be distinguished by different sets of cytokines secreted after activation and, consequently, mediating very different regulatory and effector functions. In other words, T-cell subsets are in a state of equilibrium (Coffman 2006). Upon the activation of one subset, other subsets are modulated or inhibited to promote the most specific effector response in defense against threat. The Th1/Th2 concept emerged from the original observations that mouse T-helper cell populations could be subclassified, based on the cytokines they secreted (Mosmann and Sad 1996). One theory of immune regulation involves IL-12 secreted by APCs, the most upstream components of the immune response, that induce Th1 polarization and differentiation, which enables them to secrete the major pro-inflammatory cytokines TNF-β and IFNγ. Th2 secrete anti-inflammatory cytokines, namely IL-4 IL-10 and IL-13, which inhibit Th1 responses (Elenkov 2004). Th1 and Th2 responses are mutually inhibitory. The main role of Th1 is to prevent Th2 differentiation and consequently dampen their activity. Alteration of the Th1/Th2 balance has been described in several autoimmune diseases, such as rheumatoid arthritis, multiple sclerosis, type 1 diabetes and autoimmune thyroid disease. The balance is skewed towards Th1, and it is associated with an excess of IL-12 and TNF-β levels, and a concomitant decrease in Th2 activity and lowered IL-10 level (Segal et al. 1998) (Figure 2.17). Consequently, immunomodulatory treatments that would promote a shift from the Th1 to the Th2 anti-inflammatory cytokine pathway would have a beneficial effect on the clinical course of autoimmune diseases (see details in section 4.3.4).

* *Th17/Treg (T regulator)*

It was previously hypothesized that the division of $CD4^+$ was solely dominated by Th1 and Th2. This paradigm was maintained until 2005, when a third T-cell subset was identified and named Th17 as they produce interleukin 17 (IL-17), which was originally cloned in 1995. It displayed multiple inflammatory effects, such as cytokines production and recruitment of leukocytes. Mounting evidence over the past decade reveals that Th17 cells and regulatory T-cells (Tregs) also play a role in regulating autoimmunity and cancer.

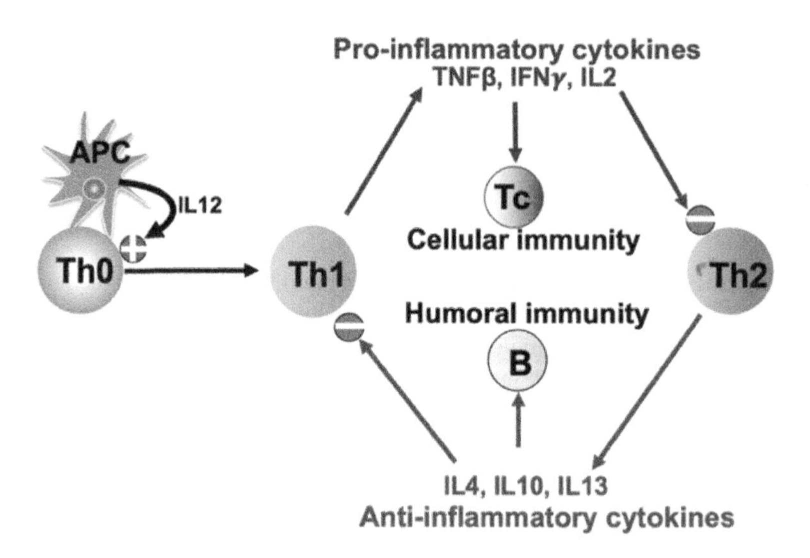

Figure 2.17. *Immune response relayed by Th1 and Th2 through the secretion of pro-inflammatory and anti-inflammatory cytokines, respectively*

Furthermore, the secretion of interleukins by Th17 cells promotes inflammation, whereas Treg cells produce anti-inflammatory cytokines. Therefore, these two cell types play opposite roles during inflammatory and immune responses (Littman and Rudensky 2010). Th17 cells cause, whereas Treg cells inhibit, autoimmunity.

Accumulated evidence suggests that the Th17/Treg balance underlies the pathogenic mechanisms driving autoimmune diseases (Noack and Miossec 2014). See an excellent review by Knochelmann et al. (2018).

Finally, cytokines have been classified as pro- versus anti-inflammatory based on global effects in animal models. This simple dichotomy may not apply to several cytokines with a pleiotropic role, such as IL-10, which has anti-inflammatory and inflammatory effects (Mocellin et al. 2003).

2.3.2.5. *Matrix metalloproteinases/tissue inhibitors of metalloproteinases*

MMPs are a group of endopeptidases that are secreted by many cells, including fibroblasts, endothelial cells and leucocytes. They degrade various protein fibers in the extracellular matrix (ECM), such as collagen, elastin and laminin. They are involved in physiological processes such as ECM remodeling, tissue invasion and vascularization, which take place during development and wound repair. On the other hand, they also participate in cancer development, progression and invasiveness. Their role is to break the cell-to-cell and cell-to-ECM adhesion, which thereby facilitates tumor invasion and metastasis.

MMPs are regulated by endogenous tissue inhibitors of metalloproteinases (TIMPs) that bind MMPs in a 1:1 stoichiometry. Tissue homeostasis is thus achieved by a tight balance of MMP proteolysis to TIMP expression. Deregulated expression and activity of both MMPs and TIMPs are key features of all human cancers. Similarly, an elevated MMP/TIMP protein level ratio is associated with many diseases of the central nervous system like Alzheimer's disease and multiple sclerosis, and it results in a net increase in the proteolytic activity of MMPs. The imbalance between MMPs and TIMPs is also a principal feature of hepatic fibrosis, and it occurs when the rate of matrix synthesis exceeds matrix degradation.

It is noteworthy that TIMPs have poor specificity for a given MMP, and each of them can inhibit multiple MMPs (Moore and Crocker 2012; Cui et al. 2017).

2.3.2.6. *Co-stimulatory/co-inhibitory signals*

Immune responses are considerably more complicated than initially thought. The model that is currently held is the following: the balance between co-stimulatory and co-inhibitory molecules decides the direction and magnitude of the immune response. These two groups of signals are the key players contributing to the immune homeostasis when the adaptive immunity is activated to eliminate pathogens, infected cells and cancer cells. Briefly, when dendritic cells (DCs), which belong to antigen-presenting cells (APC), encounter pathogens, they migrate to lymphoid organs and present antigenic peptides to naive $CD4^+$ T-cells to direct their activation, through two signals. The first signal called "antigen recognition" refers to the engagement between the T-cell receptor (TCR) and major histocompatibility complexes (MHC-II) displayed on APC. The second signal, called "co-stimulatory" signal, should be provided to effectively prime the naive T-cell into the $CD4^+$ T-helper cell (Th). This signal is delivered by CD80 in APCs to CD28 in T-cells. These two signals in combination with cytokines secreted by APCs dictate their proliferation and differentiation into different effector T-cells such as $CD4^+$ Th1 and Th2. It was therefore called the two-signal model of cell activation and it was consolidated when other co-stimulatory signals were discovered. But it turned out that this model was oversimplified when "co-inhibitory" signals and their respective receptors (CTLA4, PD-1 to name but a few) were discovered. Currently, it is clearly established that when co-stimulatory signals outweigh the co-inhibitory signals, a T-cell proliferates and differentiates into an effector T-cell. Conversely, when co-inhibitory signals overweigh the co-stimulatory signals, a T-cell enters an antigen-specific responsive state known as anergy or tolerance, which prevents autoimmunity (Azuma 2019).

It should be emphasized that, unlike $CD4^+$ T-cells, naive $CD8^+$ T-cells require additional cytokine signals IL-2, IL-12 from APCs, to proliferate and differentiate.

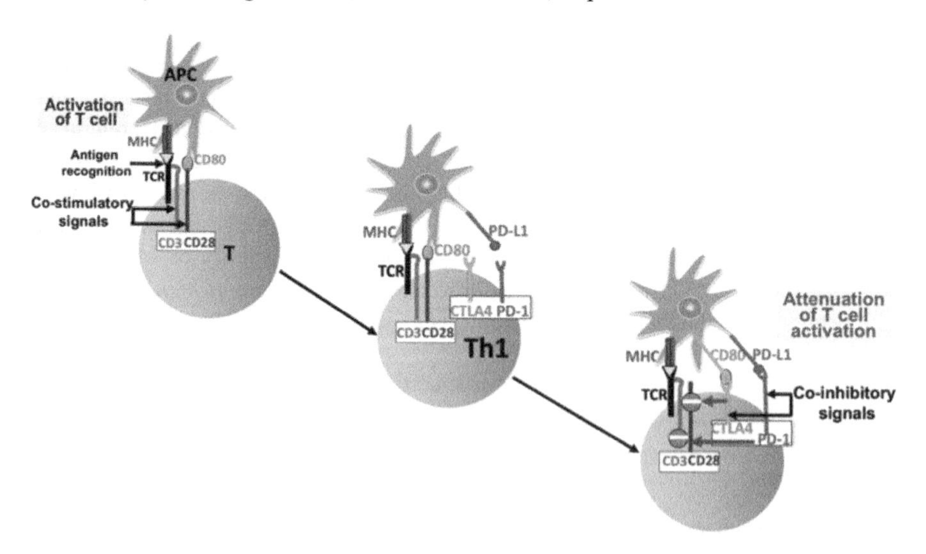

Figure 2.18. *The balance between co-stimulatory and co-inhibitory molecules decides the direction and magnitude of the immune response*

Similarly, the natural killer (NK) cell activity depends on balance from inhibitory and activating receptors. Whether an NK cell will attack or tolerate an infected cell or a tumor target cell depends on the net balance of signals transduced by these two groups of receptors. Healthy cells express self-molecules which are recognized by inhibitory NK cell receptors and as such prevent autoimmune responses (Nash et al. 2014)

2.3.2.7. *Oncogenes/tumor suppressor genes*

The oncogene concept emerged from the tumor virus field. In 1909, P. Rous described a transmissible tumor (sarcoma) of the chicken with extreme malignancy and the disease could be transmitted by a filtrate free of the tumor cells. These observations indicated that the tumor was due to a "filterable agent" (Rous 1911). Other chicken tumors with distinct pathologies also proved to contain filterable agents. He was able to show that there were distinct agents, each of which produced tumors of the same type as the tumor from which it had been derived. This strain specificity was the first hint that the ability to induce tumors was a genetic property of the virus. The discoveries were met with reactions ranging from indifference to hostility and were unrelated to human diseases. Anecdotally, Rous recounted a comment from C. Andrewes, a famous British oncologist: "But, my dear fellow, don't you see, this can't be cancer because you know its cause" (Andrewes 1971). The idea that viruses could contain genetic information only emerged some 30 or 40 years later. By the 1950s, it had been established that there are many different types of mammalian viruses that could cause cancer, such as the Hepatitis C virus (HCV), Human papillomavirus (HPV), Hepatitis B virus (HBV), Epstein–Barr virus (EBV) and Kaposi's sarcoma-associated herpesvirus (KSHV)/human herpes virus 8 (HHV8). Their names are often eponymous and a consequence of their species tropism and the type of cancer they caused. Fortunately, Rous saw that his ideas were progressively accepted, and in 1966, at the age of 85, he received the Nobel Prize in Physiology or Medicine.

In 1969, R.J. Huebner and G.J. Todaro proposed that RNA tumor viruses could be transmitted genetically (Huebner and Todaro 1969). They postulated that these viruses contained transforming genes or "oncogene" and that the activation of these viruses could cause cancer. In the early 1970s, investigators tried to identify the product of the newly discovered viral Src gene and speculated that it may be of cellular origin. In other words, rather than viruses inserting transforming genes into the genome, they proposed that transforming viruses may have captured cellular genes. In 1976, the sequence of the gene provided evidence that it was of cellular origin rather than viral origin (Stehelin et al. 1976). The term proto-oncogene was coined to describe this new type of gene to emphasize that, although representing the precursor of the viral transforming gene, it is not itself transforming unless mutated and/or overexpressed. The viral and cellular genes were named v-Src (oncogene) and c-Src (proto-oncogene), respectively. In 1989, the discovery earned J.M. Bishop and H.E. Varmus the Nobel Prize in Physiology or Medicine. These years saw the discovery of the Src homology (SH) domains and the development of the concept of modular protein interaction domains (section 3.3.3). It is worth noting that v-Src was also the first tyrosine kinase that was discovered as described in a seminal paper by T. Hunter and B.M. Sefton (1980b).

Following the discoveries that cancer is caused by malfunctioning genes that normally regulate major processes, including cell growth, proliferation survival, the genes were classified into two categories based on whether they functioned to promote or inhibit tumorigenesis. Hence, a first category of genes promoting tumorigenesis was termed "oncogenes" and the majority of genetic changes results from mutations in "proto-oncogenes", leading to a gain of function (GOF) of encoding proteins. These proto-oncogenes normally stimulate cell growth, proliferation and survival, but these cellular processes are out of control after mutations of these genes.

Like the idea of the oncogene, the concept of the tumor suppressor is over a century old (Hansford and Huntsman 2014). The first evidence for tumor suppression came from cell fusion experiments that showed that normal cells fused with cancer cells could suppress cancer formation (Harris et al. 1969). Then, A. Knudson developed the concept further when he explained the late-onset retinoblastoma, which required inactivation of both alleles to initiate the tumor process (Knudson 1971). This second category of genes restraining tumorigenesis was called "tumor suppressor genes", and mutations in these genes result in a loss of function (LOF) of encoding proteins. They normally restrain cell growth, promote DNA repair and arrest the cell-cycle progression, but after mutations in these genes, the encoding protein functions are impaired. Due to these characteristics, they are also referred to as antioncogenes and their protein products as tumor suppressor proteins.

It should be emphasized that cancer progression involves an accumulation of multiple mutations within a cell, resulting in the activation of oncogenes and impaired function of tumor suppressor genes. There are excellent reviews describing genes involved in cancer progression, as well as the mutations that lead to a GOF or LOF of proteins (Liu and Weissman 1992; Levine 1993; Weinberg 1994; Bashyam et al. 2019). It is worth noting that mutated genes encode proteins of growth factor signaling pathways, thus explaining uncontrolled cell proliferation.

An unexpected aspect of tumorigenesis has been discovered when studying the response of a tumor cell to TGFβ. TGFβ is a well-known inhibitor of cell growth in adult epithelial cells. During tumor progression, cancer cells escape the suppressive and growth inhibitory effects of TGFβ by accumulating mutations in genes encoding components of the TGFβ signaling cascade. It therefore turns TGFβ into an oncogenic factor (Seoane and Gomis 2017). These genes regulate cell migration, invasion and immunosuppression. This highlights that TGFβ exerts dual opposite roles in oncogenesis, which can be explained by the pleiotropic nature (see the pleiotropy concept in the next section) of TGFβ.

It should also be mentioned that several cancer-related genes exhibit a dual nature, functioning as both an oncogene and a tumor suppressor gene and vice versa in different contexts. For instance, the switch can occur at the protein level. Thus, the p53 tumor suppressor gene is a transcription factor mutated in over half of all human cancers and the majority of p53 mutations are missense mutations affecting its DNA binding domain. The protein loses its tumor suppressive function and may simultaneously gain novel functions (GOF), primarily through protein–protein interactions with other transcription factors (Kim and Lozano 2018).

2.3.2.8. *Sympathetic/parasympathetic activities: excitatory/inhibitory ratio*

Classically, the autonomic nervous system (ANS) is divided into the sympathetic (SNS) and parasympathetic nervous system (PNS), and in most organs, the sympathetic tone is counterbalanced by the parasympathetic activity. A disruption of this internal balance occurs in response to short-term stress. It elicits an SNS stimulation and a secretion of adrenaline. According to W. Cannon, it prepares the organism for a rapid "fight or flight" response. On the other hand, H. Selye, known as the "father of stress", studied the effect of long-term stress. He pointed out the role of the brain and the adrenal cortex in response to stress and identified several hormones that regulate the stress response. It has been addressed in detail in section 4.3.1.2. The stimulation of the PNS can be defined as the "rest and digest" response.

These two branches of the ANS have a key role in bone remodeling, which will be discussed later (section 4.3.3.2). The dysregulation of the tone of the SNS and PNS is often reported in diseases such as chronic autoimmune diseases and rheumatoid arthritis (RA), which is a prototypic immune-mediated inflammatory disease (IMD). In general, RA patients have overly active sympathetic tone and reduced parasympathetic tone. Likewise, it was discovered that autonomic dysfunction precedes and predicts arthritis development in subjects at risk of developing seropositive RA (Koopman et al. 2017).

It should also be mentioned that neural circuits in the brain are composed of excitatory and inhibitory neurons that mainly transfer the information via neurotransmitters at chemical synapses. Glutamate is the predominant excitatory (E) neurotransmitter, mediating excitatory synaptic neurotransmission, in the brain. To balance glutamatergic excitability, inhibitory (I) synapses release the neurotransmitter γ-aminobutyric acid (GABA). These two types of neurons account in part for the normal functioning of the brain. In some pathological situations, there is a disruption of the E/I ratio, which contributes to cognitive dysfunction and neurological abnormalities; see the review by Sarnat and Flores-Sarnat (2021).

2.3.2.9. *Reactive oxygen species/antioxidants*

In physiological conditions, reactive oxygen species (ROS) are produced by living cells and represent normal metabolic byproducts generated in the oxidative reaction process of the mitochondrial respiratory chain. To avoid the accumulation of ROS, cells have evolved cellular antioxidants, which are enzymes such as superoxide dismutase (SOD), catalase (CAT), glutathione peroxidase (GPx) and non-enzymatic antioxidants such as ascorbate, glutathione and flavonoids, that undergo reversible oxidation.

Oxidative stress is a consequence of an increased generation of ROS and/or reduced activity of antioxidant defenses against ROS. In normal conditions, the balance between the generation of ROS and the activity of antioxidant defense is slightly tipped in favor of the ROS production. Despite this imbalance, there is a mild oxidative damage. In contrast, in many types of cancer cells, there is an excessive production of ROS that can contribute to the initiation of the tumoral process through activation of signaling pathways, involved in cell proliferation. To quench the ROS effects, there is an enhanced production of antioxidants, but the system cannot cope with this excessive amount of ROS. A marked imbalance between production and destruction of ROS ensues, which leads to oxidative stress. Depending on the severity of the stress, it causes DNA lesions or elicit cell death.

2.3.2.10. *ATP/AMP ratio*

The physiological meaning of this ratio will be discussed in section 3.2.2.1.

2.3.3. ***Physical signals***

Cells have the ability to sense mechanical cues of their microenvironment and appropriately respond to the diversity of stimuli. The first issue that has been addressed is the identification of the mechanical sensors that detect, interpret and respond to specific signals. These signals include not only all components of force, stress and strain, but also substrate rigidity, topography and adhesiveness, and this is called mechanosensation. So far, a relatively large number of these sensors have not been clearly identified yet (reviewed in Chalfie (2009), Martino et al. (2018) and Xu et al. (2018)). Recently, two groups of investigators have examined the molecular mechanisms underlying the perception of the temperature and physical touch in detail.

Research in the detection of temperature and touch can be traced back to 1944 when J. Erlanger and H. Gasser worked on the formation of action potentials in different types of peripheral nerve fibers. These two physiologists shared the Nobel prize in Physiology or Medicine for the "outstanding contributions they have made to the electrophysiology of the nervous system" (Kellaway 1945). They demonstrated that these fibers transmitted the sensation of pain, but did not address the following question: how does a neuron translate a thermodynamic energy or pressure into an action potential? In the 1990s, D. Julius addressed this issue by exploiting an unusual characteristic of a natural compound (capsaicin), which is perceived as having "temperature" when eating chilies. He showed that capsaicin acts on the transient receptor potential cation channel vanilloid (TRPV1) uniquely expressed in thermosensitive nociceptive neurons and elicits changes in intracellular calcium (Jordt and Julius 2002). TRPV1 is otherwise activated by temperatures above ~43°C, which underlies the sensation of heat-pain as opposed to warm, which is sensed by a different group of membrane receptors and thermosensory cells.

With respect to the search for sensors that enable neurons to respond to physical touch, it started with Patapoutian's group. They first observed that a neuroblastoma cell line was sensitive to mechanical force and suggested that a cation channel may be involved. In this cell line, they found about 70 channels or proteins of unknown function. Anecdotally, using siRNA, they systematically knocked down expression of each channel and observed no effect until their last candidate of the list, which displayed a marked reduction of inward current in response to pressure. When it was expressed in kidney cells, it conferred sensitivity to mechanical force. In 2010, this team reported the identification of the first mammalian mechanosensing ion channel. They named it PIEZO1 after the Greek term for pressure (Ranade et al. 2015). Then, in 2018, it was revealed that PIEZO channels were at the center of somatosensory systems, critical for the perception of the pain, touch, shear stress and a broad range of internal homeostatic processes.

In 2021, the Nobel Prize in Physiology or Medicine was awarded jointly to D. Julius and A. Patapoutian for their discoveries of receptors for temperature and touch.

Information about other molecules that sense the mechanical cues are presented in section 3.2.1.5.

2.4. Nuclear receptors

2.4.1. *Chemical nature of signals and functional characteristics of nuclear receptors*

In the early years, steroid hormones, thyroid hormones, retinoids and vitamin D3 were purified and characterized long before they were recognized as nuclear receptor ligands. They were identified because of their abilities to modulate a wide range of physiological processes and because of associations with human diseases. It was also clearly established that most steroid hormones were derived from the precursor steroid cholesterol and synthesized in specialized endocrine cells within specific endocrine glands. These hormones include the adrenal steroids cortisol and aldosterone, the adrenal androgen dehydroepiandrosterone (DHEA) and the reproductive steroids. The latter comprise estradiol and progesterone, which are predominantly synthesized in the ovary, and testosterone and dihydrotestosterone synthesized predominantly in the male testis. After the initial characterization of these hormone actions on their target tissues, a fundamental question was brought up: how do these lipophilic signals mediate their action? Development of radiolabeled ligands allowed us to address this issue. In 1966, Jensen's group showed with radioactive estradiol that rat uterus and other hormone-responsive tissues contain characteristic steroid-binding components with which the hormone reacts to induce growth without itself undergoing chemical alteration. This raised the question of whether these tissue binding sites had any direct role in hormonal action. The subsequent experiment demonstrated that a progressive inhibition of estradiol uptake by uterine cells occurs with increasing amounts of antiestrogen (nafoxidine), and it parallels the inhibition of uterine growth (Jensen 1966). It provided evidence that binding is implicated in uterotropic action. Hence, the binding entities could be regarded as true receptors. More support for this concept came from cell fractionation studies which showed that, before administration of labeled estradiol, the binding entity is localized in the cytosol fraction, whereas after estradiol administration, most of the hormone is localized in the nucleus (Toft and Gorski 1966). These data strongly suggested a link between transcriptional control and physiology. The subsequent identification of hormonally responsive target genes within these tissues completed the initial characterization of a steroid hormone signaling pathway. Together, these biochemical studies (reviewed in Yamamoto (1985)), provided the classical model of steroid hormone action, from ligand to target gene product. The cloning of the steroid receptors was an essential prerequisite for ultimately understanding the molecular basis of this model.

In 1985, the primary structure of a cDNA encoding a human glucocorticoid receptor was described (Govindan et al. 1985). At the same time, cDNA encoding the first human estrogen receptor was cloned (Walter et al. 1985). In less than a decade, 30 or more receptors have been identified (reviewed in Reichel and Jacob (1993)). Among the receptors identified (48 human NRs), about half had assigned ligands. Those that did not have ligands were classified as "orphan receptors", but later on and for some of them, intrinsic or xenobiotic ligands have been identified (Gustafsson 2016; Weikum et al. 2018). It should be emphasized that the accelerated understanding of nuclear receptor biology represents the impact of new technologies on modern biology. The term "nuclear receptor" (NR) is used because many of the functions mediated by these receptors depend on their translocation to the nucleus either bound to a ligand or without a ligand that is subsequently bound within the nucleus. Noteworthy, NRs are both receptors and transcription factors. Unlike the membrane receptors, they bypass the complex "second messenger", such as cAMP or Ca^{2+}, in response to their ligands and can directly modulate gene transcription.

By the mid-1990s, key concepts about nuclear receptor biology have emerged (reviewed in Mangelsdorf et al. (1995)), which are briefly reported. Despite diversity in the size, shape and charges of ligands, almost all members of the nuclear superfamily share a common modular domain structure, composed of five or six domains: A–E. Two of them play a key role: a DNA binding domain (DBD) and a ligand binding domain (LBD), which is a complex allosteric (see section 3.1.3.1) signaling domain that binds ligands and interacts with coregulator proteins (see the next section: regulation of transcription). The N terminal domain also contains the activator Function-1, which interacts with coregulator proteins.

NRs are divided into seven subfamilies, from 0 to 6, mainly based on their binding to DNA as homodimers, heterodimers or monomers (Weikum et al. 2018). Two subfamilies are reported: (1) the subfamily 1 that binds non-steroid ligands, including thyroid hormones, fatty acids, bile acids and sterols, and (2) the subfamily 3 that binds cholesterol-derived hormones, such as glucocorticoids, mineralocorticoids, androgens, estrogens and progesterone.

Some historical aspects of NR research are interesting to mention. It is worth noting that the road to the discovery of retinoic acid receptors started in the 1880s. It was postulated that essential factors other than proteins, lipids, carbohydrates or minerals are present in food. The first discoveries in the field were made by C. Eijkman and F.G. Hopkins, who found that rice polishings contain substances preventing beriberi, which was ultimately characterized as thiamine (Carpenter and Sutherland 1995). These investigators received the Nobel Prize for their work in 1929. In 1912, C. Funk identified the active fraction and this substance belonged to a class of compounds called amines, and it was named "vitamine" (vital amine) (Piro et al. 2010). By 1920, other "vitamines" were isolated and were classified as Vitamin A, Vitamin B, etc.

Subsequent studies revealed that vitamin A serves as a precursor for two active derivatives: 11-cis retinal and all-trans retinoic acid. Further experiments concluded that retinoic acid (RA) was the hormone involved in cell growth. Other vitamin A metabolites were identified and in 1976 were grouped as *retinoids*. Clues as to how RA works inside cells came from studies performed in the 1960s with estrogens and glucocorticoids.

Also at that time, it was shown that cytosolic proteins bind retinol or retinoic acid and transfer them into the nucleus to specific binding sites on the chromatin to modulate the transcription of specific genes. The first retinoic acid receptor, RARα, was identified in 1987, and subsequently a second family of nuclear retinoid receptors was cloned: the RXRs. The interesting feature of the RXR proteins is that they can activate transcription in response to RA, but are unable to bind all-*trans* RA. However, 9-cis retinoic acid has a high affinity for RXRs, but this ligand is not detected in most tissues.

The discovery of the concept of vitamin D is also worth mentioning. It started in 1922, when McCollum's group described the factor that cures rickets (McCollum et al. 2002), which was named the antirachitic factor (vitamin) or vitamin D. Then, in 1968, H. DeLuca isolated an active vitamin D metabolite (25(OH)D3) (Blunt et al. 1968), which was followed by the discovery of a second biologically active vitamin D metabolite: 1,25$(OH)_2$D3. The latter metabolite was the biologically active form of vitamin D3. In 1969, M.R. Haussler and A.W. Norman discovered the receptor which conveys the 1,25$(OH)_2$D3 signal, leading to cellular responses (Haussler and Norman 1969). Detailed discoveries of vitamin D and its metabolites are reviewed in Deluca (2014).

2.4.2. *Molecular mechanisms underlying regulation of gene transcription*

The ability of receptors to regulate gene expression in a selective manner is mediated by the recognition of specific DNA sequences known as hormone response elements (HREs), and they use a common consensus sequence separated by variable of non-consensus nucleotides. The common consensus sequence can take the form of a palindrome (Gronemeyer and Moras 1995). The spacer region that varies in length has been shown to allosterically modulate NRs, resulting in a variety of transcriptional outputs (Schwabe and Rhodes 1991). Depending on the subfamily of NR, they function as homodimers, heterodimers or monomers (Figure 2.19). Once DNA-bound, NRs possess the ability to both repress and activate target gene expression, reflecting the diverse biological roles of these receptors.

To explain the diversity of NR actions, it was postulated that they interact with non-DNA "nuclear acceptor" molecules (Spelsberg et al. 1971) that were also called intermediary factors. Then, our knowledge drastically expanded in this field due to advancements in new methodologies. In 1994, two groups provided the first evidence that ligand binding resulted in the recruitment of associated molecules by estrogen receptor (Cavaillès et al. 1994; Halachmi et al. 1994). Within the same decade, 30 of these molecules were cloned and organized into a coherent model of their biological significance. These auxiliary proteins were collectively named coregulators of NRs functions and among them (approximately 200), two distinct classes were identified: coactivators and corepressors.

To modulate the gene transcription, NRs recruit these coregulators. To activate transcription, NRs recruit coactivators, which are large protein complexes that harbor histone acetyltransferase (HAT) or histone methyltransferase (HMT) enzymes (Bulynko and O'Malley 2011). Once these enzymes are activated, it facilitates the opening of chromatin, allowing the transcriptional machinery to turn on the transcription.

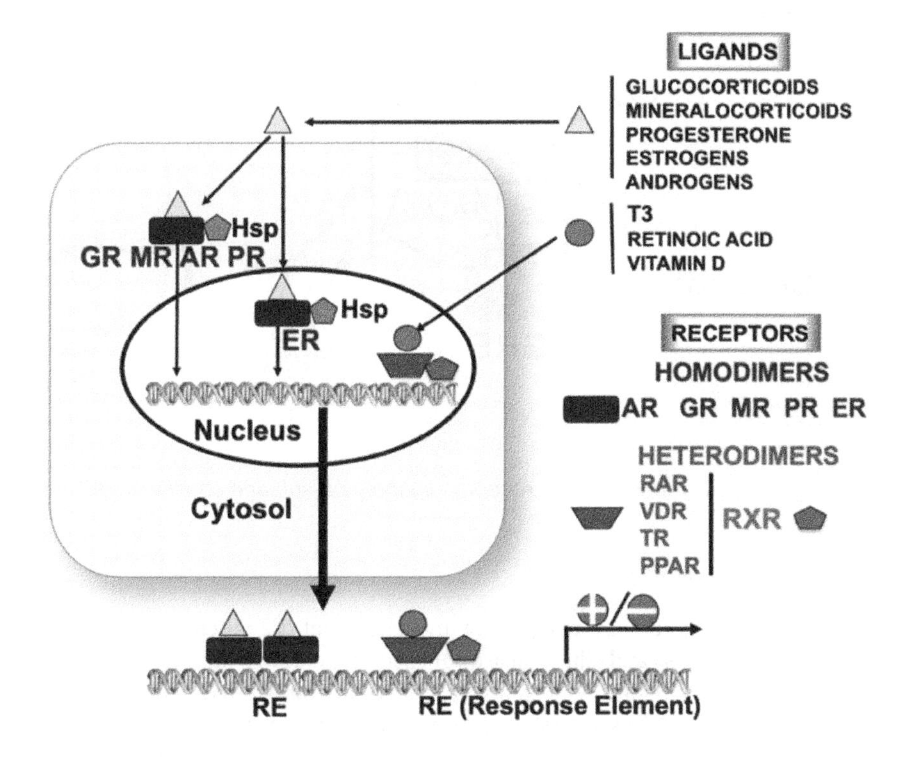

Figure 2.19. *Characteristics of nuclear receptors and their ligands*

There are two mechanisms to repress transcription: (1) NRs recruit corepressor proteins such as histone deacetylases (HDACs), acting in opposition to HATs and therefore restricting the chromatin opening which blocks the transcriptional machinery (Glass and Rosenfeld 2000) and (2) NRs interact with a "negative DNA response element" (Surjit et al. 2011).

To date, advances in understanding the mechanism of action of NRs have been made by imaging structural features of ligand-binding domains with ligands. But this approach has some limits; it does not capture conformational and allosteric (see section 3.1.3.1) effects transmitted by other domains within the receptors.

The next goal is to decipher the dynamic interactions of NRs with their coregulators and chromatin at a spatio-temporal resolution to understand how NRs control gene expression (Lenstra et al. 2016). This requires single-molecule studies in live cells to simultaneously track the 3D spatial distribution over time and monitor 3D enhancer organization (Presman et al. 2017).

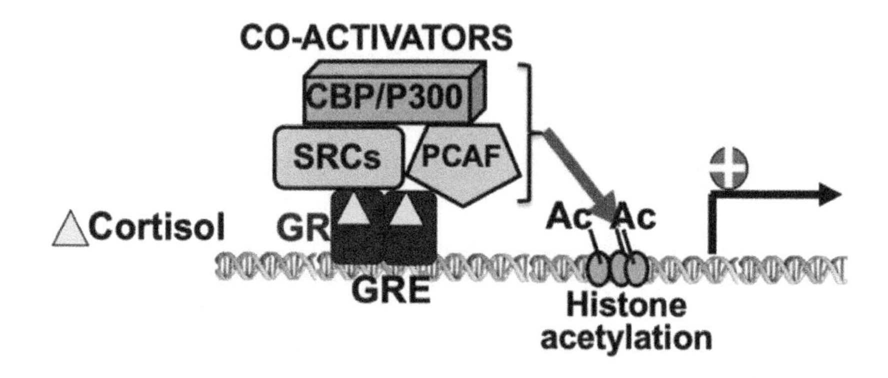

Figure 2.20. *Glucocorticoid receptor, ligand (cortisol) and coactivators*

Non-genomic action of NRs

In the 1970s, there was compelling evidence that E2 altered the firing of hypothalamic neurons and the release of neuropeptides within seconds. This observation was not in line with genomic effects of this steroid. The concept of "rapid" non-genomic effects for estrogen was unthinkable for neuroendocrinologists. This changed in the 1990s when membrane localization of receptor ER α was documented in pituitary cells, and in neuropeptide Y (NPY) and pro-opiomelanocortin (POMC) hypothalamic neurons. Thus, E2 can act on these neurons by eliciting a combination of rapid changes in membrane excitability accompanied by slower alterations in gene expression. This has been reviewed in detail in Stincic et al. (2018).

3

Intracellular Events in Response to Signals

3.1. Signaling pathways

3.1.1. *General overview*

The development and maintenance of homeostasis of multicellular organisms require communication between cells within and between tissues through extracellular signals. These signals provide a message that is relayed by a receptor which in turn activates flux of intracellular signals regulating key cellular events such as proliferation, differentiation and survival. But the identification and characterization of these signaling pathways have led to the surprising conclusion that only a few exist. Multiple mechanisms fine-tune and orchestrate the duration, intensity and nature of these signals. Hence, the stimulation of a given pathway does not have a predefined outcome. For instance, two RTK ligands, NGF and EGF, cause different signaling outputs from the Ras transduction cascade in PC12 cells. Stimulation with EGF results in proliferation, whereas NGF leads to a neuronal differentiation (outgrowth of neurites) and these differences have been linked to transient and sustained activity of ERK, respectively (Marshall 1995). Thus, the nature of the signal-decay phase can define the signaling outcome (Ryu et al. 2015).

Other experiments with variable duration and periodicity of ligand exposure also led to the activation of different signaling pathways. For instance, sustained or transient stimulations with insulin in cultured cells led to sustained or transient phosphorylation-induced activation of Akt, but the two downstream components of Akt selectively responded to transient S6K phosphorylation or sustained G6Pase transcription (Kubota et al. 2012). Similarly, PTH exerts two opposite effects

---

For a color version of all the figures in this chapter, see www.iste.co.uk/gilbert/concepts.zip.

depending on the duration of the administration of the hormone (Li et al. 2007). Continuous infusion and pulsatile treatment or daily injection causes a bone resorption and a bone synthesis (anabolism), respectively. This paradoxical anabolic effect can be explained by a transient upregulation of genes classically associated with a resorptive response. This observation suggests that the PTH likely regulates differentially two signaling pathways.

Finally, in response to varying GnRH pulse frequencies, the GnRHR differentially activates distinct signaling pathways. When applied over time with high pulse frequency, it stimulated the transcription of β-LH but not β-FSH, whereas the reverse was true with lower GnRH frequencies.

It is obvious that the response to signaling pathway activation is complex and involves the regulation of many processes, and the question is how a single ubiquitous signaling molecule can effectively integrate the myriad extracellular signals and generate a diverse array of responses. In the early 1980s, it was suggested that the wide range of physiological responses initiated by a variety of stimuli that produce cAMP might be indicative of compartmentalization of cAMP molecules in the cell. Studies throughout the last four decades have identified three major types of proteins which assist in the compartmentalization of cAMP signaling and generate a wide range of cellular responses (reviewed in Robichaux III and Cheng (2018)). These systems include adenylate cyclases (ACs), phosphodiesterases (PDEs) and A-kinase-anchoring proteins (AKAPs). Detailed roles are reported in section 3.1.5.2.3. Other mechanisms can explain the diversification of cellular responses. For instance, Akt is a central node in signaling and each physiological response appears to be mediated by multiple targets. Furthermore, some Akt substrates control more than one cellular function, such as GSK-3 which regulates metabolism, proliferation and survival (Manning and Cantley 2007).

Taken together, these findings point out that the pattern and the dynamics of cell signaling are determined by different mechanisms that contribute to signal termination initiated by the receptor activation. The question is whether, how and when to put a brake(s) on the response. This theme is now developed, but it is interesting to briefly discuss how these regulatory mechanisms took place and have been applied to the regulation of signaling pathways.

3.1.2. *Signal termination*

Historically, we can trace interest in regulated systems back to Ctesibius (250 BCE), a Greek scientist who used a simple float system to maintain a constant supply of water to a water clock. This is the first example of the use of negative feedback in a man-made device. Other historical insights can be found in the review by S. Bennett (1996). With respect to the history of biological feedback, this goes back to observations by E. Pflüger in the 1870s that organs and other living systems "satisfy their own needs". The concept was then introduced in the 1950s by the engineering mathematician N. Wiener (2019) who defined the feedback as "the chain of the transmission and return of information". Evidence for negative feedback was mentioned in 1956 by H.E. Umbarger after his discovery of feedback inhibition in the isoleucine biosynthesis pathway. He stated: "in the living organism, processes are controlled from one or more feedback loops that prevent any one phase of the process from being carried to a catastrophic extreme" (Umbarger 1956). These feedback loops became central to investigators studying the kinetics properties of sequential enzyme reactions, including metabolic end-products, inhibition and transcriptional self-repression (Umbarger 1956; Monod and Jacob 1961), but it illustrates only one aspect of how homeostasis works.

This concept of feedback – negative or positive – also applies to molecular mechanisms that control signal activation and termination. Cellular responses to extracellular signals are, in general, transient; they last from seconds up to a few minutes, despite continued presence of the ligand in the pericellular space. To avoid inappropriately strong cellular response, several molecular mechanisms exist that terminate the cell signaling, thus reducing cellular or tissue response. In this way, the molecular mechanisms that are involved bring the system closer to a state of stability or homeostasis.

It should also be emphasized that all negative regulatory events should be reversible. In certain circumstances, terminating the cell signaling can be a response to a new cellular state or fate, such as apoptosis, commitment to differentiation and cell-cycle entry. Readers are referred to an excellent review by M.A. Lemmon (Lemmon et al. 2016).

3.1.2.1. *Negative and positive feedback loops*

Feedback loops are processes that connect output signals to their inputs. Feedback became an influential concept that led to W. Cannon's theory of physiological homeostasis (Cannon 1929), and it can be defined as follows: it is a property of a control system to use its output as (a part of) its input. This concept is useful as a framework for understanding how intracellular signaling systems elicit specific cell responses. It is now well established that the magnitude, duration and oscillation of cellular signaling pathway responses are often limited by negative feedback loops.

Signaling with a single negative feedback loop

A negative feedback loop is defined as sequential regulatory steps that feed the output signal inverted back to the input. Upon stimulation, the output signal rapidly increases but is attenuated once it passes a threshold. In other words, a stimulus triggers a response inside the cell and, at the same time, it also induces a process that inhibits the response to a future exposure to the same stimulus (Dohlman 2002).

The simplest feedback loop is observed when the blood glucose level rises, hepatic glucose production decreases and vice versa. A second feedback loop involves the α- and β-cells and glucose. The latter inhibits glucagon secretion, whose role is to stimulate hepatic glucose production. Glucose also stimulates insulin release, whose role is to lower glucose production (Cherrington 1999). This dual control of glucose on glucagon and insulin release leads to a quick adjustment of its circulating level, and it operates through two feedback loops.

A negative feedback loop also operates in the immune system. To regulate an immune response, the inhibition of T-cell activation occurs through PD1. This response is also observed in tumor cells that then evade the immunosurveillance.

Note that other examples of feedback loops are given in section 3.1.2.2.

Signaling with a single positive feedback loop

This type of feedback occurs when a deviation in the controlled quantity is further amplified by the control system, or activation of one system tends to activate the other as well. It is inherently unstable.

Functions of positive feedback in biological systems were described in the 1910s (Lotka 1910). This feedback is defined as a set of regulatory steps that feeds the output signal back to the input signal. If signaling output activity increases, positive feedback will further increase the responses to subsequent stimulations, thereby enhancing the output signal (Mitrophanov and Groisman 2008). Here are three examples: (1) increasing levels of circulating estradiol (E2) during the follicular phase of the ovarian cycle act on the brain to trigger a massive release of gonadotropin-releasing hormone (GnRH) that evokes the pituitary luteinizing hormone (LH) surge responsible for ovulation in mammals (Herbison 2008) (see further details in section 4.3.1.2); (2) increased prolactin secretion in response to suckling (see details in section 4.3.1.2); and (3) extracellular matrix binding promotes integrin clustering and association with the cytoskeleton, which in turn promotes further integrin clustering and matrix organization in a positive feedback system (Giancotti and Ruoslahti 1999) (details are provided in section 2.3.1.6).

Positive feedback of estrogen on the hypothalamic–pituitary axis is essential for ovulation in the female. The ratio of LH to FSH secretion rises as the frequency of pulsatile GnRH release increases during the late follicular phase of the normal menstrual cycle. Increased LH secretion stimulates estrogen production from the ovary, which through positive feedback leads to the midcycle LH surge that causes ovulation. This feedback response is a sexually differentiated phenomenon, male rats do not respond to elevated estradiol with a surge of LH. Elegant studies carried out in rats showed that neonatal castration of males allows them to respond to estradiol in adulthood with an LH surge. Conversely, testosterone treatment of neonatal females defeminizes the brain such that they can no longer respond to estradiol treatment in adulthood with a surge of LH (Handa et al. 1985). Thus, sensitivity of the post-pubertal female brain to respond to elevations in estradiol with a surge of LH is primed by the absence of testosterone in neonatal females.

It must be mentioned that the sympathetic nervous system (SNS) and the hypothalamic–pituitary–adrenal (HPA) axis are implicated in a mutual positive feedback loop and activation of one system activates the other as well (Elenkov et al. 2000).

Multiple lines of evidence indicate that biological systems often feature positive and negative feedbacks that are active in only part of the cell rather than the whole cell. This results in distinct functions such as local pulses, waves and cell polarization.

Experimental and clinical data provide clear evidence that accurate signaling between the nervous endocrine and immune systems should lead to proper feedback (timing, magnitude, duration, etc.) and their defects may result in uncontrolled stimulation and disease.

Autoregulatory feedback loop

This feedback circuit is a mechanism by which the activator induces expression or activity of the inhibitor. Here are two examples: p53 and NF-κB.

In the absence of cellular stress, the intracellular level of p53 is low due to its interaction with the protein mdm2 (ubiquitin ligase), which targets it towards the proteasome for degradation (Wu et al. 1993). In response to a stress (intracellular or extracellular), p53 undergoes post-translational modifications, resulting in disruption of its interaction with mdm2. It leads to stabilization of p53 which in turn activates several stress response programs (cell-cycle arrest, DNA repair, etc.). It creates an autoregulatory feedback loop that regulates both the activity of p53 and the expression of the mdm2 gene.

This feedback loop is also initiated after the stimulation of the nuclear factor kappa B (NF-κB) signaling pathway (see Figure 3.10). In the absence of extracellular signals, IκB sequesters NF-κB in the cytoplasm, thus preventing its transcriptional activity. In response to cytokine signals there is a dissociation of

NF-κB from IκB that leads to a translocation of NF-κB into the nucleus, to activate transcription of target genes, including IκB itself. Upon IκB resynthesis, NF-κB is subsequently inactivated and redirected into the cytoplasm. It terminates the activation of the signaling pathway (Figure 3.10B).

3.1.2.2. *Termination of plasma membrane receptor signaling*

In many systems, signal termination is a result of a combined action of several different mechanisms that operate at different levels of signaling.

3.1.2.2.1. Signal attenuation: ligand, receptor and post-receptor levels

Ligand

Removal of the receptor ligand is the simplest way of signal termination. It operates after a release of a neurotransmitter (ligand) in the synaptic cleft, and it is achieved by a reuptake and/or degradation of the ligand (Figure 3.1).

* *Reuptake*

The reuptake is mediated by transporters. They are localized on presynaptic neuronal terminals as well as on astrocytes. Many different types of transporters exist, including monoamine transporters (e.g. for dopamine, noradrenaline and serotonin), amino acids and their derivatives (e.g. glutamate, glycine, GABA).

The importance of these transporters is best illustrated by the effects of their inhibitors. Inhibition of reuptake into presynaptic neuronal terminals prolongs effects of neurotransmitters on the post-synaptic receptor.

The astrocyte reuptake of glutamate prevents excitotoxicity (Figure 3.1).

* *Extracellular degradation*

– Degradation (hydrolysis) of acetylcholine (ACh) by the acetylcholinesterase (AChE). Hydrolysis of the ester bond produces choline and acetate. Choline is subsequently taken up by the presynaptic, cholinergic, neuronal terminal, where it can be reused for the synthesis of ACh. Inhibition of AChE prolongs action of ACh. For instance, nerve gas sarin blocks activity of AChE thus initially leading to overstimulation of post-synaptic cholinergic receptors in the neuromuscular junction (motor endplate). Ultimately, cholinergic receptors in skeletal muscle fibers are desensitized (see below) which results in respiratory arrest.

– Degradation of neuropeptides by neutral endopeptidase (NEP), also known as the endopeptidase. NEP degrades several peptides, such as substance P, somatostatin and bradykinin.

* *Intracellular degradation*

– Degradation of monoamines (noradrenaline, dopamine, serotonin) by the monoamine oxidase (MAO), which is in the external mitochondrial membrane.

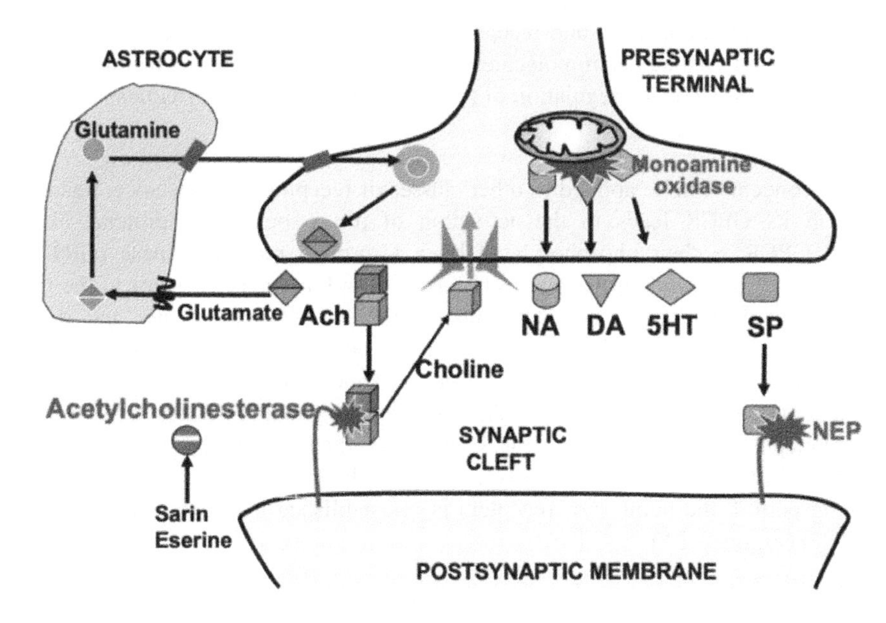

Figure 3.1. *Signal attenuation of neurotransmitter signal through degradation and reuptake of the ligand*

Receptor

The most well-known mechanism for switching off signaling is to reduce the levels of the receptor at the plasma membrane, which is called downregulation. It has been well studied for GPCRs and RTKs and other classes of receptors. There are two key steps, briefly described.

* *Endocytosis and uncoupling*

Endocytosis was first described in phagocytic cells in the 1930s, and the process was firmly established by the electron microcopy studies of Palade in the 1950s. It was considered as a nonspecific process that transported fluid into cells. In 1975, Pearse discovered that LDL (Low-Density Lipoprotein) receptors clustered and are destined for endocytosis. After internalization, the receptors dissociate from their ligand in endosomes, and they are recycled back to the cell surface.

This work on the LDL receptor introduced three concepts: the concepts of receptor-mediated endocytosis and receptor recycling, which provide a mechanism by which cells internalize macromolecules (proteins, hormones, growth factors etc.), and the concept of feedback regulation of receptors (see the review by Goldstein and Brown (2009)).

These concepts can be applied to other classes of receptors. For instance, ligand binding to its GPCR leads to the activation of the associated G protein. The activated GPCR is then phosphorylated by a G-protein receptor kinase (GRK) followed by arrestin binding that prevents further coupling of the receptor to the G protein. The GPCR-arrestin complex initiates the endocytosis process and this mechanism represents a direct (short-loop) feedback regulation. Following internalization, a fraction of receptors is deactivated and recycled back to the plasma membrane and the rest is targeted to the lysosomal degradation (Goh and Sorkin 2013; Irannejad and von Zastrow 2014). In this case, the number of active receptors in the plasma membrane is temporarily decreased (downregulation). When it lasts for a long period, the number of receptors is re-established by resynthesis (Böhm et al. 1997) (Figure 3.2A).

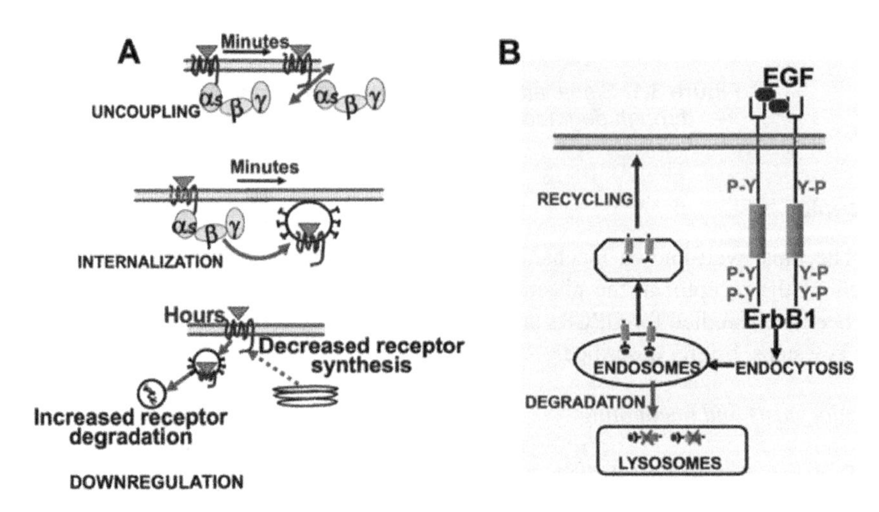

Figure 3.2. *Signal termination of activated GPCRs: uncoupling, internalization and downregulation. Signal termination of activated RTKs: endocytosis, degradation and recycling of receptors*

Regarding RTKs, ligand-induced activation causes a trans-autophosphorylation within receptor dimers, which serves to induce the endocytosis of the complex. It is then delivered to protein tyrosine phosphatases (PTPs) located in internal membranes in order to promote its inactivation by dephosphorylation (Haj et al. 2002). It is then targeted towards the endosomes and one fraction of receptors is degraded into the lysosomes and the other one is recycled to the plasma membrane (Figure 3.2B).

Ubiquitylation also plays a role in initiating GPCR and RTK endocytosis. It should be emphasized that the cycle of ubiquitylation/deubiquitylation share conceptual similarities with that of phosphorylation/dephosphorylation.

Post-receptor: GPCR, RTK cytokine and RSTK signaling pathways

Any receptor that activates intracellular pathways has multiple negative feedback loops and negative regulators in the signal networks that can keep a signal to terminate, the signaling, which ensures transient activation of the pathway and downstream transcription factors. The termination of these four activated signaling pathways is discussed.

– GPCR signaling pathway

Role of RGS (regulator of G-protein signaling)

The duration of GPCR signaling depends on the rate of Gα-GTP hydrolysis. Although the Gα subunit possesses intrinsic GTPase activity to promote conversion of the Gα subunit from an active to an inactive state (Gα-GDP), the role of RGS proteins is to enhance the intrinsic GTPase activity. Once GTP is hydrolyzed to GDP, the Gα-GDP subunit reassociates with the Gβγ subunit, which rapidly inactivates the signaling pathway. This process differs from that of GRK and arrestins reported above, by acting at the G-protein instead of the receptor.

Role of phosphodiesterases (PDEs)

PDEs hydrolyze cAMP, which inactivates the signaling pathway triggered by the elevation of intracellular cAMP concentration. There are several PDE isoforms. For instance, PDE3 is responsible for short-term inactivation of the signaling pathway. Intracellular cAMP concentration is therefore a result of equilibrium between the activities of AC and PDE (Conti et al. 2003) (Figure 3.3).

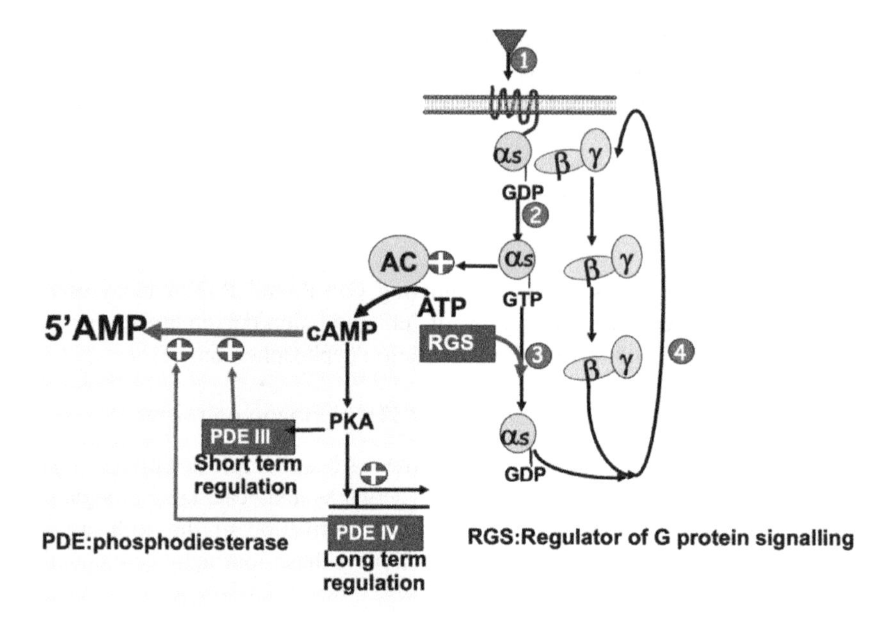

Figure 3.3. *Attenuation of the GPCR signaling pathway*

– RTK signaling pathway

Two signaling pathways are initiated from RTK and the mechanisms of negative feedback occurring at the post-receptor level include GTPase activation, dephosphorylation and inhibitory phosphorylation events.

1) Raf/MEK/ERK signaling pathway

**GTPase activating proteins (GAPs)*

In response to RTK activation, the downstream molecule Ras binds GTP (Ras-GTP), which in turn activates the signaling Raf/MEK/ERK. The deactivation of this pathway is mediated by GTPase Activating Proteins (GAPs) which bind to Ras-GTP and increase its intrinsic GTPase activity, resulting in GTP hydrolysis and consequently a deactivation of downstream events, that is, Raf/MEK/ERK signaling (Figure 3.4A).

**Phosphatases and protein tyrosine phosphatases (PTPs)*

The Raf/MEK/ERK signaling pathway is terminated by dephosphorylation of their components by several key negative key regulators of cell signaling. They are classified into two categories: p-Tyr-specific phosphatases and dual specificity phosphatases (p-Tyr and p-Ser/Thr phosphatases). The PTPs terminate activation of RTKs by reversing the autophosphorylation of the receptor.

* *Phosphorylation of signaling pathway proteins*

Another mechanism of negative feedback involves Sos (Son of sevenless), which is a Ras-GEF (guanosine exchange factor). Activation of ERK in the signaling pathway causes a phosphorylation of Sos, which can no longer interact with Grb2, preventing the conversion of Ras-GDP to Ras-GTP required for the activation of downstream events. This is the mechanism of negative feedback, which is shared by numerous signaling pathways: the most distal molecule of the signaling pathway inhibits function and/or activity of a molecule that is upstream in the signaling pathway.

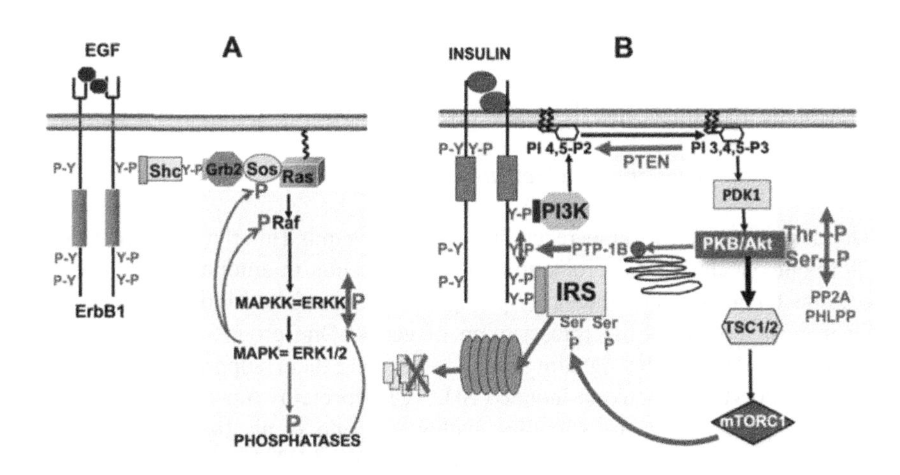

Figure 3.4. *Attenuation of RTK signaling pathways. Ras/MAPK signaling (A) and PI3K/Akt/TORC1 signaling (B)*

2) PI3K-Akt signaling pathway

This is also subjected to negative feedback to assure a transient response to stimulatory signals. For instance, activation of insulin and IGF1 receptors leads to PI3K-Akt activation which in turn activates mTORC1. The latter promotes IRS degradation through phosphorylation of multiple serine residues on this protein, thereby dampening PI3K activation (Tzatsos and Kandror 2006). Inversely, mTORC1 inhibition increases IRS stability, resulting in more robust and prolonged insulin and IGF1 signaling.

In this pathway, Akt is a key master regulator of a wide range of physiological functions, such as metabolism, proliferation, survival and growth. It is activated in response to the binding of a growth factor to its receptor. Briefly, there ensues an activation of PI3K that generates PIP2 and PIP3 and enables an interaction of Akt with the 3' polyphosphoinositides. This relocalization to the plasma membrane promotes a conformational change that unmasks the catalytic kinase core and enables its phosphorylation at specific sites, Thr308 and Ser473, which are required for its activation (Andjelkovic et al. 1997). To terminate Akt signaling, three phosphatases are involved: (1) the protein phosphatase 2 (PP2A) which dephosphorylates Thr308; (2) the PH domain and leucine-rich repeat protein phosphatase (PHLPP); and (3) the protein phosphatase and tensin homologue (PTEN). In addition, the activation of the protein tyrosine phosphatase-1B (PTP-1B) by Akt results in dephosphorylation of IRS, leading to an attenuation of the signaling pathway (Elchebly et al. 1999) (Figure 3.4B).

– Cytokine and RSTK signaling pathways

* *Transcription of genes encoding negative regulators*

The activation of cytokine receptors by their ligands (interleukins, interferons and hormones such as GH, prolactin and leptin) results in autophosphorylation of JAK (Janus Kinase), which thereafter phosphorylates the transcription factor STAT, which in turn modulates the transcription of genes. One group of genes encodes proteins which dampen the signaling pathway. Among them, suppressor of cytokine signaling (SOCS) and cytokine-induced SH2 (CIS) proteins function in a negative feedback loop to inhibit the signaling pathway (Endo et al. 1997). Each SOCS protein contains an SH2 domain that compete with the SH2 domains of STATs for binding to activated cytokine receptors, which causes a suppression of the signal (Palmer and Restifo 2009) (Figure 3.5A).

Likewise, activation of RSTK signaling (TGFβ-stimulated signaling) results in the expression of inhibitory proteins I-Smads: Smad 6 and 7. The latter interacts with the type I receptor and blocks recruitment of R-Smads (Smad 2, etc.) to the receptor (Figure 3.5B).

These proteins SOCS and I-Smad play a role of decoy molecules (see section 3.3.5).

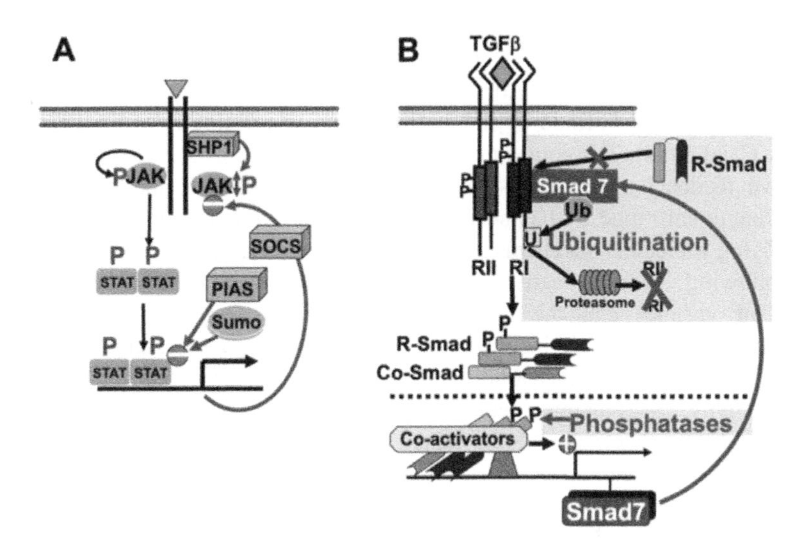

Figure 3.5. *Attenuation of signaling pathways. The JAK/STAT cytokine signaling pathway (A) and the Smad/Co-Smad/R-Smad of RSTK signaling pathway (B)*

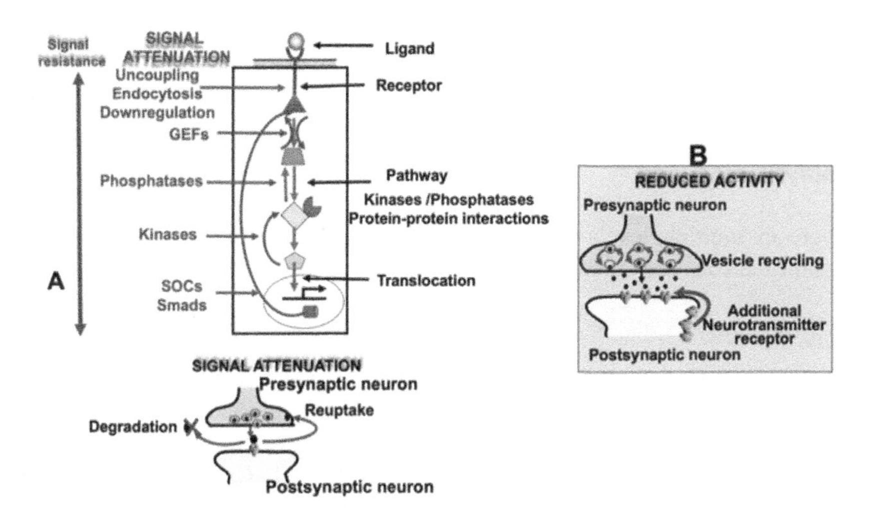

Figure 3.6. *Schematic illustration of cell signaling dynamics in time and space. Stimulation and attenuation of signaling pathway, through a membrane receptor (A), and neurotransmission and attenuation of the response, through degradation and reuptake of the neurotransmitter (A). Homeostatic synaptic plasticity at an excitatory synapse (B). Neurons offset changes in network activity by adjusting their pre- and postsynaptic strengths. Thus, reduced activity is offset presynaptically by enhancing the recycling of the vesicles and postsynaptically through incorporation of additional receptors at the synapse*

3.1.2.2.2. Homeostatic synaptic plasticity

The idea of synaptic plasticity first emerged in 1894. S. Ramón y Cajal proposed that memories must be formed by the strengthening of existing neural connections (Ramón y Cajal 1894). The concept of homeostatic synaptic plasticity refers to a set of negative feedback mechanisms that are used by neurons to maintain activity within a functional range by balanced synaptic excitation and inhibition. Although the underlying molecular mechanisms are not fully elucidated, the proposed model is the following: reduced activity is offset presynaptically by enhancing the recycling of vesicles and the release probability. Post-synaptically, additional neurotransmitter receptors are incorporated at the synapse by a mechanism involving lateral diffusion from extrasynaptic sites. Inversely, to compensate an increased network activity, presynaptic neurons decrease their neurotransmitter release probability while post-synaptic cells reduce the number of post-synaptic receptors by endocytosis (Pozo and Goda 2010) (Figure 3.6B).

3.1.2.2.3. Homeostatic hormone and ion levels

The homeostatic hormone level is discussed in section 4.3.1, which illustrates the negative feedback concept.

Ca^{2+} homeostasis: increased blood level of Ca^{2+} leads to a negative feedback on the parathyroid glands, thereby inhibiting further PTH release (Hannan et al. 2018).

3.1.3. *Control of protein activities: allostery, covalent modifications and proteolytic cleavage*

There are multiple levels of protein control that function in response to different environmental cues and signals.

3.1.3.1. *Allostery*

This biological phenomenon was first introduced over a century ago by C. Bohr, who observed positive cooperativity of oxygen binding to the hemoglobin protein (Bohr et al. 1904). Since then, numerous approaches have emerged to explain this phenomenon, and the concept was initially proposed by J.P. Changeux while he was struggling with the interpretation of the inhibitory mechanism mediated by bacterial regulatory enzymes. The data could not be accounted for by the classical Michaelis scheme of a competitive inhibition, and it was suggested that substrate and regulatory effector bound at topographically distinct sites. The model of "no-overlapping" sites was presented at the 1961 Cold Spring Harbor Symposium on Quantitative Biology (Changeux 1961), and the word allosteric was subsequently coined in the General Conclusions of the same meeting by J. Monod and F. Jacob. Then, the Monod-Wyman-Changeux (MWC) concerted model was introduced to describe this cooperativity, and it is the leading allosteric model (Monod et al. 1963). This mechanism enables a modulation of protein properties and regulation of activity. In contrast with the classical mechanism of competitive interaction between ligands for a common site, allosteric interactions take place between distinct sites and are mediated by a reversible conformational change of the protein. It can be broadly defined as the binding of a ligand at one site causing a change in affinity or catalytic efficiency at a distant site. Allostery is composed of two Greek roots expressing the difference (allo) in (stereo) specificity of the two binding sites; namely, that regulatory and active sites are distinct and non-overlapping.

In recent years, an increasing number of reports have pointed to common mechanisms governing the allosteric modulation of membrane proteins, including conformational selection, oligomerization and the modulation of allosteric sites.

The concept has been extended to membrane receptors for neurotransmitters and shown to apply to the signal transduction process, which, in the case of the nicotinic acetylcholine receptor (nAChR), links the ACh-binding site to the ion channel. The transition to an open active state of the channel is triggered by the occupation of agonist-binding site, which changes the structure from a tense (closed) state towards a more relaxed (open) state. These allosteric properties have been successfully engineered into proteins for drug design or the development of novel biosensors.

Examples of coordinated allosteric regulation in metabolism

– Fatty acid synthesis versus oxidation: malonyl-CoA coordinates the synthesis and oxidation of fatty acids to prevent these two opposite pathways (anabolic and catabolic) to occur simultaneously (Figure 3.7). Thus, in the presence of glucose availability, citrate is shuttled out of the mitochondria to act as an allosteric activator of acetyl-CoA carboxylase, resulting in increased malonyl-CoA level that fuels the lipogenesis pathway. In addition to being the first intermediate of fatty acid biosynthesis, malonyl-CoA is also an inhibitor of the β-oxidation (catabolic pathway). It is an allosteric inhibitor of the carnitine palmitoyltransferase-1 which regulates the entry of acyl-CoA for β-oxidation in the mitochondrial matrix. Hence, these two pathways are not active at the same time but rather they are segregated in time as well as space (cytosol vs. mitochondrial matrix).

– Control of glycolytic flux: PFK1 (6-phosphofructo-1-kinase) is the site of the most complex control over the glycolytic flux and allosteric regulation. A total of six binding sites are found: the catalytic ATP and F6P-binding sites, activator-binding sites for adenine nucleotides and fructose-2,6-bisphosphate (F2,6P) and inhibitor-binding sites for ATP and citrate. Citrate seems to play a major role in downregulation of PFK1. In fact, citrate build-up is a sign of anaplerotic conditions in the cells that are characterized by elevated TCA cycle intermediates and by the need to slow down glycolysis (Kemp and Foe 1983).

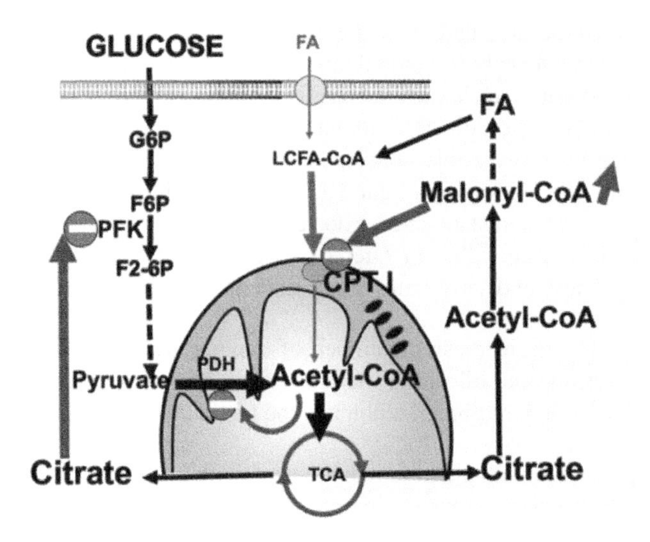

Figure 3.7. *Example of coordinated allosteric regulation in metabolism. To promote fatty acid synthesis, malonyl CoA inhibits fatty acid oxidation through an allosteric inhibition of the CPT1*

3.1.3.2. *Post-translational modifications*

In a broad sense, post-translational modifications of proteins confer novel properties to the modified proteins, including changes in enzymatic activity, subcellular localization, interaction partners, protein stability, etc. Given that about 5% of the genome is dedicated to enzymes that carry out these modifications and include phosphorylation, glycosylation, methylation, acetylation, amidation sumoylation, ubiquitination, and many other types, numerous studies have been focused on the role of these changes in the enzymatic activity (for more detailed vocabulary, see: http://www.uniprot.org/docs/ptmlist).

In accordance with cellular requirements, an interconvertible enzyme is shifted between covalently modified and unmodified forms by two converter enzymes that catalyze opposing reactions. In other words, this interconvertible enzyme system is in fact a continuous cyclic process consisting of two tightly coupled opposing cascades. A theoretical analysis of cyclic cascades reveals that: (1) they possess a capacity for signal amplification, (2) they modulate the amplitude of the maximal response and the sensitivity to changes in the concentration of primary allosteric effectors, (3) they serve as biological integration systems which can sense simultaneous fluctuations in the intracellular concentrations of metabolites and adjust the specific activity of the interconvertible enzyme accordingly and (4) they serve as rate amplifiers that are capable of responding within millisecond range to changes in metabolite levels (see the review by Chock et al. (1980)).

Protein phosphorylation

It is the most studied post-translational modifications and occupies a central role in regulating signal transduction, metabolism and other cellular processes. It was discovered in 1959 (Fischer et al. 1959), and it is an addition of phosphoryl group from ATP to serine, threonine and tyrosine residues. Glycogen phosphorylase was the first cellular protein recognized as undergoing a reversible covalent modification: phosphorylation–dephosphorylation (Fischer and Krebs 1955). Since elucidation of this process other covalent modifications have been reported such as acetylation-deacetylation and methylation-demethylation. To serve a protein function, these processes must be tightly regulated. For instance, the phosphorylation–dephosphorylation of enzymes requires a tight control of the protein kinases and/or the protein phosphatases. For instance, the regulation of the glycogen phosphorylase phosphatase reaction in the liver is substrate-directed. Thus, the binding of glucose or glucose-6-P to the phosphorylase provides a better substrate for the phosphorylase (Hers 1976). The activities of more than 30 enzymes – involved in carbohydrate, protein, lipid, and nucleic acid metabolism – are modulated by a phosphorylation/dephosphorylation cycle. The reader is referred to an excellent review (E.G. Krebs and J.A. Beavo 1979) for information with respect to extensive work that has been carried out on the phosphorylation of enzymes.

The process of phosphorylation–dephosphorylation also plays a key role in response to the activation and termination of a signaling pathway by a growth factor. The Akt phosphorylation and dephosphorylation which occurs after PI3K stimulation is reported in Figure 3.4B.

Protein acetylation

Protein lysine acetylation was first reported in 1963 (Phillips 1963). In 1964, V. Allfrey reported the identification of histone acetylation and proposed a regulatory role for this protein modification in transcription regulation. Subsequently, histone acetyltransferases (HATs) and histone deacetylases (HDACs) were identified, and Allfrey proposed a "dynamic and reversible mechanism for activation as well as repression of RNA synthesis" by reversible post-translational histone acetylation (Allfrey et al. 1964). But it took another 30 years to validate this hypothesis. Since then, proteomic analyses on protein acetylation revealed that many acetylated proteins play a key role in regulating gene expression and metabolic enzymes. Like regulation by phosphorylation, acetylation appears to have divergent effects on metabolic enzymes, including inhibition and activation (Wang et al. 2010). It should be pointed out that ≈30 acetyltransferases and ≈ 18 deacetylases have been identified in the human genome and ≈2,200 proteins undergo acetylation, and they display a great functional diversity. By comparison, more than 500 kinases and nearly 150 phosphatases have been identified, and they outnumber acetyltransferases, which likely provide a higher degree of substrate specificity for phosphorylation. Nearly all enzymes involved in glycolysis, gluconeogenesis, the tricarboxylic acid (TCA) cycle, fatty acid oxidation, the urea cycle and nitrogen metabolism and glycogen metabolism are acetylated (Guan and Xiong 2011). These findings revealed a new mechanism that might provide a means of coordinating different metabolic pathways. However, it raises several questions such as: how do a limited number of acetyltransferases and deacetylases control the activity of so many metabolic enzymes? Is there a mechanism for recruiting specific substrates? To name but a few.

3.1.3.3. *Protein ubiquitylation and sumoylation*

The process of ubiquitination is the addition of ubiquitin to lysine residues in order to target the protein to the proteasome, which degrades it. Sumoylation is the addition of a small ubiquitin-like modifier (SUMO), and it controls many aspects of nuclear function.

3.1.3.4. *Zymogen activation*

Many enzymes are synthesized as inactive precursors, which are called zymogens or proenzymes. They are subsequently activated by the cleavage of one or a few specific peptide bonds. Many proteases are produced as proenzymes and the first "hints" on regulation of protease activity date back to 1935 from studies on trypsinogen and chymotrypsinogen (Kunitz and Northrop 1935). A common mechanism of activation is the presence of an amino acid sequence which ranges between 1 and 200 amino acids. These prodomains or propeptides interact with the protease active site and hamper the interaction with substrates. In contrast with allosteric control and covalent regulation, the process does not require ATP, and it is irreversible. Here are two examples of protease activation:

– MMPs: matrix metalloproteinases are the proteinases that participate in extracellular matrix (ECM) degradation. Their activities are regulated at the level of the proenzymes and by endogenous inhibitors, called tissue inhibitors of metalloproteinases (TIMPs), which bind MMPs in a 1:1 stoichiometry (Visse and Nagase 2003). In general, proMMPs results from the disruption of a prodomain ranging in length from 66 to 120 amino acids.

– Caspases: caspases are the main drivers of apoptosis or programmed cell death. Their activity is extremely important to maintain cellular homeostasis and, therefore, their activation mechanisms are tightly controlled. The primary structure of all caspases is similar; they are intracellular proteases expressed as zymogens.

3.1.3.5. *Bifunctional enzyme*

PFK2 (6-phosphofructo-2-kinase)/FBPase-2 (fructose-2,6-bisphosphatase) is the homodimeric bifunctional enzyme that catalyzes the synthesis and degradation of F-2,6-P. The latter is a signal molecule that controls glycolysis, through allosteric stimulation of the phosphofructokinase (PFK).

3.1.3.6. *Isoenzymes*

Isoenzymes catalyze the same chemical reaction, but typically differ with respect to their primary structure, intracellular location and physiological role. They represent multiple forms of enzymes that permit the fine-tuning of metabolism to meet the particular needs of a given tissue. For instance, there are four distinct forms of pyruvate kinase, named L (liver), R (red blood cells), M (muscle) and K (kidney). There are four isoenzymes of hexokinase referred to as Type I, Type II, Type III and Type IV, the latter commonly called "glucokinase". They differ with respect to their catalytic and regulatory properties as well as subcellular localization, which is likely an important factor in determining the pattern of glucose metabolism in tissues/cells (see the review by Wilson (2003)).

3.1.4. *Impaired cellular responses to extracellular signals*

3.1.4.1. *The concept of signal resistance*

Two components of signal resistance: reduced sensitivity and responsiveness

Under pathological conditions, a biological response to an extracellular signal may be reduced. Such resistance to extracellular signals is characterized by reduced sensitivity and/or responsiveness.

Reduced sensitivity means that an extracellular signal achieves its effects at higher concentrations than under normal conditions. Its concentration (dose)–response curve is therefore shifted to the right towards higher concentrations (Figure 3.8). If only sensitivity is reduced, the extracellular signal may still achieve its maximal effect (E_{max}), but the concentrations at which it achieves its actions are increased, which is best characterized by an increase in EC50 (i.e. concentration at which a half-maximal effect is triggered).

A reduction in sensitivity is typically caused by a reduction in the number of receptors. Importantly, cells and tissues usually have more receptors than are needed to achieve a maximal biological response. A corollary to this is that often only a minor fraction of receptors needs to be occupied by a ligand to trigger a full effect, while the remaining receptors can be regarded as a reserve population of receptors. Hence, if the total number of receptors in a certain tissue is reduced but is still higher than the number needed for a maximal response, a full response can be achieved, but it will typically occur only at increased concentrations of the ligand. This is a consequence of the chemical equilibrium between ligand (L) and receptor (R): if we take the simplest model, the formation of the active ligand–receptor complex (LR), which is proportional to intensity of the biological effect, is dependent on the concentration of both ligand and receptor ($L + R \leftrightarrow LR$). Thus, if the number of receptors is reduced, more ligand is needed to "push" the reaction towards the formation of the ligand–receptor complex.

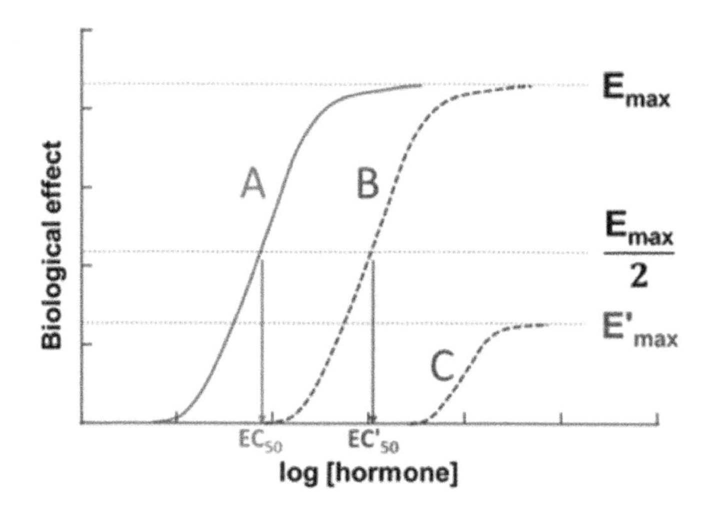

Figure 3.8. *Concentration–response curves for a hypothetical hormone. A: Normal concentration-response curve. B: A shift to the right due to reduced sensitivity, characterized by an increase in EC50 (i.e. concentration at which half-maximal effect ($E_{max}/2$) is achieved). C: The curve indicates resistance due to a decrease in sensitivity as well as responsiveness. Reduced responsiveness is characterized by a lower maximal effect (E'_{max})*

Responsiveness refers to the maximal response that can be achieved by an extracellular signal. Reduced responsiveness is therefore characterized by a reduction in E_{max}. This may occur due to a reduced number of effector molecules. For instance, if the number of GLUT4 transporters is markedly reduced in skeletal muscle, insulin is not able to increase the uptake of glucose to the same degree as under normal conditions.

Historical development of the resistance concept

The first observation of signal resistance/insensitivity is usually considered to be a report by F. Albright in 1942 (Arnold 1942). In a cohort of patients, he described a marked hypocalcemia and hyperphosphatemia despite elevated serum PTH levels (Albright and Reifenstein 1948). Given that the normal biological response would have been the opposite, it was coined pseudohypoparathyroidism (PHP) and later it has been explained by a decreased ability of the PTH to increase calcemia and lower phosphatemia. The molecular defect resides in GNAS, the gene encoding the α-subunit of the stimulatory GTP-binding protein (Gαs). This mutation leads to reduced protein level and cellular activity. Interestingly, in PHP, hormones (glucagon, vasopressin) that rely on Gαs mediate their action. It therefore suggests that Gαs mRNA is transcribed in certain cells only from one parental allele. Detailed investigations of GNAS have recently provided new insights into Gαs expression and the mechanisms underlying hormone resistance (see the review by Bastepe and Jüppner (2003)).

With the benefit of the hindsight, we can actually consider a report concerning the resistance to the effects of vitamin D as the first description of hormone resistance. In 1937, Albright and colleagues reported that rickets is not always responsive to standard doses of vitamin D (Albright et al. 1937). In such cases it was possible to achieve a therapeutic effect only with extreme doses of vitamin D. Clearly, in those days, vitamin D was not considered a hormone and molecular mechanisms of its action were not understood. Nevertheless, Albright and colleagues were able to show that the lack of antirachitic effects of vitamin D was not caused by malabsorption, highlighting that rickets in this case was not due to deficiency of vitamin D, but due to intrinsic resistance to its effects. The condition may be referred to as a pseudodeficiency of vitamin D or a pseudovitamin D deficiency because pathological consequences mimic those of a real deficiency (e.g. due to low oral intake or malabsorption of vitamin D). We now know that the pseudovitamin D deficiency can be caused either by mutations in 25-OH-1α-hydroxylase, the kidney enzyme that converts 25-(OH)-vitamin D, an inactive precursor, into 1,25-$(OH)_2$-vitamin D, the physiologically active form, or by mutations of the vitamin D receptor (VDR). Although we cannot establish with certainty the underlying cause for the apparent resistance to vitamin D actions in Albright's 1937 report, this case established the concept that resistance may underlie deficient action of a nutritional factor or a hormone (Sinding 1989).

Another description of complete signal insensitivity was reported by Z. Laron in 1966 (Laron et al. 1966) in two siblings with the classical appearance of growth hormone (GH) deficiency (short stature), but presenting elevated levels of GH. The GH insensitivity was associated with a lack of IGF1 in response to GH treatment. It was not until 1989 that the molecular defect was characterized. Subsequently, several genetic defects have been reported to explain impairment of GH and IGF1 actions (see the review by Domene and Fierro-Carrion (2018)).

In most cases, resistance to an extracellular signal is a state in which a given concentration of signal produces a subnormal biological response. The molecular mechanisms underlying the phenomenon have been investigated and the potential defects lie between the receptor which binds the signal and the last molecular event of the signaling pathway responsible for the cellular response.

3.1.4.2. *Receptor defects*

The target cell cannot respond to an extracellular signal without an appropriate receptor. Moreover, it is the receptor, not the ligand *per se*, that defines the response. For instance, while a cell may be responsive to catecholamines, such as adrenaline and noradrenaline, there are different types as well as subtypes of adrenergic receptors, meaning that the same ligand may trigger different effects. For instance, if a cell predominantly expresses the α1 adrenoceptor, which is coupled to $G_{q/11}$, the cell will respond by activating phospholipase C (PLC), thus leading to formation of DAG, which activates protein kinase C (PKC), and IP3, which triggers the release of Ca^{2+} from endoplasmic/sarcoplasmic reticulum. On the contrary, the activation of α2 adrenoceptor, which may be coupled to different G-proteins, may lead to the inhibition of adenylate cyclase with the attendant reduction in cAMP concentrations or activation of PLC or even PLA2. In contrast to α2 adrenoceptor, activation of β adrenoceptors stimulates adenylate cyclase, thus leading to an increase in cAMP levels. Clearly, even in the presence of the same ligands, such as adrenaline, the cell may respond in many different ways depending on the population of receptors that it possesses.

Given the importance of receptors for biological actions of hormones, it is not surprising that mutations leading to a loss of function (LOF) of receptor are the main causes of insensitivity to a signal. Here are some examples:

*Mutations of the melanocortin type 2 receptor MC2R cause resistance to adrenocorticotropic hormone (ACTH), a pituitary hormone that stimulates the secretion of cortisol, the major glucocorticoid in humans, from the cortex of the adrenal gland. Patients therefore suffer from glucocorticoid deficiency despite having elevated plasma concentrations of ACTH. An increase in ACTH levels occurs in a compensatory manner because deficiency of cortisol relieves negative feedback inhibition of ACTH secretion. Although ACTH cannot trigger its effects via the mutated MC2R it can still act via the related types of melanocortin receptors, which remain functional. For instance, high concentrations of ACTH stimulate melanocortin receptors in the skin, resulting in pronounced hyperpigmentation, which again highlights that it is the receptor that defines the final response. It also highlights that deficiency of hormone action can coexist with exaggeration of some of its effects if these effects are mediated via a different set of receptors.

*Defects of androgen receptors lead to a condition referred to as androgen insensitivity syndrome, also known as testicular feminization. In this condition, cells are resistant to the effects of testosterone and dihydrotestosterone. As a result, a person who is a chromosomal male (i.e. has the X and Y sex chromosome), (dihydro)testosterone is unable (if resistance is complete) or is less able (if resistance is partial) to effect masculinization. This results in a female phenotype or at least undermasculinization, depending on the extent of the defect (i.e. complete vs. partial resistance to androgens).

*Resistance to thyroid hormones (i.e. thyroxine and T3) is another important example. These hormones regulate various cellular functions via nuclear receptors TRα and TRβ (with splice variants TRβ1 and TRβ2), which dimerize with RXR to form functional transcription factors. Mutations of TRβ may affect the ligand-binding domain, thus leading to LOF. However, the mutated TRβ retains the ability to form dimers with RXR, which prevents the binding of functional TRα as well as TRβ. Since the mutated TRβ opposes the function of normal receptors, it acts in a dominant-negative manner, meaning that resistance to thyroid hormones occurs even in heterozygous carriers.

Other notable examples include the previously mentioned vitamin D resistance and Laron syndrome, which is characterized by a defect of growth hormone receptor, as well as defects of leptin receptor and melanocortin type 4 receptor (MC4R), which mediates effects of α-MSH in the anorexigenic pathway in hypothalamus downstream of the leptin receptor. Both types of defects lead to hyperphagia and consequently obesity. It is important to note that although such monogenic causes of obesity are rare (they altogether account for ~5% of all cases of obesity), mutations of MC4R are the most common reason for monogenic obesity.

3.1.4.3. *Suppression of the signaling pathway*

**Resistance to hormonal action may arise even though the receptor itself is fully functional*

One interesting example of this type is resistance to cortisol and other glucocorticoids. Cortisol exerts its action by binding to the glucocorticoid receptor α (GRα). In the absence of cortisol, GRα is located in cytoplasm, where it is bound to Hsp90 and other proteins. Upon binding of cortisol, GRα is released from the complex, is translocated into the nucleus and forms a GRα/GRα homodimer, which then regulates gene expression in positive as well as negative manner. Interestingly, this is not the only way GRα mediates effects of cortisol. For instance, GRα is able to bind to NF-κB and other transcription factors, thus modulating their action. In this way, GRα suppresses NF-κB-driven gene expression, including expression of inflammatory cytokines.

In addition to GRα, cells express GRβ, a truncated form of GR which is unable to bind glucocorticoids and is not transcriptionally active. However, GRβ can bind to GRα as homodimers and suppress its transcriptional activity. This dominant-negative effect of GRβ resembles the one described above for the mutated TRβ, the difference being that GRβ is a physiological inhibitor of glucocorticoid signaling. Upregulation of GRβ suppresses glucocorticoid signaling, which may be one of the mechanisms underlying reduced effectiveness of glucocorticoids used in pharmacotherapy.

*Insulin resistance also illustrates this phenomenon. Thus, increased IRS-1Ser/Thr phosphorylation is strongly correlated with insulin resistance in insulin-responsive cells or tissues (Tanti et al. 1994). Similarly, even partial LOF mutations of Akt (Latva-Rasku et al. 2018), a kinase downstream of insulin receptor, reduce response to insulin although insulin receptor is activated normally.

It is also important to note that signaling pathways branch downstream of receptors, which explains why resistance to hormone actions may occur selectively. For instance, in skeletal muscle, the pathway via PI3K/Akt may be severely resistant to insulin, which explains deficiency of its metabolic effects. However, other pathways, such as the mitogen-activated protein kinase (MAPK) signaling via ERK1/2 may remain normally responsive to insulin. Having two signaling pathways with different responses to insulin may lead to a paradoxical situation, where one pathway may be insufficiently active, while the other one is overly active. For instance, hyperinsulinemia, which occurs in response to insufficient metabolic actions of insulin, opposes a reduction in insulin-stimulated glucose disposal, thus preventing (if insulin resistance is not yet severe and insulin secretion is sufficient) or at least blunting (if insulin resistance is severe and insulin secretion cannot compensate it anymore) hyperglycemia. However, the activation of the MAPK pathway, which is normally responsive to insulin, may be exaggerated under hyperinsulinemic conditions, which may facilitate development of some types of cancer. Finally, it should be emphasized that extracellular signals contributing to hormone resistance: GH (Møller and Jørgensen 2009) and fatty acids, antagonize the hepatic and peripheral effects of insulin on glucose metabolism.

3.1.4.4. *A special case: an inappropriately increased response to a ligand*

The above paragraphs focus on the mechanism of hormonal resistance. However, sometimes the opposite effect may occur, i.e. the response of cells may become inappropriately increased. One of the most interesting examples is an inappropriate response to glucocorticoids in a condition referred to as apparent mineralocorticoid excess syndrome.

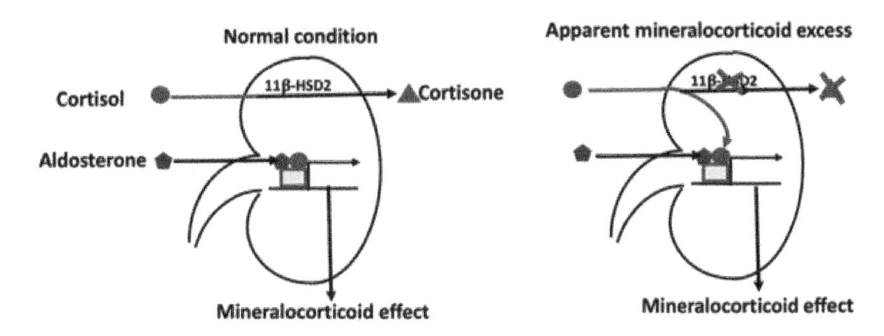

Figure 3.9. *Under normal conditions, 11β-hydroxysteroid dehydrogenase type 2 (11β-HSD2) protects the kidney cells from inappropriate activation of the mineralocorticoid receptor by converting cortisol, which can activate the receptor, to cortisone, which is inactive. Mutations of 11β-HSD2 increase cortisol levels, thus leading to activation of the receptor and an apparent excess of mineralocorticoids (i.e. aldosterone)*

The kidney is one of the major target organs for aldosterone, the main adrenocortical mineralocorticoid. Aldosterone triggers its effects via the mineralocorticoid receptor, a nuclear receptor that is not completely specific for aldosterone. Indeed, glucocorticoids, such as cortisol, can also bind to this receptor, which leads to mineralocorticoid-like effects, such as retention of Na^+ in the kidney. However, the kidney tubular cells are usually protected from an inappropriate activation of the mineralocorticoid receptor by cortisol because they express 11β-hydroxysteroid dehydrogenase type 2 (11β-HSD2), which converts the active cortisol into inactive cortisone. The specificity of aldosterone vs. cortisol actions in the kidney is therefore not achieved by selectivity of the receptor, which is responsive to both, but by intracellular inactivation of cortisol (Figure 3.9).

In individuals harboring mutations of 11β-HSD2 or in those who ingest too many sweets or cough lozenges based on the extracts of licorice (*Glycyrrhiza glabra*), which contain a natural inhibitor of 11β-HSD2, cortisol is not converted or is at least less efficiently converted to cortisone, thus leading to an inappropriate cortisol-induced activation of the mineralocorticoid receptor, which mimics an excess of aldosterone, for instance, by causing arterial hypertension. This example again highlights the importance of the receptor for the final biological effect in the cell.

3.1.5. *Subcellular localization and sequestration*

3.1.5.1. *Compartmentalization, sequestration, bioavailability and localization*

Cells are not homogeneous solutions of enzymes and metabolites. Indeed, reactions are often spatially restricted and take place in specific compartments, among which cytosol, mitochondria and nucleus are particularly prominent. Compartmentalization serves several purposes.

First, by spatially separating enzymes, opposite reactions or reaction sequences do not occur at the same time, which prevents or at least limits futile cycles. For instance, the synthesis of fatty acids occurs in the cytosol, while β-oxidation takes place in the mitochondrial matrix.

Second, compartmentalization enables the cell to concentrate enzymatic machinery and metabolites in restricted space, which increases the efficiency of reactions. If reactants and enzymes were distributed completely at random, metabolic pathways, especially those with long reaction sequences, could not proceed efficiently under conditions that usually prevail under physiological conditions. Even within a particular compartment, reactions likely do not proceed in a free, completely mixed, solution. For instance, enzymes of multistep metabolic pathways, such as the *de novo* purine synthesis pathway, which comprises 10 reactions from phosphoribosyl pyrophosphate to inosine monophosphate, form multi-enzyme complexes, that physically restrict the space, where reactions take place, thus enabling effective channeling of metabolites from one enzyme to the next.

Third, the localization of enzymes and transporters, regulatory factors, such as transcription factors, within specific compartments offers the cell the opportunity to control their localization, thus adding an additional layer of physiological regulation. For instance, the exclusion of an enzyme from a compartment, where its reaction takes place, suppresses the metabolic flux, while a membrane transporter that is sequestered in the intracellular compartment cannot participate in transmembrane transport. Similarly, exclusion of a transcription factor from the nucleus prevents its effects on gene expression. There follow some examples.

Glucokinase

This is the first enzyme of glycolysis, which catalyzes the conversion of glucose to glucose-6-posphate and is located either in the cytosol or in the nucleus. Under low glucose conditions, glucokinase is bound to a regulatory protein, which sequesters it in the nucleus, thus slowing down glycolysis. Once glucose concentration is increased, for instance, after a meal, glucokinase is translocated to the cytoplasm, which promotes glycolysis. Conversely, fructose-6-phosphate, which is synthesized from glucose-6-phosphate in the glycolytic pathway, promotes association of glucokinase with its nuclear regulator protein and thereby stimulates its translocation back to the nucleus, thus inhibiting glycolysis (Van Schaftingen et al. 1992).

GLUT4

Glucose uptake in insulin-responsive tissues, such as skeletal muscle and adipose tissue, is largely dependent on GLUT4. In the absence of insulin, GLUT4 is located in the intracellular vesicles. Upon stimulation with insulin, exocytosis is triggered, which leads to translocation of GLUT4 to the plasma membrane, and consequently, an increase in glucose uptake. Translocation of GLUT4 is also triggered by contractions, which stimulate glucose uptake in skeletal muscle independently of insulin. As well as translocation of GLUT4, insulin and contractions activate ion stimulate redistribution of Na^+/K^+-ATPase from intracellular vesicles to the plasma membrane, thus increasing the capacity of skeletal muscle to perform ion transport.

SREBP-1c

Sterol regulatory element-binding protein-1c (SREBP-1c), a transcription factor, is located in the membrane of the endoplasmic reticulum. By activating the PI3K/Akt/mTOR pathway, insulin stimulates the release and translocation of SREBP-1c to the nucleus, where it induces expression of acetyl-CoA carboxylase and fatty acid synthase, key enzymes for the synthesis of fatty acids from acetyl-CoA.

Foxo1

As a transcription factor, Foxo1 regulates the gluconeogenesis and energy intake. In the fasting state, an enhanced glucagon level triggers the activation of PKA, which then phosphorylates specific serine residues of Foxo1, leading to its nuclear localization, thus promoting the transcription of glucose-6-phosphatase (G6Pase) and phosphoenolpyruvate carboxykinase (PEPCK), which are two critical rate-determining enzymes of gluconeogenesis. In the fed state, the elevation of insulin level stimulates Akt which then phosphorylates Foxo1 on different serine residues leading to its translocation out of the nucleus and sequestration by 14-3-3. It ensues a decreased transcription of the PEPCK and G6Pase (Oh et al. 2013). Foxo1 also regulates the transcription of two major genes, NPY and AgRP, involved in the regulation of energy intake. Its role is similar to that described above for the PEPCK and the G6Pase both in the fasting and fed states (Webb and Brunet 2014) (Figure 3.10A).

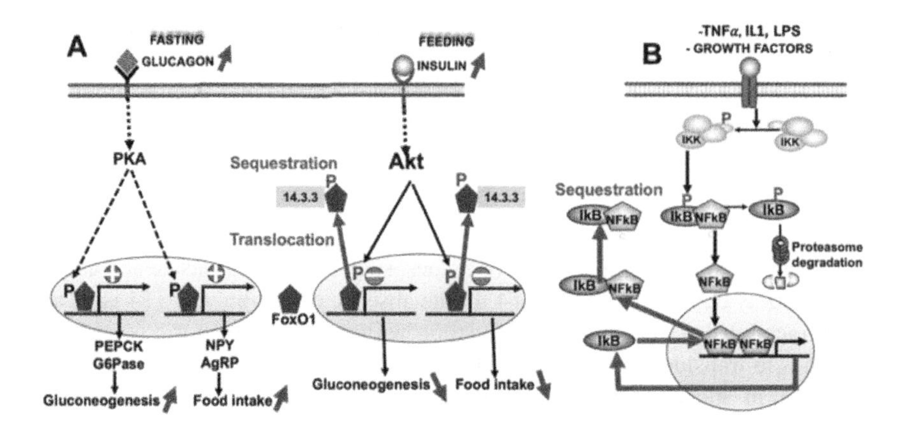

Figure 3.10. *Translocation and sequestration of the transcription factor Foxo1 (A), and translocation and sequestration of the transcription factor NF-κB (B)*

NF-κB/IκB/IKK

The transcription factor nuclear factor kappa B (NF-κB) discovered by Sen and Baltimore (1986) plays a key role in physiological and pathophysiological processes, notably in immunity and inflammation. In unstimulated cells, NF-κB binds to inhibitory IκB proteins located in the cytosol, which maintains NF-κB in an inactive state and prevents its translocation to the nucleus. Activation of NF-κB occurs in response to the pro-inflammatory cytokine tumor necrosis factor α (TNF-α). Briefly, the stimulation of the TNF-α signaling pathway results in activation of the IκB kinase (IKK) complex, which then phosphorylates IκB. Phosphorylated IκB is targeted to the proteasome, thereby enabling NF-κB to translocate to the nucleus where it binds to sequences in the promoter of target genes to stimulate their transcription (see the review by Karin and Ben-Neriah (2000)). Among these genes, newly synthetized IκBα, in the cytosol, enters the nucleus and associates with DNA-bound NF-κB, and shuttles it back to the cytoplasm, thereby rapidly repressing NF-κB activation, through cytoplasmic sequestration. Hence, it terminates the NF-κB signaling pathway and IκBα functions as regulator of the negative feedback loop (see the review by Ruland (2011)) (Figure 3.10B). More detailed information is presented in section 3.3.1.

In addition to the above-mentioned examples, metabolites, enzymes or ions may be sequestered in a specific compartment to prevent their potentially toxic effects:

– Intracellular lipids (LDs)

In adipocytes, cells protect themselves from lipotoxicity by either oxidizing FAs or sequestering them as triacylglycerol (TAG) within lipid droplets (LDs). This mechanism makes it possible to keep the low concentration of lipotoxic intermediates resulting from lipid oxidation (Walther and Farese 2012).

– Proteases

Cells contain potent enzymes that may lead to serious damage if they are released or activated inappropriately. For instance, various proteases are sequestered in lysosomes, which prevents their contact with cellular components. However, in certain situations proteases may leak, thus leading to tissue damage. This may occur in inflammation when neutrophils engulf the invading bacteria in order to destroy them by producing reactive oxygen species and by releasing lysosomal contents, which may subsequently leak into the tissue.

– Intracellular Ca^{2+} concentration

Intracellular fluid contains approximately 10 times more Ca^{2+} (~1% of total body Ca^{2+}) than the extracellular fluid (~0.1% of total body Ca^{2+}); however, intracellular concentration of free Ca^{2+} (~100 nM) is ~10,000 lower than its extracellular concentration (~1 mM). In an unstimulated cell, Ca^{2+} is sequestered in the intracellular stores, such as the sarcoplasmic/endoplasmic reticulum, from where it is rapidly released upon arrival of an appropriate stimulus. The ensuing change in the intracellular concentration of Ca^{2+} is an important signal that triggers major physiological effects, including muscle contraction and exocytosis of hormones and neurotransmitters. Uncontrolled release of Ca^{2+} may lead to serious pathology, including death. For instance, some people harbor a mutation of a ryanodine receptor, a Ca^{2+} channel in the sarcoplasmic reticulum, that makes it particularly sensitive to anesthetics such as halothane. During anesthesia, mutated ryanodine receptors open in an uncontrolled manner, leading to massive release of Ca^{2+}, muscle contractions, rigidity, acceleration of metabolic rate and heat production that leads to a life-threatening increase in body temperature. This condition, known as malignant hyperthermia, is one of the most serious complications of general anesthesia.

– β-Catenin/E-Cadherin/Wnt

Some proteins sequester, i.e. they act as scaffolds to control assembly of signaling complexes and their localization, serve as molecular switches or regulate signaling networks through substrate or enzyme sequestration.

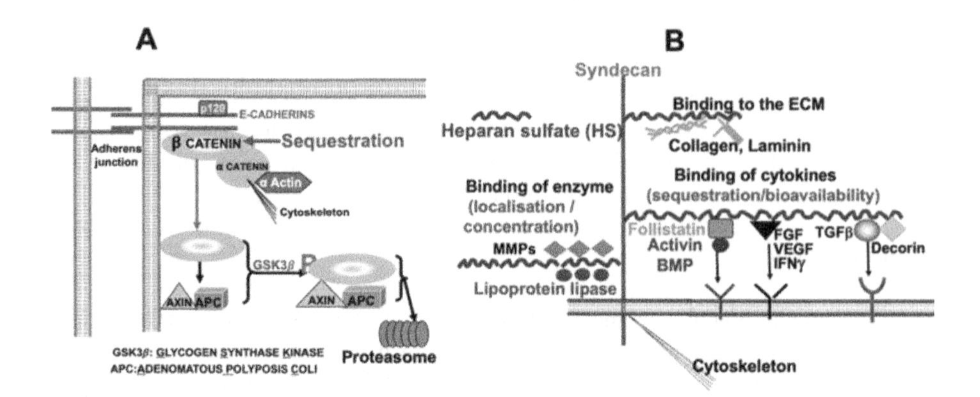

Figure 3.11. *Sequestration of the β-catenin by the E-cadherin (A), and bioavailability, sequestration of growth factors by the syndecan (proteoglycan) (B)*

* *Bioavailability*

This has been briefly described in section 2.3.1.8 (GFs-HSPGs-GFRs).

The supramolecular networks of extracellular matrix (ECM) play a crucial role in the phenomenon of bioavailability of peptide growth factors. It is a key event in regulating their cellular actions as it prevents aberrant signaling events which may result in dysregulated growth and differentiation processes and also in vicious feed-forward cycles of tissue destruction (Zimmermann et al. 2021). The main role of ECM proteins is to target a great variety of ligands (IFNγ, FGF, EGF, BMPs, TGFβ, avidin, follistatin, decorin, etc.) to their cellular receptors and regulate their release (Figure 3.11B). Functional interactions and mechanisms that control the bioavailability of these ligands have recently been studied and the concept has been established that multidomain ECM proteins integrate GF signaling in a spatio-temporal and context-dependent manner. Here are some examples:

– TIMPS inhibit MMP (section 2.3.2.5), which results in decreased bioavailability of ligand to its receptor. A transgenic model provided the first example that signaling of insulin-like growth factor receptor (IGF1-R) was blocked by hepatic TIMP1 overexpression. TIMP1 inhibited the MMP-mediated release of IGF2 from IGF-binding protein 3 (IGFBP3).

– The regulation of TGFβ signaling by ECM binding is another example. The TGFβ is secreted as part of a latent complex bound to latency-associated peptide (LAP), called small latency complex (SLC). Then, LAP binds to the latent TGFβ-binding proteins (LTBPs) and forms large latent complexes (LLCs) which represent the form of TGFβ secretion. In turn, the LTBPs bind to ECM proteins, thereby incorporating TGFβ into ECM in latent form. There are several mechanisms to activate this growth factor, which are described in detail in the review by Hynes (2009).

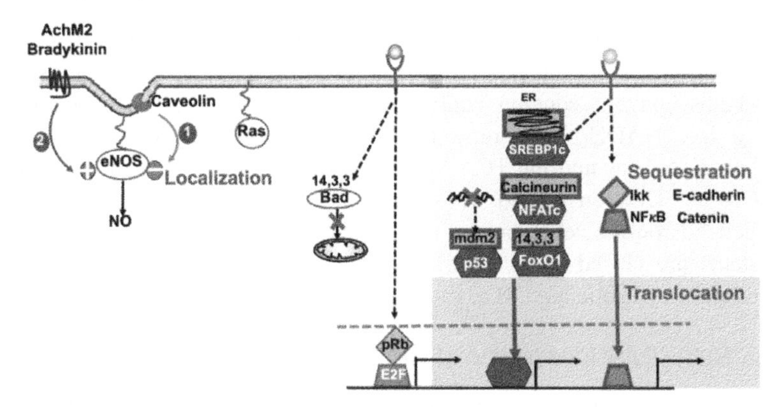

Figure 3.12. *Schematic illustration of different concepts discussed in section 3.1: localization, sequestration and translocation. The symbol ● represents: p53, SREBP1c, NFATc and FoxO1*

3.1.5.2. *Compartmentalization concepts in metabolism and signaling*

Intracellular compartmentalization is a fundamental feature of eukaryotic cells. It enables the physical segregation and simultaneous execution of distinct biochemical processes within the same cell. With the advances of live cell imaging techniques and the generation of fluorescent recombinant proteins, numerous studies have clearly shown that the distribution of proteins inside the cell is a dynamic process which is critical for cellular homeostasis.

3.1.5.2.1. Concept of "shuttle systems"

The compartmentalization of metabolites and enzymes between and within cytosol and mitochondria plays an important role in the regulation of energy metabolism. Given that these two compartments have distinct requirements for NAD^+ a tight control of reducing equivalents NAD^+/NADH is crucial for maintenance of cellular homeostasis. Biochemical reactions create an oxidizing environment in the cytosol, while the opposite prevails in mitochondria.

In the absence of a direct mode for NAD^+ transport, cells rely on the compartmentalized flux of metabolites to support the balance of reducing equivalents. Cytosolic NADH generated during glycolysis must be transferred into the mitochondrial matrix to enter the electron transport chain to achieve maximal ATP production. Simultaneously, cytosolic NAD^+ must be regenerated to maintain glycolytic flux. Hence, the concept of the "shuttle system" was introduced and seven systems have been proposed, but only three have received the most attention, the α-glycerophosphate, the malate-citrate/fatty acids and malate-aspartate shuttles. These systems convey reducing equivalents from cytosol into mitochondria and maintain a low $NADH/NAD^+$ ratio for continuation of glycolysis. Thus, these shuttles link glycolysis and the TCA cycle, and detailed molecular mechanisms are reviewed by A. Dawson (1979). It is worth noting that the operation of shuttle systems depends on the cell type and the metabolic status of the cell. In summary, these systems are closed cyclic pathways by which reducing equivalents can be transferred from cytosolic NADH to mitochondrial oxygen.

3.1.5.2.2. Subcellular localization of GPCR signaling

There has been a considerable resistance to accepting a non-random distribution of signaling molecules in the plasma membrane, likely due to the well-established fluid mosaic model previously described (Singer and Nicolson 1972). The concept of signal compartmentalization is now widely accepted, but it has been a challenge to demonstrate, signaling domains in living cells. Recently, the development of methods based on single-molecule microscopy has allowed probing the organization and dynamics of GPCR signaling nanodomains with unexpected spatiotemporal resolution. It has enabled us to revisit the dynamic aspects of this signaling pathway, which is briefly reported.

Investigators have been able to track, with the aforementioned technique, the signaling dynamics and activity in living cells (Ni et al. 2018). There is now evidence supporting the concept that GPCR-mediated activation of G proteins occurs not only from the plasma membrane but also from internal membrane locations, such as endosomes and Golgi apparatus. Likewise, β-arrestin-dependent signaling can be transduced from the plasma membrane by β-arrestin trafficking to clathrin-coated pits after dissociating from a ligand-activated GPCR (Eichel and von Zastrow 2018).

3.1.5.2.3. Signaling pathways

In this section, three molecules, cAMP, Akt and Bcl2, illustrate the concept of compartmentalization.

Compartmentalization of cAMP signaling (reviewed by Robichaux III and Cheng (2018))

Three major types of proteins are known to assist in the compartmentalization of cAMP signaling in cells and generate the diverse array of physiological responses. These regulatory systems include ACs, PDEs and AKAPs (see Figure 3.3).

Adenylyl cyclase

The first group of molecules responsible for regulation of the cAMP signal is cAMP manufacturing enzymes, the AC family. Mammalian ACs are not randomly distributed within the cell, but targeted to discrete subcellular loci such as centrioles, mitochondria or nuclei (Steegborn 2014). The regulation, as well as the level of each AC isoform, differs in specific tissues (Hanoune and Defer 2001).

Phosphodiesterase

Opposing the action of ACs, another group of molecules that are also responsible for cAMP compartmentalization is PDEs. These enzymes degrade cAMP to 5-AMP, restoring the basal cAMP state after activation of ACs (Houslay 2010) The activation of PDEs is vital for regulating the strength and duration of the cAMP signal within a cell.

Scaffolding proteins: a-kinase anchor proteins

Investigators discovered a third set of molecules, the structurally diverse family of AKAPs that tether relevant signaling components to specific subcellular organelles or regions and form discrete multiprotein signalosomes for efficient biological responses (Kapiloff et al. 1999; Feliciello et al. 2001). This scaffold family coordinates the assembly of key regulators of cAMP effectors, ACs, PDEs.

Importantly, as cAMP is generated, the second messenger acts on downstream effectors such as PKA, EPAC (exchange proteins directly activated by cAMP) or ion-gated channels with exquisite spatiotemporal precision promoting the appropriate signal response.

Compartmentalization of Akt signaling

In response to specific extracellular signals, Akt interacts with phosphoinositides and PDK1, resulting in the activation of its kinase activity. Akt isoforms redistribute throughout many cellular compartments and associate with specific organelles. They have been localized to virtually all cellular compartments, including the cytoskeleton, cytoplasm, the nucleus, the Golgi and mitochondria membrane (Martelli et al. 2012; Toker and Marmiroli 2014).

Compartmentalization of Bcl-2 signaling

Bcl-2 proteins are the main regulators of apoptosis through a control of the mitochondrial outer membrane permeability. Once they are synthetized, they are targeted to multiple subcellular compartments from where they remotely control the apoptosis (Kaufmann et al. 2003). However, part of these proteins can translocate and change their subcellular localizations depending upon the cellular status and apoptotic stimuli. The molecular mechanisms behind these localization changes currently remain unexplained, but an interesting question has been raised: is there a difference between the subcellular distribution of Bcl-2 proteins between normal and cancer cells. If so, it could dampen their apoptotic role and thereby contribute to cancer progression.

It should be emphasized that Bcl-2 proteins are also involved in other cellular processes, such as calcium homeostasis, cell migration and cell-cycle control. This is another example illustrating the concept of multifunctional role of certain proteins.

3.1.5.2.4. Intracellular ceramide partitioning

Ceramides are sphingolipids that are generated by ceramide synthases (CerSs) and, depending upon the fatty acyl chain lengths integrated in the molecules, they exert different effects. It was initially demonstrated that CerSs reside at the cytosolic side of the endoplasmic reticulum. However, recent studies have reported that these enzymes are also detected in other intracellular compartments, such as Golgi apparatus, mitochondria and nucleus, and generate ceramides with different acyl chain lengths (reviewed in Turpin-Nolan and Brüning (2020)). Given that increased levels of ceramides in obese individuals contribute to cellular dysfunction, such as insulin resistance in metabolic tissues, there is an emerging concept that postulates that alterations in ceramide synthesis in specific cellular compartments could have consequences on metabolic homeostasis.

3.1.6. *Crosstalk*

3.1.6.1. *GPCR-EGFR crosstalk*

Two distinct major mechanisms of GPCR-EGFR crosstalk have been reported: the EGFR ligand-dependent and the EFGR ligand-independent transactivation (Cattaneo et al. 2014; Köse 2017; Di Liberto et al. 2019).

* The EGFR ligand-dependent mechanism has been discussed in section 2.3.1.2 and Figure 2.12A. It is characterized by three transmembrane signal transmission events: (1) extracellular ligand-mediated GPCR activation, which results in an intracellular signaling that induces MMP activation; (2) subsequent processing of EGFR ligand precursor and release of the mature form, which extracellularly binds

and activates the RTK; (3) activation of RTK intracellular signaling, Ras/Raf/MEK/ERK and PI3K/Akt/mTOR (see Figure 2.12A).

* The ligand-independent mechanism that occurs in the absence of RTK ligands involves the stimulation of GPCR with a selective ligand which leads to the activation of several second messengers such as Src, Ca^{2+}. The latter may in turn induce tyrosine phosphorylation of RTK, enabling a stimulation of the signaling pathway.

It is noteworthy that RTKs can also recruit components of the G protein signaling cascade, creating a bidirectional intricate interplay that provides complex control over multiple cellular events.

There are also other examples that illustrate that a cellular response is rarely determined by isolated signaling units. For instance, there is an extensive crosstalk between FGFRs and other cell surface receptors, integrins and adhesion molecules. Some binding partners of FGFR1 are listed: FGFR1/GPCR (CB1R, mACh)/RTK (EphR)/Integrin (αv/β3). They likely modulate cellular responses.

3.1.6.2. *Crosstalk between signaling pathways*

This is a process in which the cellular signaling machinery is shared among distinct ligand-activated pathways. It amplifies and dampens the activity of extracellular and intracellular signals. For instance, interactions between GF, polypeptide hormones, cytokines, work together in a complex but coordinated fashion to lead to an integrated control of both cell and organ cell physiology.

Crosstalk between PI3K–Akt and Ras–ERK pathways

Growth factors, hormones that bind to RTKs, activate these two signaling pathways that are involved in proliferation, differentiation and apoptosis. Studies have revealed a crosstalk between these pathways, and it has been shown that RAS proteins seem to play a critical role. The first Ras-effector pathway to be identified was the Raf–MEK–ERK pathway (Moodie et al. 1993), and Ras also activates PI3K through an interaction with the kinase to promote its catalytic activity (Vivanco and Sawyers 2002). Note that PI3K activity is controlled by RTKS and Ras. It should also be mentioned that the PI3K pathway activates Akt, which in turn phosphorylates Raf, and inhibits it, resulting in decreased Raf–MEK–ERK activity (Moelling et al. 2002) (Figure 3.13A).

Points of crosstalk between the PI3K-Akt and Ras-Raf-ERK pathway are well established, but these signaling pathways also interact with AMPK signaling (see below), which brings another layer of complexity in the crosstalk mechanisms.

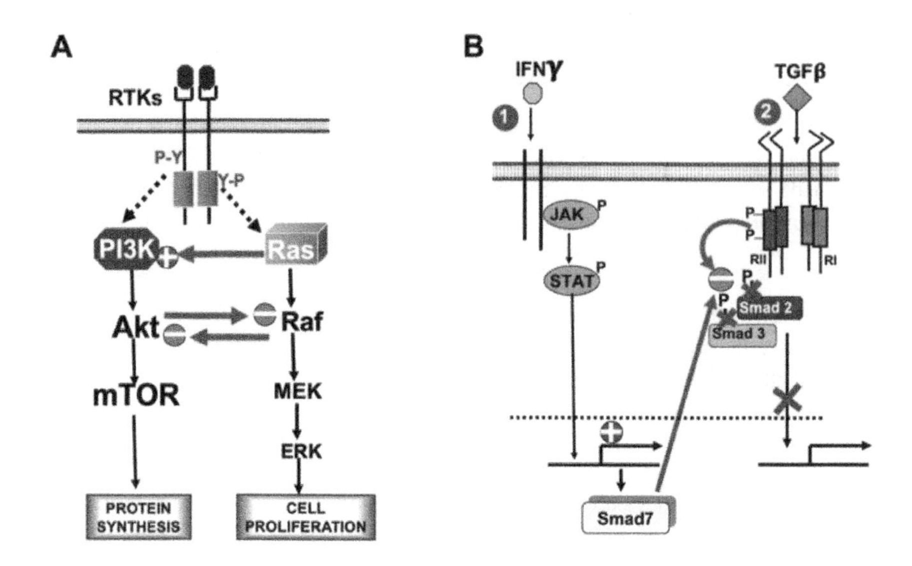

Figure 3.13. *Crosstalk between PI3K/mTOR and Ras/ERK signaling pathways (A), and crosstalk between IFNγ and TGFβ1 signaling pathways (B)*

Crosstalk between IFNγ and TGFβ1 pathways

In the wound healing process, collagen deposition is one of the most crucial events that depends on the stimulation of TGFβ1. As IFNγ is considered to be a negative regulator of collagen biosynthesis, it has been hypothesized that a crosstalk between these signaling pathways would explain a reduced collagen deposition which often occurs during the healing process. The molecular mechanisms can be summarized as follows: (1) TGFβ1 mediates its signals mainly by phosphorylating stimulatory Smads, Smad2 and 3; and (2) IFNγ mediates its signals through Stat1 that increases the transcription and expression of Smad7. The latter prevents the phosphorylation of Smad2 and 3, thereby inhibiting the actions of TGFβ1. Thus, the blockade of the IFNγ signal transduction pathway might enhance TGFβ1 signaling in a positive feedback manner and would be an important strategy to accelerate the wound healing process (Figure 3.13B) (Ulloa et al. 1999; Ishida et al. 2004).

Crosstalk between integrins and growth factor/cytokine receptors

Although different receptor types can be activated separately by their own ligands, some signaling pathways are regulated by both integrins and growth factor/cytokine receptors, providing key nodal points at which the two systems can interplay (Giancotti and Tarone 2003).

– Crosstalk between integrin-EGFR

Mechanical forces, growth factors and the extracellular matrix all play crucial roles in cell adhesion. Recent studies suggest that epidermal growth factor receptor (EGFR) impacts the mechanics of adhesion through a regulation of integrin tension and the spatial organization of focal adhesions (FA) (Rao et al. 2020). A model highlighting the role of EGFR signaling in the mechanics adhesion is briefly depicted. In the absence of EGF, integrin engagement of RGD results in low direct activation of the integrin signaling pathway. This results in the formation of a limited number of immature FAs with unorganized cytoskeleton, leading to relatively small cell spread areas. In the presence of EGF stimulation, activated EGFR acts as a mechano-organizer, facilitating integrin tension, cytoskeletal rearrangement and FA maturation. Consequently, cells with organized FAs and cytoskeleton enhance cell spreading. This allosteric regulation of cell mechanics is probably generalizable across the RTK family. Overall, these results bridge the gap between microenvironment sensing and intracellular signaling.

ECM adhesion can trigger GF receptor activation and vice versa. Moreover, in many cases, engagement of both adhesion and GF receptors is required for an optimal output of sustained synergistic signaling.

* *GF regulation of integrins*

GFs can influence integrin function. One mechanism is by altering the expression of integrin α/β pairs of adhesome components. For example, FGF2 increases α5/β1 integrin expression in endothelial cells, whereas TGFβ1 elevates β1 and β5 integrin levels.

* *Integrin regulation of GF receptors*

GFs direct cell-fate decisions during development. However, the signaling responses of GF receptors (i.e. RTKs) are intricately controlled by adhesion to the ECM, where synergy leads to an optimal response (Streuli and Akhtar 2009). Integrin interception of GF signaling can occur at the receptor level, where integrins may influence expression, localization or post-translational modification of RTKs or levels of GFs themselves. One mechanism for activating GF receptors is through integrin clustering, in which integrin–RTK complexes can trigger RTK activity in the absence of the GF (Miyamoto et al. 1996; Moro et al. 1998). Several RTKs can be stimulated simply by adhesion to the ECM, including PDGFR, EGFR, bFGFR (FGFR1) and VEGFR.

– Crosstalk between β1 integrin–IL-3 β receptor

IL-3 secretion by infiltrating T-cells promotes angiogenesis, and the β1 integrin is involved in this process. Upon fibronectin adhesion, the integrin associates with the IL-3 β receptor and independently of IL-3 it leads to a transient activation of JAK2/STAT5 signaling. In the presence of IL-3, the β1 integrin–IL-3 β receptor enables a sustained activation of JAK2/STAT5 signaling which results in cell-cycle progression in adherent cells. Expression of an inactive STAT5 inhibits cell-cycle progression upon IL-3 treatment, identifying integrin-dependent STAT5 activation as a priming event for IL-3-mediated cell-cycle entry (Defilippi et al. 2005) (Figure 3.14A).

– Crosstalk between β1 integrin–prolactin receptor

A similar situation occurs for prolactin signaling, where cells need to be in contact with the correct ECM environment in order to respond to instructive signals from this endocrine hormone. β1 integrin adhesion to laminin (LM) provides the spatial signals controlling this cell-fate process, and prolactin provides the temporal signals (Katz and Streuli 2007). Prolactin activates the JAK2/STAT5 cascade when mammary cells adhere to LM, but not to collagen I, being inhibited on the latter ECM by a PTP, which is likely to be SHP-2. This PTP can be suppressed chemically or with constitutively activated Rac (V12Rac), restoring prolactin-dependent STAT5 signaling. The mechanism by which V12Rac regulates SHP-2 is unknown, but one possibility is via ROS (reactive oxygen species), which inhibit PTPs (Edwards et al. 1998) (see Figure 3.14B).

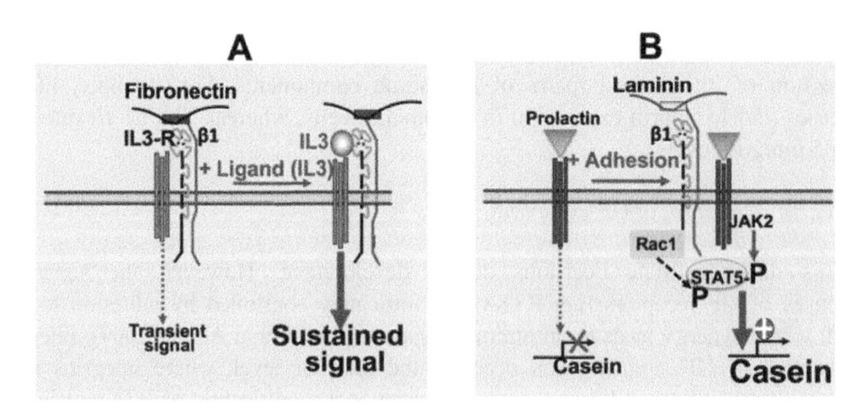

Figure 3.14. *Crosstalk between the β1 integrin and the IL-3 β receptor and their respective ligands, fibronectin and IL-3 (A). Crosstalk between the β1 integrin and the prolactin receptor and their respective ligand laminin and prolactin (B)*

– Crosstalk between integrins and interferons (IFN)

Integrin-mediated adhesion also regulates IFN responses, which in this case is via an indirect mechanism involving PKCε. In cells responsive to IFNγ, signaling to STAT1 is much stronger in cells that are attached to either FN or the basement membrane collagen IV than in detached cells or those adhering to fibrillar collagen I (Ivaska et al. 2003). PKCε is also required for the ability of IFNγ to activate STAT1, with the input occurring at a point downstream of JAK1/2. Since integrin-mediated adhesion leads to the phosphorylation and activation of PKCε, it appears that PKCε facilitates the coupling of JAK1/2 to STAT1.

Crosstalk between PI3K-mTORC1, Ras-ERK and AMPK signaling pathways

The molecular links between these signaling pathways have been reviewed in detail by Yuan et al. (2013) and Carroll and Dunlop (2017), and are summarized in this section. It is clearly established that AMPK interacts with PI3K-Akt-mTOR and Ras–Raf–MEK–ERK signaling pathways. The former pathway which is activated by the insulin and IGF1 represents an anabolic route, whereas AMPK signaling represents a catabolic function. Given that these pathways act in opposition to each other, it requires an appropriate balance between these two systems (Figure 3.15).

For instance, when the energy status of cells is compromised, they do not grow or divide, even though nutrients are still available. Under these conditions, and to optimize ATP synthesis, it is essential for AMPK to switch off PI3K-Akt-mTORC1 signaling that stimulates ATP-consuming anabolic processes. Thus, in response to ATP depletion, AMPK activation results in inhibition of mTORC1, thereby reducing protein synthesis (Buttgereit and Brand 1995).

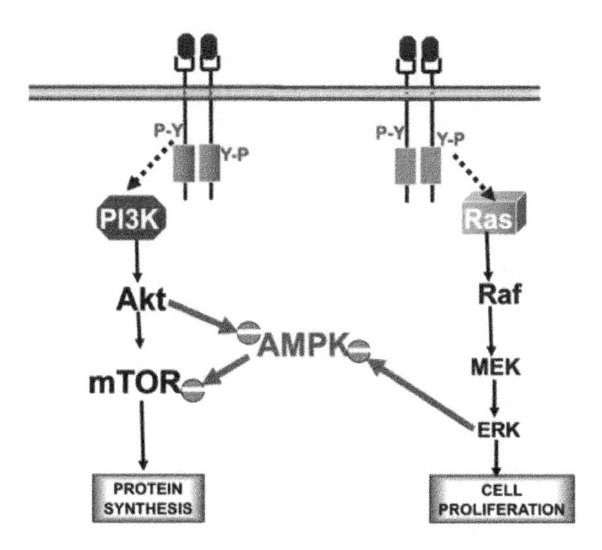

Figure 3.15. *Crosstalk between PI3K–mTORC1, Ras–ERK and AMPK signaling pathways*

Conversely, when anabolic processes take place, AMPK signaling should be switched off. This is achieved by Akt which phosphorylates AMPK and inhibits its activity (Hawley et al. 2014). Notably, when a negative energy status prevails, AMPK phosphorylates mTOR to switch off the PI3K pathway (anabolic route). Similarly, when the mitogenic Ras–Raf–ERK pathway is activated by the chemokine receptor, CCR7, induced the binding of ERK to AMPK and through its phosphorylation inhibited it (López-Cotarelo et al. 2015). Given that Ras-Raf-ERK is a mitogenic pathway, it seems reasonable to postulate that ERK can inhibit the anti-mitogenic AMPK. This crosstalk between Ras–Raf–ERK and AMPK signaling pathways is binary, as AMPK can phosphorylate Raf and alter its activity. Moreover, AMPK signaling inhibits mTOR activity, thereby suppressing protein synthesis initiated by PI3K-Akt (Mihaylova and Shaw 2011).

Taken together, these findings illustrate the concept that a positive signal efficiently stimulates a cellular response when it simultaneously switches off the influence of the negative signal working in the opposite direction.

3.2. Sensing of extracellular and intracellular cues

Historically, the discovery of the lactose operon in the 1950s clearly demonstrated that bacteria (*E. coli*) actively sense and respond to their environment, often by detecting changes in nutrient levels and down- or upregulating expression of selected genes. This discovery revealed the concepts of gene regulation across the living world (Jacob and Monod 1961). Similarly, yeast (*S. cerevisiae*) switches the expression of a family of genes depending on its nutrient source (glucose or galactose). The switch involves signaling pathways that sense environmental changes and send a signal to specific domains of the genome.

In multicellular organisms, the whole body, tissue and cells also have the ability to sense external perturbations. As mentioned earlier (Figure 2.1), in the whole organism, sensors (endocrine cells and sensory neurons) constantly monitor the values of regulated variables. When perturbations occur, the system engages in a specific response that aims at restoring homeostasis. In cellular homeostasis, the sensors are signaling proteins that detect alterations in various key processes, such as nutrient availability.

3.2.1. *Sensing of extracellular cues*

The metabolic and energy status of the organism mainly depends on the supply of nutrients, which fuels the metabolic pathways so as to meet ATP production requirements for maintaining physiological functions. The glucose is the nutrient which plays a central role in metabolism as it provides the bulk of cell energy requirements of all living organisms. Hence, most of the research in the field of metabolism has been focused on the intracellular glucose fate, and this information is displayed on maps hanging on the wall in all laboratories. In the past, a question often brought up was, how do cells sense their immediate environment?

3.2.1.1. *Glucose sensing*

3.2.1.1.1. Glucose-sensing neurons

Since the initial observation by C. Bernard that a "piqûre" on both sides of the floor of the fourth ventricle of the rabbit induces diabetes (Bernard 1849), the brain has been recognized as an important regulator of glucose homeostasis. This seminal observation is reported in his notebook the "Cahier rouge", and details of the experimental approach as well as a drawing are displayed on the document below. Note that the experiment was performed on April 30, 1858. A second experiment on May, 3 1858, concluded that when the "piqûre" was performed on one side it did not induce diabetes. This document has been kindly provided by the Collège de France.

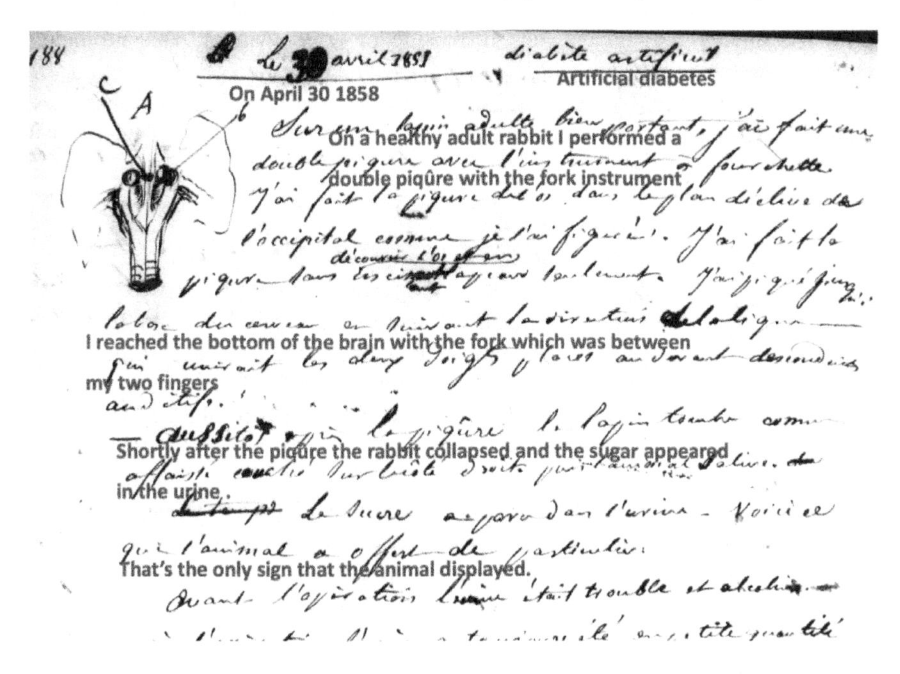

In 1953, J. Mayer predicted that arterio-venous differences in blood glucose levels were sensed by "glucoreceptors" localized in the hypothalamus as a means of controlling food intake (Mayer 1953). Then, in 1964, Y. Oomura (Oomura et al. 1964) and B. Anand (Anand et al. 1964) identified neurons in the lateral hypothalamus (LH) and ventromedial hypothalamus (VMH), which used glucose as a signaling molecule to regulate their firing rates. It took an additional 32 years before such neurons were shown to use an ATP-sensitive K^+ sensitive channel (K_{ATP}) to regulate glucose sensing (Rowe et al. 1996). It is similar to the one in the pancreatic β-cell, which is the most downstream molecular event for glucose-stimulated insulin release. Today, we know quite a lot about the neurons which sense and respond to ambient glucose levels. Unlike most neurons that use glucose and other metabolites to meet the metabolic demands associated with their activity, glucose-sensing neurons use the glucose as a signal to alter their activity. It was demonstrated that an increase in glucose levels led to increased activity in some neurons (glucose excited; GE) and decreases in others (glucose inhibited; GI) (Levin 2006).

**Glucose-excited neurons (GE)*

These sense an elevation of the glucose level by a mechanism similar to that used by pancreatic β-cells in response to hyperglycemia. This relies on three main components: (1) glucose transporter GLUT2, (2) glucokinase (GK) and (3) K_{ATP} channel. As in the β-cell, glycolysis initiates a sequence of steps leading to the generation of the ATP, which results in closure of K_{ATP} channel. This promotes membrane depolarization, an influx of Ca^{2+}, which triggers the secretion of γ-aminobutyric acid (GABA) neurotransmitter in the VMH, which then decreases the secretion of glucagon and epinephrine.

**Glucose-inhibited neurons (GI)*

The activity of GI neurons increases when extracellular glucose levels fall, which results in a reduced availability of ATP generated in the glycolytic cycle and deactivates the Na^+/K^+ ATPase pump. This leads to an intracellular accumulation of Na^+, resulting in depolarization of the plasma membrane and a release of glutamate and noradrenaline neurotransmitters. These neurons are distributed throughout the LH, and they have certain similarities with GE neurons, sharing the expression of GK and GLUT2.

Inversely, in the presence of increased extracellular glucose concentrations, GI neurons enhance their glucose uptake. This causes an increase in ATP concentration and leads to a stimulation of Na^+/K^+ ATPase pump. It results in hyperpolarization of the neuron, ultimately leading to a decrease in neuronal firing.

It is worth mentioning that two groups of neurons, neuropeptide Y (NPY) and proopiomelanocortin (POMC), in the hypothalamic arcuate nucleus are prototypic metabolic sensors. Both use glucose as signaling molecules and both have receptors for peripheral hormones, including insulin and leptin. More detailed information is presented in section 2.2.3.

– Role of GLUT2

Recently, a subpopulation of GI neurons has been identified in the nucleus of the solitary tract of the brainstem (NTS). They are GLUT2-expressing neurons and are excited when glucose levels drop. Like the GI neurons mentioned above, they increase parasympathetic nerve firing and glucagon secretion, thus defining a neuronal circuit linking hypoglycemia detection to counter-regulatory response (Lamy et al. 2014).

– Role of glucokinase

Several lines of evidence indicate that glucokinase is critical for neuronal glucose sensing, as its inhibition decreases GE neuronal activity, whereas it increases that of GI neurons in VMH neurons. The most clearly defined role of neuronal glucokinase in the VMH is for the regulation of glucose homeostasis (Thorens 2012). McCrimmon stated that glucokinase plays a pivotal role in inducing the counter-regulatory response (CRR) to hypoglycemia (McCrimmon 2009). Injections of the glucokinase inhibitor alloxan into the third ventricle impaired the CRR to hypoglycemia in rats (Sanders et al. 2004).

Induction of glucopenia in the VMH elicits a significant increase in plasma glucagon and catecholamines despite systemic normoglycemia. These pioneering studies of the physiology of glucose counter-regulation have established the classic concept that the secretion of both these hormones represent the first line of defense against hypoglycemia (Borg et al. 1995).

3.2.1.1.2. Glucose-sensing by β-cells

Like glucose sensing neurons, β-cells are capable of sensing variations in glucose levels, which in response triggers insulin secretion. Pancreatic GLUT2 transporter and glucokinase play a key role in the process as genetic inactivation of GLUT2 in mice (Thorens et al. 2000). GLUT2 is less prominent in human β-cells, which express GLUT1, but mutations in the glucokinase gene lead to abnormalities in glucose homeostasis, highlighting its essential role in β-cell function (Osbak et al. 2009).

3.2.1.1.3. Portal vein glucose sensing

During feeding, glucose is absorbed from the gut, causing the hepato-portal glucose concentration to rise. The portal vein glucose sensor monitors this change and generates a neural signal, which in turn increases liver glucose uptake and glycogen storage (Moore et al. 2012). It simultaneously induces muscle glucose utilization, independently of insulin action (Burcelin et al. 2003). This sensor also appears to block the secretion of counter-regulatory hormones (Mithieux 2014). Together these actions might cooperate to blunt the post-prandial glycemic excursion. However, following oral glucose ingestion, there was no evidence to support the involvement of hepatic portal vein glucose sensors in the regulation of incretin hormone secretion (Edgerton et al. 2019).

Although many studies have explored the effects of the portal glucose signal (e.g. on hepatic glucose uptake), the outcomes are controversial.

3.2.1.2. *Metabolite-sensing GPCRs*

Cells need a timely and accurate perception of dynamic changes in intracellular and extracellular metabolites, especially the concentration of nutrients. Historically, the discovery of the lactose operon in the 1950s clearly demonstrated that bacteria actively sense and respond to their environment, often by detecting changes in nutrient levels and down- or upregulating expression of selected genes, the availability of carbon sources and modulate gene transcription to tightly control protein synthesis. Over the years, we have learnt that several different nutrient metabolites, such as long-chain fatty acids, oligopeptides, amino acids and the ketone body β-hydroxybutyrate (βOHB) are sensed by and signal via specialized GPCRs. Interestingly, the βOHB itself, signals into our cells' nuclei to regulate gene expression and chromatin structure (Newman and Verdin 2014). The concept of key metabolites, generated from nutrients or by the gut microbiota, as ligands for specific GPCR is quite recent and they target primarily enteroendocrine, neuronal and immune cells in the intestinal epithelium of the GI tract. In accordance with their function as sensors of "signaling metabolites", these receptors constitute approximately 50% of the receptors controlling hormone secretion in enteroendocrine cells (see detailed review by Husted et al. (2017)).

3.2.1.3. *Ca^{2+} sensor*

The existence of an extracellular Ca^{2+} sensor was suggested in 1966 from a study correlating serum Ca^{2+} concentration and parathormone (PTH) secretion (Care et al. 1966). Subsequently, a link between extracellular Ca^{2+} concentration changes, PTH secretion and intracellular Ca^{2+} signals was definitively demonstrated. The molecular identification of the Ca^{2+} sensor was established in 1993 and named calcium-sensing receptor (CaSR) (Brown et al. 1993). It is a peculiar G protein-coupled receptor of the C family. The main physiological function of the CaSR is to sense changes in plasma Ca^{2+} levels and maintain Ca^{2+} homeostasis via changes in PTH secretion which in turn regulates Ca^{2+} handling via its action on bone, kidney and intestine. Once activated, CaSR leads to an increase in intracellular Ca^{2+} concentration, which suppresses both PTH synthesis and secretion. Regulation of PTH via CaSR therefore represents a negative feedback loop that prevents hypercalcemia. CaSR also controls calcitonin secretion from thyroid parafollicular (aka C) cells (see section 4.2.3 and the review by Brown and MacLeod (2001)).

The CaSR is also allosterically activated (in the presence of Ca^{2+}) by amino acids and small peptides (Chavez-Abiega et al. 2020).

3.2.1.4. *Pathogen-sensing systems*

The ability to sense infections is primordial to preserve organisms, and the innate immune system is the first line of defense against pathogens. In the mid-1990s, a key advance in this field was the discovery of molecular sensors that enable eukaryotes to detect and eradicate microbes. This system is non-specific and rapid in its response. Innate immune cells, including macrophages and dendritic cells (DCs), play a major role, non-professional cells such as epithelial cells, endothelial cells also contribute to the innate immunity. These professional and various non-professional immune cells express receptors, collectively termed pattern recognition receptors (PRRs), capable of sensing a wide variety of molecular entities. Three main families of sensing molecules have been identified, including Toll-like receptors (TLRs), retinoid acid-inducible gene I (RIG-I)-like receptors (RLRs) and other cytosolic nucleic acid sensors, and nucleotide-binding and oligomerization domain (NOD)-like receptors (NLRs). They are localized both on cell surfaces and within intracellular compartments (Baccala et al. 2009). Their primary function was described earlier in section 2.3.1.9.

3.2.1.5. *Mechanosensors*

Cells have the ability to sense and respond to mechanical forces, such as shear, compressive, and tensile forces on endothelial cells, chondrocytes and in myocytes, respectively. This is critical for numerous biological processes such as tissue homeostasis, development and embryogenesis. We also sense physical contact, gravity, sound waves and air flow. The understanding of the molecular pathways involved in mechanosensing is still in progress. Some mechanosensing elements have been identified, such as, adhesion complexes, cell-cell junctions, cytoskeletal components, stretch-activated ion channels, and G protein-coupled receptors. In response to physical signals, these sensors located at the plasma membrane activate cellular signaling pathways resulting in the expression of mechanoresponsive genes. The question of whether the nucleus itself can also sense mechanical stimuli, and therefore could act independently of cytoplasmic mechanotransduction pathways, remains challenging. Recent findings support nuclear mechanotranduction which may modulate the effect of cytoplasmic signals (reviewed by Alonso and Goldmann (2016) and Kirby and Lammerding (2018)).

**Mechanosensitive ion channels*

Because mechanical or auditory stimuli generate electric signals quickly (with a latency of less than 1 ms), transduction is thought to be too rapid to involve chemical intermediates. This has therefore led to the assumption that mechanosensory transduction involves ion channels that are directly gated by mechanical stimulus. They have been termed mechanosensitive (MS) ion channels and they are directly activated by stresses applied to the lipid bilayer or its associated non-membrane components (Árnadóttir and Chalfie 2010). The key question is how mechanical stimuli are transduced from the environment to the MS channel from a closed to an open conformation. Two models have been proposed and one is called "the membrane tension model" in which force applied to the lipid bilayer generates membrane tension to gate the channel. The second one is "the tether model" in which force applied to the cell is transmitted to the cell through a tether connecting the channel with ectomembrane components to gate the channel (reviewed in Jin et al. (2020)).

To understand the force transduction and gating mechanisms of MS channels, it is important to know the physical property of the channel and the channel conformations at different states. The ion channel property is usually characterized by measuring channel activity using patch-clamp electrophysiology. Recent developments in single-particle cryogenic electron microscopy (cryo-EM) technology have enabled structural studies of many membrane proteins allowing the visualization of a 3D protein structure at near-atomic resolution. Over the past decade, new MS channels have been identified and it will be important to ask whether there is a general mechanism underlying channel gating.

The notion that bone tissue is responsive to mechanical influences has been known for well over 100 years (Wolff 1892). Defective bone mechanosensing due to immobility and/or sarcopenia results in decreased osteoblast-mediated bone formation and increased adipogenesis, leading to a decline in bone mass (osteoporosis) (Moerman et al. 2004). Osteocytes are thought to be the major cell type responsible for sensing mechanical stress in bone and show greater sensitivity than osteoblasts. They are mechanically stimulated by bone loading, but the physical stimulus to which osteocytes respond is a matter of ongoing research. The identity of the primary mechanosensory apparatus in osteocytes has been of great interest in the field, but a consensus on the exact mechanism(s) is lacking. However, integrin complexes and ion channels are considered likely candidates for osteocyte mechanosensors. Although the signal reception mechanism in the osteocyte is unclear, downstream events are better characterized. Mechanical stimulation of the osteocyte induces the release of ATP, prostaglandins E2 (PGE2), nitric oxide (NO) insulin-like growth factor 1 (IGF1) (Robling and Bonewald 2020). Overall, the identities of the channel and gating linker remain elusive.

* *Integrins*

Because they connect the cell cytoskeleton to the microenvironment, integrins are continuously subjected to forces transmitted between cells and the ECM. As such, they are ideally positioned to serve as sensors of mechanical signals over a sustained period of time. Tissues *in vivo* are continuously subjected to mechanical forces generated by cells, through the actomyosin cytoskeleton and by indirect factors, such as blood flow in endothelia. Such forces lead to complex stresses and a large fraction of forces is transmitted from the extracellular matrix to cells through integrins. The latter are considered essential mechanosensors within tissues (see the review by Kechagia et al. (2019)).

Measuring traction force and assessing downstream effects, such as signaling and ECM remodeling, provide information about numerous biological processes and disease progression. Recently, a high-resolution sensor has been designed, and it can report single- and multiple-cell forces in 3D ECM over a long period of time with a resolution of 1 nN (nanonewton) and quantify the change of stiffness of the tissue remodeled by the ECM, which in turn influences cell functions and forces.

It should be mentioned that circadian clocks and photoreceptors are also mechanisms for sensing and responding to the light environment. More detailed information is presented in section 5.2.

3.2.2. *Sensing of intracellular cues*

To maintain cellular homeostasis, specialized sensors constantly monitor the values of regulated variables. They detect variations in core processes, such as ATP depletion, genotoxic stress, ER stress, hypoxia, ion concentrations, glucose deprivation and oxidative stress (Chovatiya and Medzhitov 2014). These sensors are mostly signaling proteins.

3.2.2.1. *Energy sensors*

Inside each cell, there are thousands of metabolic reactions that are displayed on maps in many laboratories and teaching institutions. They describe metabolic fluxes that lead to the production and utilization of energy, but they are not indicative of the mechanisms that regulate these fluxes when an intact organism should cope with the marked fluctuations in nutrient availability and energy needs. The first law of thermodynamics indicates that the energy linked to an isolated system remains constant over time. This basic principle also applies to living organisms, from bacteria to humans. Every living cell obtains chemical energy from nutrients which are oxidized to release energy, which is stored as ATP. Then, to meet the cellular energy requirements, there is a decrease in the ATP concentration associated with an increase in ADP concentration. Therefore, an organism requires a tight control of catabolic and anabolic processes that produce and use ATP, respectively. Thus, it seems logical that proteins should sense the cellular energy status and restore intracellular energy balance, i.e. when there is a decrease in cellular energy. In such situations, it is accompanied by an increase in AMP, resulting in a greater AMP:ATP ratio. This ratio is directly sensed by some metabolic enzymes, including the muscle glycogen phosphorylase and phosphofructokinase, which are activated by an increasing AMP/ATP ratio, while fructose-1,6-bisphosphatase is inhibited. However, it is now clearly established that another kinase, also activated by AMP and named AMP-activated protein kinase (AMPK), represents the principal energy sensor in eukaryotic cells. It is unquestionably one of the most important discoveries in Cell Biology, and the protein was first described by D. Carling and G. Hardie in 1987 (Carling et al. 1987).

AMPK detects changes in cellular energy status that occur during hypoglycemia, hypoxia, muscle contraction and food deprivation. It is characterized by an increase in AMP/ATP and ADP/ATP ratios which activates the kinase after the binding of AMP or ADP to AMPK. Inversely, binding of ATP antagonizes the activating effects of AMP and ADP. As a consequence of this activation, a global metabolic counter-regulatory response is achieved and AMPK restores intracellular energy balance by switching off ATP-consuming processes (protein, glycogen, fatty acid and sterol synthesis) while turning on catabolic processes, favoring ATP production (lipolysis, fatty acid oxidation and mitochondrial biogenesis). From the standpoint of energy homeostasis, ATP generation needs to remain in balance with ATP consumption.

Over the last two decades, several other molecules have been identified and characterized as central energy gauges. JNK1/2 (c-Jun N-terminal protein kinase), IKKα (inhibitor of nuclear factor kappa-B kinase subunit alpha), IKKβ (inhibitor of nuclear factor kappa-B kinase subunit beta), iNOS (inducible nitric oxide synthase), PI3K (phosphatidylinositol-3 kinase), mTOR (mechanistic (or mammalian) target of rapamycin) (Haissaguerre et al. 2014), sirtuins (SIRTs) (Quiñones et al. 2014), HIF-α (hypoxia-inducible factor) (Palmer and Clegg 2014), K-(ATP) channels, FTO (fat mass and obesity related) (Yeo 2014), PKC (protein kinase C), FAS (fatty acid synthase) (Duca and Yue 2014) and CPT1 (carnitine palmitoyl transferase 1), among others, have been demonstrated to be expressed in the hypothalamus and to control, not just the whole body energy balance, but also almost all the neuroendocrine axes, reproductive function, growth and inflammation.

3.2.2.2. *Nutrient sensors: mTORC1/Sestrin2/CASTOR/GCN2*

mTOR ("*m*ammalian *t*arget *o*f *r*apamycin", but now officially "mechanistic TOR") is a serine/threonine protein kinase that is shared by two multiprotein complexes called mTORC1 and mTORC2. The story of rapamycin reinforces the notion that discoveries may lead to unanticipated outcomes in other disciplines. In 1964, a streptomycete was isolated from a soil sample from the island of Rapa Nui (Easter Island). Studies following its discovery showed that it exhibited multiple properties, including antifungal, immunosuppressive and anticancer. The antifungal principle was named "rapamycin" in reference to its place of origin. The mTOR signaling pathway is critical to cell growth, proliferation and survival and rapamycin inhibits these processes of cancer. Of note, only mTORC1 is sensitive to rapamycin, and it controls cell growth by maintaining the balance between anabolic and catabolic processes (see the review by Laplante and Sabatini (2012)). mTORC1 is an important molecule that senses, quantifies and responds to basic nutrients such as glucose, amino acids and fatty acids.

Extensive efforts have been taken to elucidate the mechanisms through which mTORC1 senses intracellular amino acids. Rag GTPases have been shown to be key transducers between amino acids and mTORC1 activation (Kim et al. 2008). Studies argue that amino acid signaling to mTORC1 begins within the cell instead of at the plasma membrane, and the lysosome appears to be a key player, as Rag GTPases are localized to the lysosomal surface.

It should be pointed out that leucine, arginine and glutamine are among the most effective amino acids for the activation of mTORC1. Prior to its activation, the intracellular availability of leucine and arginine is transmitted to mTORC1 through the amino acid sensors sestrin2 and castor, respectively. In response to its activation, mTORC1 increases protein synthesis, lipogenesis and energy metabolism. The downstream targets of TORC1 are eukaryotic initiation factor 4E-binding protein-1 (4EBP1) and S6 kinase (S6K aka p70S6K) and their phosphorylation leads to a stimulation of protein synthesis. Consequently, mTOR and AMPK pathways regulate the phosphorylation and activity of S6K and 4EBP1, indicating a convergence of these pathways. However, these two sensors function coordinately and act in opposition to each other, as activation of TOR pathway promotes anabolism (protein synthesis) while AMPK activation switches off the anabolic processes (Xu et al. 2012).

With respect to TORC2, it is activated by growth factors and regulates cell survival and metabolism through phosphorylation of Akt, serum-and glucocorticoid-induced protein kinase 1 (SGK1) and PKCα.

GCN2 (General Control Non-derepressible-2) kinase functions as an amino acid sensor in eukaryotic cells. Under amino acid deficiency, uncharged aminoacyl-tRNA accumulates leading to the activation of GCN2 kinase (Hinnebusch 2005). Activated GCN2 is able to phosphorylate eukaryotic translation initiation factor 2-alpha (eIF2α) and reduces general translational initiation.

3.2.2.3. *Cellular stress sensor: ER*

The ER coordinates many diverse cellular processes, such as maintaining intracellular calcium levels and synthesizing lipids. The vast network of the ER membrane physically and functionally interacts with every other membranous structure in the cell. Thus, this structure is well positioned to sense cellular perturbations, integrate them and adjust signaling pathways to restore homeostasis. It plays a major role in maintaining protein homeostasis (proteostasis) and is responsible for folding and processing nearly all polypeptides destined for secretion (Frakes and Dillin 2017). It should be emphasized that a stress can arise from an excess or a lack of upstream signals or as a result of genetic perturbations in upstream effectors of the pathway.

ER stress is emerging as a major contributor to the pathogenicity of age-onset metabolic disease, such as obesity, insulin resistance and type 2 diabetes. Type 2 diabetes is characterized by increased levels of blood glucose due to insufficient insulin secretion from pancreatic β-cells and insulin resistance in adipose tissue, muscle and liver (Hotamisligil 2006).

3.2.2.4. *Nutrient-sensing nuclear receptors: PPAR FXR*

The liver undergoes major changes in glycolytic, lipogenic and gluconeogenic fluxes over the daily feeding and fasting cycle. Recent studies have found that two nutrient-sensing nuclear receptors, peroxisome proliferator-activated receptor alpha (PPARα) and farnesoid x receptor (FXR), are central modulators of these metabolic pathways, through regulation of gene transcription. Furthermore, the two nutrient sensors regulate each other's expression. For instance, in the fed state, FXR activation induces PPARα mRNA expression and in the fasted state PPARα activation induces FXR mRNA expression (Preidis et al. 2017).

The two nutrient sensors exert complementary but opposing regulatory effects on glycolysis, gluconeogenesis and fatty acid oxidation. These effects would participate in the liver energy balance. However, this functional dichotomy does not apply to lipogenesis, since both appear to suppress this pathway.

3.2.2.5. *Intracellular NAD$^+$ levels sensor: SIRT1*

As a protein deacetylase, SIRT1 (Silent Information Regulator T1) has a wide range of protein substrates, and in response to certain environmental stimuli it shuttles between nucleus and cytosol. Its activity is tightly controlled by the cellular levels of one of its substrates NAD$^+$ which is an essential coenzyme. In metabolism, NAD$^+$ is involved in redox reactions as it can readily switch from electron accepting form (oxidizing) to electron donating form (reducing) NADH and vice versa. Given that numerous metabolic reactions rely on NAD$^+$, any changes in its availability, which reflect cellular energy status, should modify SIRT1 activity. Hence, a low-energy status that increases the NAD$^+$/NADH ratio, including starvation and exercise, stimulate SIRT1 activity (Rodgers et al. 2005) and one of its main roles is to promote the deacetylation of proteins. Inversely, in the presence of high nutrient availability, the acetyl-CoA/CoA ratio increases and promotes the acetylation of proteins, and in parallel it reduces SIRT1 activity (Yoshino et al. 2011). Taken together, these findings demonstrate that SIRT1 is an essential metabolic sensor through intracellular NAD^+ levels. Similarly, we can conclude that acetylation may act as an indicator of nutrient availability (abundance or deficiency) (Figure 3.16).

Notably, NAD^+ levels fluctuate substantially in a circadian manner, linking the peripheral clock to the transcriptional regulation of metabolism by epigenetic mechanisms involving SIRT1 (Sassone-Corsi 2012). Details regarding these mechanisms are provided in section 5.2.4.2.

3.2.2.6. *Oxygen sensor: HIF*

Oxygen was first isolated by a Swedish chemist C. Scheele in 1772, but officially discovered by J. Priestley an English chemist (because he published first in 1772) (Priestley 1772). A. Lavoisier, a French chemist who also discovered oxygen in 1775, was the first to recognize it as an element, and coined its name “oxygen” – which comes from a Greek word that means “acid-former”. The work of Priestley and Lavoisier disproved and overthrew the phlogiston theory that speculated that a substance “phlogiston” was present in air which after combustion became dephlogisticated. The discovery of oxygen in the late 1770s was a seminal event when chemistry transitioned from a largely “black art” (alchemy) to science.

As a follow-up to this discovery, biologists pointed out in the 1930s that a sufficient supply of oxygen to the tissues is an absolute requirement for cellular homeostasis in any living organisms. Given its pivotal role, cellular and molecular strategies have evolved to adjust/maintain the supply of O_2 to prevent irreversible cellular damage which can occur under acute and chronic conditions of reduced O_2 availability (hypoxia). The question which then arises is how cells/organisms detect O_2 deficiency?

To acutely (in seconds) adapt to systemic hypoxia (0.5–2% oxygen in the tissue), like high altitude, mammals elicit diverse homeostatic responses – hyperventilation, tachycardia – that in a few seconds increase O_2 uptake in deprived tissues. These rapid reflexes are mediated by O_2-sensitive chemoreceptor cells (CB) located in specialized sensory tissues/organs tissues. The first organ of the homeostatic O_2-sensitive system was described by anatomists (Von Haller and Taube) in the mid-20th century. In 1868, E. Pflüger, a German physiologist, reported that hypoxia stimulates breathing (Pflüger 1868). Then, studies between the 18th and 20th centuries tried to identify the structures that “sense” systemic O_2 levels and trigger physiological responses. A structure resembling a “ganglion” or a “gland” was identified between the external and internal carotid arteries, but no physiological function was assigned to this organ. It was named the carotid body (CB) or *glomus caroticum* and its detailed study started in the first quarter of the 20th century. In 1925, F. de Castro, a brilliant disciple of S. Ramón y Cajal, studied the structure and innervation of the CB and in a series of scientific papers published between 1926 and 1929 he considered the CB as a sensory receptor or chemoreceptor to detect chemical changes in the blood (de Castro 1929). Thereafter, C. Heymans brought the physiological demonstration of de Castro's hypothesis regarding chemoreceptors and was awarded the Nobel Prize in Physiology or Medicine in 1938 (de Castro 2009). Among the organs of the homeostatic O_2-sensitive system (Weir et al. 2005), the CB is of particular importance as it assumes the main responsibility for the detection of changes in blood O_2 levels and activation of the respiratory center to elicit the proper adaptive ventilatory response.

When hypoxia is maintained for long periods (hours or days), a powerful and generalized genetic program is activated, increasing non-aerobic ATP synthesis, erythropoiesis and the generation of new blood vessels. The discovery of HIF started with the observation that the levels of erythropoietin (EPO), a glycoprotein hormone synthetized in the kidney and liver that stimulates production of red blood cells, were upregulated several hundred-fold at low oxygen conditions (Fisher 1983). Then, in the early 1990, two groups studied regulatory regions in the gene encoding erythropoietin (EPO). A DNA sequence was identified and was coined "hypoxia response element" (HRE) and was also present in other genes not related to erythropoiesis or angiogenesis (now >2.500). Then, the protein that binds to HRE was purified and called "hypoxia-inducible factor" (HIF) (Wang et al. 1995), which comprises an oxygen-sensitive HIF-α subunit (HIF-1α or HIF-2α) and a constitutive HIF-1β (aka ARNT) subunit.

The next issue to address was how HIF-α is regulated by O_2 levels. A third group, working on kidney tumors generated by mutations in the von Hippel–Lindau (VHL) gene, reported that the protein produced (pVHL) is part of a ubiquitin ligase complex that targets HIF-α for its degradation by proteasomes. Given that pVHL is deficient due to VHL mutations, it prevents HIF-α degradation, resulting in constitutive expression of HIFα-dependent genes even in the presence of normal O_2 levels. Another piece of information was added to the puzzle, when it was shown that HIF-α hydroxylation in specific proline residues is required for pVHL recognition to target it for degradation. A family of prolyl hydroxylases (PHDs) was identified, and the enzymatic activity is regulated by O_2 levels. In normoxic conditions, the activity of PHDs is high, and the HIF-α level is low because hydroxylated HIF-α is rapidly targeted for degradation. In hypoxia, the opposite situation prevails, the activity of PHDs is low, resulting in accumulation of HIF-α which binds to HREs in the regulatory regions of O_2-sensitive genes. Collectively, this tripartite mechanism ensures that the system responds to small decreases in O_2 levels in a very sensitive manner. Given its ubiquitous localization, HIF-α functions as a master regulator of the expression of several thousand genes coding for a wide variety of proteins involved in myriad cell functions (Figure 3.16).

The elucidation of the molecular mechanisms that regulate response to hypoxia in multicellular organisms was awarded the Nobel Prize in Physiology or Medicine in 2019.

In the section dealing with the Warburg effect (section 4.3.6.4), other aspects of HIF-α activation are discussed with a special emphasis on its role in metabolic homeostasis, depending on the cause and duration of hypoxia. For instance, HIF-α induces expression of glycolytic genes, allowing cells to use oxygen-independent glycolysis, to produce energy to maintain cellular activity.

3.2.2.7. *Glucose sensing in hepatocytes, adipocytes and pancreatic β-cells: ChREBP*

It has long been recognized that glucose is the principal energy source for most cells, and recently studies have also revealed that glucose acts as a signaling molecule in liver and adipose tissue. In these tissues, glucose stimulates the transcription of genes encoding glycolytic and lipogenic enzymes and these genes share a conserved consensus sequence, called the carbohydrate response element (ChoRE). Thereafter, the discovery of the transcription factor carbohydrate responsive element-binding protein (ChREBP) which binds to ChoRE revealed the molecular link between glucose metabolism and transcriptional reprogramming induced by glucose. After this discovery ChREBP has emerged as a key mediator of glucose sensing in hepatocytes, adipocytes and pancreatic β-cells. By regulating genes involved in glycolysis (LPK) and lipogenesis (ACC, FAS), ChREBP mediates metabolic adaptation to changing blood glucose levels, resulting in the activation of the fatty acid synthesis (see reviews by Filhoulaud et al. (2013) and Ortega-Prieto and Postic (2019)).

The molecular mechanisms underlying ChREBP activation are complex and still poorly understood. They are briefly reported. At low glucose levels, ChREBP is repressed by an intramolecular interaction with the N-terminal segment. When glucose metabolism increases, ChREBP is modified, leading to a relief of this interaction, thereby resulting in its activation. Other mechanisms have been investigated and its activation also depends on O-GlcNAcylation, acetylation and/or glucose metabolites: (1) xylulose-5-P (Xu-5P), an intermediate in the pentose phosphate pathway, (2) glucose-6-phosphate (G6P), the first intermediate in glucose metabolism, and (3) fructose-2,6-P_2 (F2,6P_2), the major regulator of glycolysis and gluconeogenesis (Baraille et al. 2015).

3.2.2.8. *DNA lesions sensing: ATM/FOXO3a/ATR*

The objective of all living organisms is to transmit an intact and unchanged copy of its genome to the next generation. This must be achieved despite the fact that each human cell is exposed to approximately 70,000 DNA lesions per day (Lindahl and

Barnes 2000). DNA damage-causing agents have two origins: (1) an external origin, represented by chemical and physical agents, and (2) internal origin, such as spontaneous hydrolysis of the bases and biological events (DNA replication, chromosome segregation, etc.). After the discovery of the double helix, F. Crick wrote, "We totally missed the possible role of enzymes in repair although … I later came to realize that DNA is so precious that probably many distinct mechanisms would exist. Nowadays, one could hardly discuss mutation without considering repair at the same time" (Crick 1974). We now know that despite a great number of lesions of the DNA, most of them are efficiently repaired which prevents or delays genetic instability and tumorigenesis. Damaged DNA elicits a cellular response, called the DNA-damage response (DDR), to detect the lesions and recruit specific damage sensor protein complexes associated with the sensing kinases that are members of the phosphatidylinositol-3-kinase-like family: the Ataxia Telangiectasia Mutated (ATM), the ATM and Rad3-related (ATR) kinase and the catalytic subunit of the DNA-dependent protein kinase (DNA-PKCs). The best characterized effectors of ATM and ATR are CHK2 (Checkpoint kinase 2) and CHK1, respectively. CHK2 phosphorylates effector proteins which are involved in mediating both G1/S and G2/M cell-cycle arrest through the p53 transcription factor. When the DNA lesion is not repairable, p53 induces the apoptosis (Colombo et al. 2020). ATR also influences DNA repair processes and CHK1 affects progression through the S phase of the cell cycle (Tubbs and Nussenzweig 2017). The elucidation of three major DNA repair pathways have been identified and were awarded with the Nobel Prize in Chemistry in 2015.

Notably, both ATM and ATR provide a barrier to tumor progression by inducing cell-cycle arrest and apoptosis. One future challenge is to understand how the DNA damage response impacts on cellular functions and how the repair and checkpoint pathways are connected.

3.2.2.9. *Neuronal calcium sensors*

In neurons, changes in intracellular free calcium Ca^{2+} concentration $[Ca^{2+}]i$ result in a wide range of cellular responses, including neurotransmitter release, modulation of ion channels, gene expression and effects on neuronal survival and apoptosis. The diversity of effects of $[Ca^{2+}]i$ depend on the magnitude, duration and location of the Ca^{2+} signal. This is a classical concept that has already been mentioned in other sections and which is often encountered in biology. In neurons, these different effects are mediated through various Ca^{2+}-binding proteins acting as Ca^{2+}-sensors. These proteins undergo a conformational change on Ca^{2+} binding, enabling them to interact and regulate different target proteins. Therefore, specific neuronal Ca^{2+} signals are likely to be decoded by various Ca^{2+} sensor proteins.

The examples of neuronal Ca^{2+} sensor proteins are as follows:

– Calmodulin (CaM)

As its name suggests, calmodulin is a CALcium MODULated proteIN. It has ubiquitous expression and contains a structural motif known as the EF-hand. An E–F hand consists of an N-terminal helix (the E helix) immediately followed by a centrally located, Ca^{2+}-coordinating loop and a C-terminal helix (the F helix) (Chin and Means 2000). It senses large Ca^2 oscillations due to Ca^2 influx from the host channels. CaM is known for regulating several classes of proteins and enzymes in a Ca^{2+}-dependent manner. The mechanisms underlying the coupling CaM-effectors are transduced into a change in the affinity of CaM for the effector and/or a change in the effector's function. Many of the most highly characterized effectors (e.g. the CaM-dependent adenylyl cyclases, phosphodiesterases, protein kinases and the protein phosphatase calcineurin) are directly or indirectly involved in protein phosphorylation (Figure 3.16). CaM also regulates the activities of the plasma membrane, a Ca^{2+} pump and various ion channels (Marcelo et al. 2016).

– Neuronal calcium sensor proteins (NCS proteins)

These have limited similarity to calmodulin (≈ 20%). There are five classes of NCS proteins (A–E), which are cytosolic at resting Ca^{2+} concentrations and constitutively associate with the plasma membrane through palmitoyl or myristoyl groups when Ca^{2+} is elevated. NCS-1 was discovered originally in *Drosophila melanogaster* and was designated "NCS-1" (class A protein) as it was thought to be expressed only in neuronal cell types. It is implicated in the regulation of neurotransmitter release. Recoverin (class C protein) is believed to have a role in the regulation of phototransduction by preventing the downregulation of rhodopsin due to its phosphorylation, and thereby prolonging the light response.

The NCS proteins regulate many cellular events in neurons and retinal photoreceptors (see reviews by Burgoyne (2007) and Burgoyne et al. (2019)). The question which is often brought up is: why do multiple NCS proteins exist?

– Calcium-binding proteins (CaBPs)

These represent another example of a diverse family of Ca^{2+}-sensors capable of regulating discrete processes in the nervous system (see the review by Burgoyne et al. (2019)). It is worth noting that two members of this family CaBP1 and CaBP2 are amino terminally myristoylated, which allows localization to the plasma membrane and Golgi apparatus. This subcellular localization confers specific function(s) to these proteins, and it gave rise to a concept which is discussed in section 3.1.5.2.

3.2.2.10. *Endoplasmic reticulum calcium sensor*

The control of calcium influx at the plasma membrane by endoplasmic reticulum (ER) stores is central to physiological calcium signaling and cellular calcium balance. Two key proteins, ORAI a calcium-selective channel in the plasma membrane (PM) and stromal interaction molecule (STIM), are involved in the regulation of Ca^{2+} fluxes. STIM is a calcium store sensor localized to the ER membrane which senses an ER-luminal calcium depletion that occurs in response to an elevation in cytosolic IP3 which then binds to its receptor in the ER membrane. The calcium-store depletion relocates STIM to ER-PM junctions and recruits and gates ORAI channels (see reviews by Grabmayr et al. (2020) and Gudlur et al. (2020)). Note that STIM is a sensor of ER-luminal Ca^{2+} depletion, whereas CaM is a sensor of increased Ca^{2+} influx (Figure 3.16).

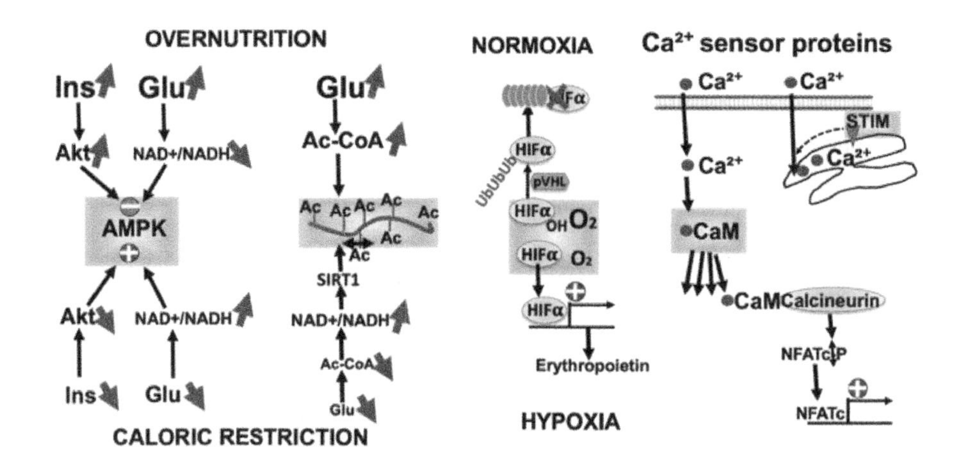

Figure 3.16. *Schematic illustration of four types of sensors. AMPK (energy), protein acetylation (nutrient availability), HIF (oxygen availability) and CaM, STIM (Ca^{2+} sensors)*

3.3. Functional diversity of proteins

3.3.1. *Multifaceted "master regulators"*

A key feature of these proteins is their functional diversity and versatility.

3.3.1.1. *Transcription factors*

MyoD family

In 1987, Davis, Weintraub and Lassar reported that overexpression of the transcription factor myogenic differentiation 1 (MyoD), normally expressed in skeletal muscle cells, converts mouse embryonic fibroblasts (MEF) to myoblasts (Davis et al. 1987). This finding revealed that a single transcription factor could induce a direct reprogramming, also known as transdifferentiation, without transitioning through an intermediary pluripotent state (indirect reprogramming).

Subsequently, three other transcription factors, myogenin (MyoG), myogenic factor 5 (Myf5) and myogenic regulatory factor 4 (Mrf4) were identified. In mice, it was shown that MyoD, Myf5 and Mrf4 function as myogenic determination factors, controlling entry into the myogenic program (Moncaut et al. 2013). With respect to MyoG, it acts as a differentiation factor, controlling the differentiation of myoblasts into skeletal muscle fibers. Interestingly, NeuroD a closely related transcription factor drives a neuronal program when it is introduced into P19 cells and MyoD introduces into MEFs (mouse embryonic fibroblasts) converts them into muscle (Fong et al. 2012). However, NeuroD cannot convert MEFs into neurons and MyoD cannot convert P19s into muscle. Such unexpected data could be explained by the fact that the binding of both transcription factors is constrained by chromatin accessibility, and thus sites that are open are determined epigenetically in a lineage-specific fashion.

In differentiating myoblasts, MyoD binds to approximately 25,000 sites throughout the genome, while only approximately 2,000 genes display modified expression during myogenesis (Cao et al. 2010). The limited number of expressed genes observed in response to extensive MyoD occupancy at genomic loci suggests that this transcription factor must itself be regulated to coordinate the temporally ordered muscle-specific transcriptional program.

During the past 30 years, cellular reprogramming has exploded, notably indirect reprogramming. Thus, Yamanaka's group found that four transcription factors reprogram somatic cells into induced pluripotent stem (iPS) cells (Takahashi and Yamanaka 2006), which subsequently can produce target differentiated cells.

Taken together, these observations have led to a reevaluation of the molecular aspects of the concept of cell differentiation.

p53/HIF-α/Myc

p53 was identified in 1979, in complex with the simian virus 40 T antigen in transformed rodent cells (Lane and Crawford 1979). It was suggested that the protein might act as a cellular oncogene (DeLeo et al. 1979), but was later identified as a tumor suppressor protein (Finlay et al. 1989). Several studies have confirmed this role: (1) many individuals affected by a germline mutation of p53 display an abnormally high incidence of tumor development (Li–Fraumeni syndrome); and (2) the transformation of cells from normal to cancerous initiates its activation due to stress signals which accompanied malignant progression. Its activation prevents tumor progression by its ability to induce apoptosis or senescence (Zuckerman et al. 2009). Broadly defined, stress is the state when cellular homeostasis is disrupted due to environmental changes or fluctuations in environmental factors. A multitude of different types of stress often encountered during tumor development initiate its activation, such as DNA damage, oncogene activation, hypoxia, telomere erosion and others (Vousden and Lu 2002). In response to these stresses, p53 undergoes a series of post-translational modifications which contribute to its activation. Then, p53 stimulates (or in some cases represses) the expression of a large network of genes involved in many cellular processes aiming at restoring cellular homeostasis to the former state or to cope with the new environment. Stress responses are mediated via multiple mechanisms, depending on the type, severity and duration of stress encountered. These specific responses to any stress exert a protective effect on the organism or the cell. For instance, in response to a severe DNA damage, p53 induces apoptosis which eliminates cells with mutated genes. In response to mild DNA damage, p53 can temporarily arrest the cell cycle (Sherr and Roberts 1999), which allows for the repair of any damaged DNA, preventing mutations from being passed on to daughter cells. It is still an open question on how p53 integrates the extent and type of DNA damage to either activate cell-cycle arrest and DNA repair or apoptosis. Because of its central role in the DNA damage response, p53 is often referred to as the "guardian of the genome". Oncogene activation also leads to p53 activation, resulting in senescence, thus limiting the oncogenic potential of preneoplastic cells.

During the past decade, it has been reported that p53 is a master regulator of many biological processes, including aging (Poyurovsky and Prives 2010), innate and adaptive immunity (Menendez et al. 2013), development (Danilova et al. 2008), reproduction (Levine et al. 2011) and neuronal degeneration (Chang et al. 2012). There is now growing evidence that the cancer metabolic reprogramming observed in most cancer types is predominantly driven by impaired p53, hypoxia-inducible factor (HIF-1) and c-Myc activation. Their respective impact on glucose metabolism is mentioned in the section dealing with the "Warburg effect" (see section 4.3.6.4).

NF-κB

One of the primary regulators of the inflammatory response is the ubiquitously expressed, inducible transcription factors of the NF-κB (nuclear factor kappa-light-chain-enhancer of activated B-cells) family. The NF-κB transcription factor family includes a collection of proteins conserved from (at least) the phylum Cnidaria to humans. It is composed of two subfamilies, NF-κB proteins and Rel proteins (c-*Rel* was identified through its oncogenic derivative, *v-rel,* found in the avian leukemogenic reticuloendotheliosis retrovirus). The activated form of NF-κB is a heterodimer which often consists of a p65 subunit (also called Rel A) and a p50 subunit that enhance transcription of target genes (Hoffmann et al. 2006). The activity of NF-κB is regulated by interaction with an inhibitor of κB (IκB) proteins (Baeuerle and Baltimore 1988). Binding of IκB to NF-κB dimers blocks their nuclear localization sequence (NLS) function, and the dimer is thereby sequestered in the cytoplasm. Upon activation of the NF-κB signaling pathway, IκB kinases (IKK-α, IKK-β) target IκBs for degradation. This allows NF-κB to translocate and accumulate in the nucleus, where it binds to DNA, resulting in the expression of more than 100 target genes involved in both the innate and adaptive immune responses (Vallabhapurapu and Karin 2009).

Much attention has been focused on stimuli that trigger the activation of NF-κB, such as exogenous agents (viruses, bacteria, fungi, etc.) or by endogenous signals released by damaged cells, and all of these molecular patterns are sensed by the abovementioned receptors (TLRs, NLRs, RLHs). In response to these exogenous or endogenous stimuli, receptors activate diverse signaling pathways, converging in the activation of NF-κB which stimulates the transcription of pro-inflammatory genes acting to defend against pathogens or damaged cells (Takeuchi and Akira 2010). It is noteworthy that one of the genes which is firstly transcribed by NF-κB is that encoding IκB. Newly synthesized IκB binds to NF-κB and attenuates the pathway of response to multiple stimuli, thereby creating a negative feedback loop within this signaling pathway (Napetschnig and Wu 2013) (Figure 3.10B).

The NF-κB signaling pathway should be tightly regulated to prevent excessive immune and inflammatory responses. One candidate for this physiological balancing is the glucocorticoid receptor (GR), which has the ability to interfere with NF-κB and reduces the expression of many of the same cytokines and cytokine-induced genes that are activated by NF-κB (McKay and Cidlowski 1999). Several hypotheses have been proposed to explain the anti-inflammatory mechanisms of GR. It was suggested that glucocorticoids increase expression of IκB, which interacts with NF-κB, thus sequestering it in the cytosol. By sequestering NF-κB in the cytosol, IκB blocks its translocation to the nucleus and thereby induction of pro-inflammatory genes. GR could also interact with different NF-κB subunits or alter NF-κB heterodimer and/or homodimer, thus regulating gene expression through different NF-κB signaling pathways. This mode of regulation represents a crosstalk between these two transcription factors.

Finally, clinical data point to the importance of chronic inflammation in human diseases. Hence, prevention or reduction of inflammation can be obtained with inhibitors of IKK kinase, such as curcumin, aspirin and ibuprofen (Karin et al. 2004).

E2F/pRb

E2F is a transcription factor, and the best understood function of E2F is its ability to regulate the G1/S phase transition and S-phase entry during the cell cycle. Recent studies have shown that the E2F family is growing steadily, and members have distinct mechanisms of action and regulation, but they are not discussed. For a comprehensive coverage of the E2F field the reader is referred to a review (Dimova and Dyson 2005). pRb (Retinoblastoma) interacts with E2F, but the molecular mechanisms behind the wide-ranging roles of this complex are not yet fully elucidated. Nonetheless several conclusions can be drawn regarding the E2F/Rb pathway. In the absence of growth factor signals, pRb binds to the transactivation domain of E2F and thereby represses the transcription of genes, required for transition from G1 to S of the cell cycle. This growth-inhibitory function of pRb only applies to its hypophosphorylated form which predominates in quiescent cells (Narasimha et al. 2014). In response to growth factors, pRb is hyperphosphorylated by the complex cyclin D/Cdk4, which is activated by upstream events of the signaling pathway. Thus, it relieves the inhibitory effect of pRb on E2F, resulting in stimulation of transcription of genes required for a quiescent cell to enter and progress through the cell cycle. pRb can be functionally inactivated by mutations. As a direct consequence, E2F is liberated by control of pRb and induces cell-cycle deregulation. Almost all human cancers carry abnormalities in the E2F/pRb pathway components.

It is worth noting that phosphorylation events during mitosis are reversible. To ensure that mitotic cells do not reverse to G2, there is a system that maintains the mitotic kinases and suppresses the phosphatases that counteract the kinases' actions (Hunt 2013). This is another example illustrating that the net outcome of a biological response controlled by two opposite signal(s) (positive and negative signal(s)) is optimized when the effects of either one are switched off.

Mechanistic studies have also revealed that pRb has a multitude of molecular functions that are not linked to gene transcription and cell-cycle regulation. Recent studies demonstrated that pRb is distributed across the genome in a sequence-independent manner. Thus, pRb interacts with over 300 proteins and most of them are chromatin modifiers, suggesting that it impacts the organization of heterochromatin. As a result, these protein complexes confer new functions to pRb and some of them are briefly discussed. Readers should refer to an excellent review (Guzman et al. 2020).

**Histone acetylation*

This involves histone acetyltransferases (HATs) and histone deacetylases (HDACs). These modifications of the acetylation state are facilitated by pRb which interacts both with HDACs to repress gene transcription and HAT to facilitate its inactivation, thereby switching off the opposite action which counteracts the HDACs. This type of regulation by pRb is a mechanism which is often encountered in biology and mentioned in other sections.

**Histone and DNA methylation*

Histone methylation is regulated by methyltransferases (HMTs) and demethylases (KDMs). The latter have the property to bind and interact with pRb (transcription factor). Depending on the type of KDM recruited by pRb, it either represses or activates its transcriptional activity.

**DNA methylation*

The DNA methyltransferase family (DNMTs) can form a stable complex with pRb and E2F and repress E2F-driven transcription.

Other studies have demonstrated interactions of pRb with numerous effectors that serve to repair DNA, replicate DNA and ensure genomic stability (see the review by Dick et al. (2018)). Overall, this highlights that pRb has functions that extend beyond the regulation of cell proliferation, thereby indicating its multifunctional properties that often apply to numerous proteins.

AP-1

The transcription factor AP-1 is encoded by proto-oncogenes and regulates various aspects of cell proliferation and differentiation. AP-1 can be composed of either homodimers or heterodimers between members of the Jun (c-Jun), Fos (c-Fos), activating transcription factor (ATF) or Maf (c-Maf) families. The protein products of the fos and jun gene families, that is, the so-called immediate-early genes that are directly activated, are transcription factors that activate and repress other genes. Enhanced AP-1 activity occurs in response to various stimuli, such as growth factors, mitogens, polypeptide hormones, cell-matrix interactions, inflammatory cytokines, bacterial and viral infections and cellular stress (ultraviolet or ionizing radiation). These stimuli mostly activate p38, JNK and ERKs which then phosphorylate AP-1 and enhance its transcriptional activity.

3.3.1.2. *Signaling pathways*

ERK signaling

Since the discovery of ERKs in 1991 (Boulton et al. 1991), a tremendous amount of literature has been generated that has revealed the complexity and pleiotropism of

ERK signaling. Their broad repertoire of substrates put them on the forefront of the regulation of most vital cellular functions. A detailed discussion of this signaling is beyond the scope of this book. Readers should refer to an excellent review by H. Lavoie (Lavoie et al. 2020).

Glycogen synthase kinase

Glycogen synthase kinase-3 (GSK-3) is one of the few protein kinases (serine/threonine kinase) that is active under basal conditions and requires extracellular signals for its inactivation. The latter occurs through different mechanisms: inactivating phosphorylation by other kinases, changes in subcellular localization, changes in tyrosine phosphorylation, through inactivation of kinases that act to prime GSK substrate proteins (Patel and Woodgett 2017). GSK lies downstream of several major signaling pathways, including the phosphatidylinositol 3' kinase pathway which causes Akt activation, resulting in phosphorylation and inactivation of GSK-3 (Cross et al. 1995). It is noteworthy that activated GSK-3 recognizes and phosphorylates substrates that are previously phosphorylated by a priming kinase. It should be stressed that distinct signaling pathways are responsible for the priming of individual GSK-3 substrates. Thereby, it adds an extra layer of complexity that allows the integrated regulation of these targets by multiple signaling inputs. Since the initial discovery, nearly 100 proteins have been added to the list of putative GSK-3 targets, ranging from regulators of cellular metabolism to molecules that control growth, differentiation and survival (Kaidanovich-Beilin and Woodgett 2011).

PTEN

Immunohistochemical staining studies revealed the cytoplasmic and nuclear localization of PTEN. It was first demonstrated that cytoplasmic PTEN acts as phosphatidylinositol-(3,4,5) and -(3-4) phosphatase and opposes the activity of the oncogenic PI3K signaling from the plasma membrane. Thus, this mechanism could explain the function of PTEN in tumor suppression. However, it was noted that anti-PI3K agents have failed to counter the growth of PTEN-deficient tumors, implicating non-PI3K aspects of PTEN function in tumor suppression (Juric et al. 2015). It was therefore hypothesized that PTEN may also act as a lipid phosphatase in the nucleus. However, different studies did not confirm this potential role, and it was discovered that PTEN is instead involved in the maintenance of genome integrity when cells are exposed to genotoxic agents. Its role is to coordinate a cellular response that determines the balance between "DNA repair and survival" and "DNA damage and apoptosis" outcomes (see excellent reviews by Fan et al. (2020) and Ho et al. (2020)).

In summary, over the last decade, studies have focused on the complex implication of PTEN in a variety of processes, including genome maintenance and

DNA repair, cell-cycle control, gene expression and DNA replication. These processes are achieved through multiple signaling pathways that require PTEN's multifaceted functions in both the nucleus and cytoplasm to enable accurate transmission of genetic materials.

Akt/PKB

Akt is a multifunctional protein that is activated by PI3K in response to the stimulation of receptor tyrosine kinases (RTK) or G protein-coupled receptor (GPCR) that integrates extracellular signals, such as availability of nutrients, energy and growth factors. It is beyond the scope of this book to list and discuss the upstream regulatory inputs into Akt, the mechanisms of its activation (for a detailed review on Akt substrates and functions see Manning and Toker (2017)).

Studies carried out on Akt signaling suggest a wide diversity of downstream Akt substrates. Likewise, Akt signaling has several points of cross-regulation with AMPK. For instance, Akt signaling promotes glucose uptake, glycolysis and ATP production, which indirectly prevents AMPK activation. Of note is that both kinases stimulate glucose uptake, but it occurs in response to distinct stimuli: activation of Akt and AMPK occurs in response to insulin and ATP depletion, respectively. Furthermore, Akt signaling stimulates ATP-consuming anabolic processes, through mTORC1, whereas AMPK activation favors ATP-producing catabolic processes, thereby blocking anabolism (Dibble and Manning 2013).

A variety of stimuli can activate the PI3K–Akt pathway and in response to them Akt phosphorylates several direct downstream targets involved in multiple biological processes. Briefly, activation of Akt by insulin results in phosphorylation of transcriptional factor Foxo1, leading to its translocation towards the cytoplasm, thereby causing an attenuation of gene transcription. This explains the decrease in the transcription of neuropeptide Y (NPY) and agouti-related peptide (AgRP), in the arcuate nucleus, leading to decreased food intake (Kim et al. 2006). Similarly, in the liver, activation of Akt by insulin results in decreased gene transcription of PEPCK and G6Pase enzymes, leading to lowered gluconeogenic flux and thereby a decreased hepatic glucose release (Matsumoto et al. 2007). Of note is that Akt controls other metabolic processes in the liver, including glycogen synthesis, glycolysis and lipid synthesis.

Akt directly phosphorylates and activates ATP-citrate lyase, which generates cytosolic acetyl-CoA, the precursor of *de novo* lipogenesis, thus linking Akt signaling to lipid synthesis (Berwick et al. 2002).

mTOR complex1 (mTORC1)

mTORC1 integrates information about nutritional abundance and environmental status to tune the balance of anabolism and catabolism in the cell. Its activation initiates a downstream anabolic program that enhances the production of proteins, lipids, nucleotides and other macromolecules while inhibiting catabolic processes, such as autophagy and lysosome biogenesis. The repression of these catabolic processes prevents a futile cycle in which newly synthesized cellular building blocks are simultaneously broken down and recycled. This dual control of anabolism and catabolism has been also mentioned with insulin which simultaneously stimulates glycogenesis, lipogenesis, protein synthesis and inhibits glycogenolysis, lipolysis and proteolysis, respectively.

In the early 1990s, major regulators of mTORC1 activity were identified, which were reviewed by Saxton and Sabatini (2017). Among them were growth factors, including insulin and insulin-like growth factor 1, which stimulate mTORC1, whereas low ATP level inhibits mTORC1, due to AMPK activation.

Full activation of mTORC1 requires signaling induced by amino acids. They promote its translocation from the cytoplasm to the lysosomal surface via the RAS-related GTP-binding proteins (RAGs).

3.3.1.3. *Glycolytic enzymes*

Recent studies have provided evidence that some glycolytic enzymes have acquired additional non-glycolytic functions, which changes the concept of simple roles in the glycolysis pathway. Unexpected functional roles of several enzymes have been identified and they are briefly summarized. These new functions include transcriptional regulation (hexokinase-2 = HK2, lactate dehydrogenase A = LDH-A, glyceraldehyde-3-phosphate dehydrogenase = GAPD and enolase = ENO 1), apoptosis (HK, GAPD) and cell mobility (glucose-6-phosphate isomerase).

GAPD is encoded by a so-called “housekeeping gene”, commonly used for gene expression control, and LDH are unique because of their ability to bind NAD^+ or NADH and to DNA and RNA. Hexokinase 4 (called glucokinase) expressed in the liver and β-cell is present in the mitochondrial complex and is involved in oxidative phosphorylation and apoptosis processes in response to various extrinsic stimuli and metabolic states (readers should refer to the review by J. Kim (Kim and Dang 2005)).

3.3.2. *Molecular motor proteins*

The ability of a cell to respond and adapt to changing physiological cues relies on continual reorganization of the contents of its cytoplasm. This is accomplished primarily through active transport along cytoskeletal filaments by molecular motor proteins. Three superfamilies of motor proteins have been described: the dynein and the kinesin were discovered in 1963 by I. Gibbons and in the mid-1980s by R. Vale, respectively. The first myosin, M2, was discovered in 1864 by W. Kühne (Kuhne 1864). They all have in common the capacity to hydrolyze ATP and the hydrolysis products drive a cycle of interactions with the track (either an actin filament or a microtubule), resulting in force generation and directed movement.

Their functions have been reviewed recently (Barlan and Gelfand 2017; Sweeney and Holzbaur 2018). Briefly, their primary function is to deliver cargoes – diverse membrane organelles, messenger RNA transcripts, protein complexes and viruses, among others – to discrete cellular locations in response to various physiological stimuli. They also play a role in facilitating molecular exchanges and chemical interactions between membrane organelles. By tethering organelles to a cytoskeletal track, motors act to limit three-dimensional diffusion to movement in one dimension.

The properties of a given myosin transporter are adapted to move on different actin filament tracks, either on the disordered actin networks at the cell cortex or along highly organized actin bundles to distribute their cargo in a localized manner or move it across long distances in the cell. Transport is controlled by selective recruitment of the myosin to its cargo that also plays a role in activation of the motor.

Kinesin and dynein motors move on microtubules and transport a broad range of organelles and vesicles. The generation of biochemically distinct microtubule subpopulations allows subsets of motors to recognize a given microtubule identity, allowing further organization within the cytoplasm. Both transport and tethering are spatiotemporally regulated through multiple modes, including acute modification of both motor–cargo and motor–track associations by various physiological signals.

3.3.3. *Interactional domains*

3.3.3.1. *SH, PTP, PH, FYVE, BH and RHD domains*

The SH2 (Src homology region 2) and PTB (phosphotyrosine-binding) domains are small protein modules that mediate protein–protein interactions involved in many signal transduction pathways. Both domains recognize phosphorylated tyrosine in receptor tyrosine kinases and other signaling proteins.

* *SH domain (Sarcoma Homology domain = Src Homology domain)*

The SH2 domain is a protein module of about 100 amino acids, initially discovered as a regulatory region in Src kinases responsible for maintaining tyrosine kinase activity of Src kinases in an inactive configuration by mediating intramolecular auto-inhibitory interactions.

The role of the SH2 domain was revealed when studying the stimulation of the signaling pathway of the EGFR in response to its ligand. The major finding was that phospholipase Cγ (PLCγ) forms a stable complex with the activated receptor (Margolis et al. 1990). It was subsequently shown that interaction of EGFR and PLCγ was mediated by the Src homology 2 (SH2) domain of PLCγ that binds directly to a P-Tyr site in the carboxy-terminal tail of EGFR (Margolis et al. 1990). Of note is that Src, PLCγ and many signaling proteins contain an additional small protein module designated the SH3 (for Src homology 3) domain, which binds specifically to short, proline-rich regions in target proteins (Pawson 1995, 2004).

* *PTP domain*

Another mechanism for recruitment and activation of signaling molecules by activated RTKs involves docking proteins such as insulin receptor substrate 1 (IRS1) and other members of the IRS family and FGF receptor substrate 2 (FRS2). It was shown that docking proteins bind via their PTB domains to the insulin receptor or FGFRs, respectively, and become phosphorylated on numerous tyrosine. The P-Tyr sites of docking proteins provide a platform for the recruitment and activation of an additional complement of signaling molecules that bind to the P-Tyr sites of the docking proteins via their own SH2 domains (Lemmon and Schlessinger 2010) and are responsible for mediating many of the known insulin-induced cellular responses (Boucher et al. 2014).

* *FYVE and PH domains*

The Fab1p-YOPB-Vps27p-EEA1 (FYVE) domain and the pleckstrin homology domain (PH) detected as an internal repeat in pleckstrin are found in many proteins involved in signaling, membrane trafficking and cytoskeletal rearrangement. These domains bind to phosphoinositides which are highly concentrated in distinct pools localized in the plasma membrane, endosomes and nucleus. For instance, the FYVE domain-containing proteins serve as regulators of endocytic membrane trafficking, and they are recruited to the endosomal membrane and selectively bind PIP3. With respect to the PH domain, it is contained in different proteins such as son of sevenless (Sos), insulin receptor substrate 1 (IRS1) and Akt.

* *BH = Bcl-2 Homology domain*

In the 1980s, after the discovery of Bcl-2 (anti-apoptotic protein) in B-cell lymphoma, several homologous proteins have since been identified. The BH domains facilitate the family members' interactions with each other and can indicate pro- or anti-apoptotic function.

* *RHD domain*

The nuclear factor NF-κB family of transcription factors have a conserved Rel-homology domain (RHD) that contains a DNA-binding motif, a dimerization region and an interaction domain with IκBs (Hayden and Ghosh 2004).

* *Bromodomain (BRD)*

This module is found in about 40 diverse proteins that primarily recognize acetylated histones and regulate gene expression through a wide range of activities. Thus, they can act as (1) scaffolds that facilitate the assembly of larger protein complexes, (2) transcriptional co-regulators, and (3) methyltransferases (see Chapter 5, Fujisawa and Filippakopoulos 2017).

* *RGD sequence*

Proteins that contain the Arg-Gly-Asp (RGD) attachment site, together with the integrins that serve as receptors for them, constitute a major recognition system for cell adhesion. The RGD sequence is the cell attachment site of a large number of adhesive extracellular matrix.

3.3.3.2. *Domain structures of ECM proteins*

Many ECM proteins are modular and multidomain. Only two of them will be briefly discussed.

* *Thrombospondin (TSP)*

TSP is a multifunctional protein with various domains. Because of multidomain structure, TSPs can bind to multiple membrane receptors, which, in turn, can activate distinct signaling pathways eventually resulting in different cellular phenotypes and tissue specific effects. It antagonizes the pro-angiogenic function of VEGF in a number of ways. It integrates pro-and anti-angiogenic signal transduction to determine endothelial cell behavior (Kazerounian and Lawler 2018).

* *Fibronectin (FN)*

The multifunctional glycoprotein FN exists either as an insoluble multimeric fibrillar component of the extracellular matrix or as a soluble monomer. The key feature of FN domains is their ability to bind to many molecules, including other ECM proteins, cell surface receptors, glycosaminoglycans (GAGs) and other FN molecules. Given the number of FN binding sites, it is extremely difficult to determine whether they have different effects or whether some of them convey the same signal(s). FN influences several crucial cellular processes and contributes to multiple pathologies (see an excellent review by Singh et al. (2010)).

3.3.4. *Carrier proteins*

Lipolytic hormones secreted in the blood circulation bind to proteins, whose role and properties started to be investigated after the discovery in 1952 of the thyroxine-binding globulin (TBG). Investigators put forward the concept of an equilibrium between thyroxine bound to TBG and thyroxine in the unbound or "free" state (Oppenheimer 1968). Assessments of their respective plasma concentrations in humans revealed that free thyroxine represents only 0.1 percent of the total plasma thyroxine concentration. Given the low level of free hormone, the means of separating bound and free are prone to error. However, measurement of the free hormone level provides a better assessment of hormonal status than the total hormone levels. Furthermore, they need to enter cells as their principal actions involve nuclear receptors. Hence, only the free hormone is able to enter the cell and exert biological effect.

Other hormone carrier proteins have been described: vitamin D-binding protein (DBP), sex hormone-binding globulin (SHBG) and cortisol-binding globulin (CBG) that bind to their respective ligand, especially 25(OH)D, testosterone and cortisol (approximately 95% of cortisol/corticosterone in the circulation is bound to CBG) (Bikle 2021). Besides their role of carrier, they serve as a reservoir within the circulation to maintain a steady level of free hormone. Hence, they regulate the availability of free hormone to the tissue.

Other carrier proteins also have a prominent role in biological processes. Some examples are as follows:

– Transferrin (Tf). This binds two ferric iron (Fe^{3+}) and Tf-bound iron enter cells after binding to a cell surface receptor R1 (Kawabata 2019).

– IGF-binding protein (IGFBP) family. This consists of six distinct members of multifunctional proteins which bind to IGFs with an equal or greater affinity than the IGF1 receptor. Hence, they sequester IGFs and consequently limit their access to their receptors. Originally described as passive circulating transport proteins, they are now recognized as playing critical roles in the circulation, the extracellular environment and inside the cell (reviewed in Firth and Baxter (2002), Clemmons (2016) and Allard and Duan (2018)). The vast majority of IGFs in the serum are found in a ternary complex, consisting of IGF1-IGFBP3/5-ALS (glycoprotein called acid labile subunit). This complex prevents the bound IGF from leaving the vessels and prolongs the half-life, forming a long-lasting reservoir of IGFs. Another important function of circulating IGFBPs is to prevent the potential interaction of IGFs with the insulin receptor, which is crucial since IGF concentrations are high enough in the serum to cause hypoglycemic effects even given their lower affinity for the insulin receptor.

Recent studies have provided a great deal of information regarding their role, but there is no complete picture of how these proteins function coordinately with IGF1 to regulate diverse physiological processes.

3.3.5. *Decoy molecules*

The concept of decoy molecules has been introduced in the context of bone remodeling and immune modulation.

Bone remodeling: osteoprotegerin

This is a biological process that requires a crosstalk between bone-forming osteoblasts and bone-resorbing osteoclasts. Disrupting this intricate balance between synthesis and resorption of bone leads to skeletal abnormalities, such as osteoporosis (decline in bone mineral density) and osteopetrosis (increased density of bones). Imbalances between osteoclast and osteoblast activities can arise from hormonal, cytokine or growth factor changes.

In 1997, a major advance in the field of bone biology was the discovery of the osteoprotegerin OPG that became a key player in the control of osteoclast function and hence osteoporosis. Two other molecules were known to be involved: the receptor activator of NF-κB (RANK) and its ligand RANKL (reviewed in Leibbrandt and Penninger (2009)).

Here are the experiments that have defined the respective role of these molecules in bone remodeling. Regarding RANKL, it is highly expressed in stromal osteoblasts, and it binds to its cognate receptor RANK a transmembrane protein expressed in osteoclast progenitors, leading to osteoclast differentiation. Mice deficient in RANK or RANKL are phenocopies of one another, indicating the essential role of this signaling axis. With respect to OPG, secreted by stromal osteoblasts/stromal cells, it was demonstrated that transgenic mice overexpressing OPG exhibited increased bone mass (osteopetrosis) (Simonet et al. 1997), whereas targeted deletion of OPG in mice resulted in osteoporosis (Bucay et al. 1998). This clearly demonstrated that OPG had a critical role in the maintenance of post-natal bone mass. Subsequently, it was shown that OPG binds to RANKL, thereby downregulating signaling through the RANKL–RANK interaction (Kong et al. 1999). Taken together these findings clearly demonstrated that RANKL–RANK interactions positively regulate osteoclastogenesis and OPG functions as a decoy receptor binding to RANKL, negatively regulating osteoclast differentiation (Figure 3.17).

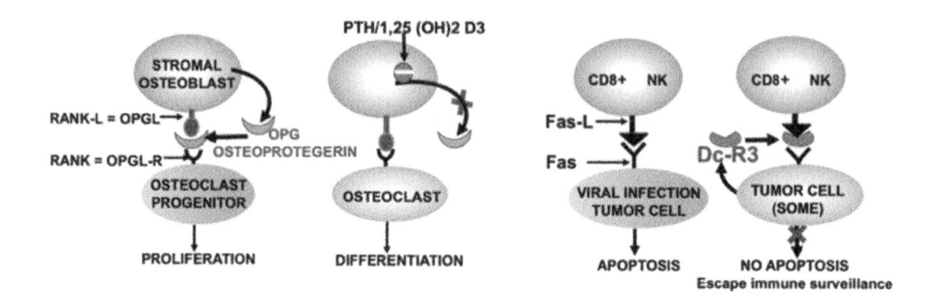

Figure 3.17. *Decoy molecules, OPG and DcR3, regulating osteoclast differentiation and tumor cell apoptosis, respectively*

Immune evasion strategy of tumor cells: DcR3

FasL was identified in 1989, which plays a pivotal role in the immune system in killing abnormal or infected cells. It is transmembrane protein, primarily expressed by cytotoxic lymphocytes, i.e. natural killer (NK) cells and $CD8^+$ T-cells. FasL has the ability to induce apoptosis (programmed cell death) in target cells through binding to its receptor Fas.

In recent studies, a new molecule has been characterized in the context of cancer. It is secreted by tumor cells and belongs to the tumor necrosis superfamily FasL and was called DcR3 (reviewed in Lin and Hsieh (2011)). It was speculated that tumor cells expressing high levels of DcR3 might be able to evade cytotoxic attack by inhibition of the FasL–Fas interaction which induces apoptosis. In support of this, it was shown in mice that injection of recombinant DcR3 attenuates lethal hepatic injury occurring after injection of cross-linked FasL and blocks FasL-induced cell death (Connolly et al. 2001). Thus, DcR3 has the ability to bind FasL and acts as a decoy receptor to help tumor cells to escape from T cell-mediated cytotoxicity. Thus, targeted reduction of DcR3 levels might be a novel therapeutic strategy in cancer (Figure 3.17).

Immune evasion strategy of viruses: MHC I-like molecule

At first, virally infected cells are normally killed by the immune system's T-cells, which recognize viral peptides that are presented by class I molecules of the MHC of infected cells. Natural killer cells (NK) are crucial for the successful control of the destruction of infected cells. NK functions are regulated by specific receptors that, upon engagement with their ligands on target cells, elicit either stimulatory or inhibitory signals. Whether an NK cell will attack or tolerate an infected cell depends on the net balance between these signals. If activating signals prevail, NK initiates its cytolytic activity (Arnon et al. 2006). However, cytomegaloviruses (CMVs) developed immune evasion strategies to evade detection and destruction by NK. To this end, throughout evolution CMVs captured genes from their hosts and used the derived proteins to evade host immune defenses. Details and mechanisms of their action are beyond the scope of this review but can be found elsewhere (Goodier et al. 2018). In brief, it has been demonstrated that an MHC I-like molecule encoded by CMV functions as a decoy molecule which engages inhibitory receptors of NK. Furthermore, CMV infection causes a downregulation of MHC-I molecules which renders infected cells invisible to $CD8^+$ T-cells.

3.3.6. *Heat shock proteins as molecular chaperones*

The discovery of heat shock proteins started with a curious incident in a laboratory in Italy in 1962. The geneticist F. Ritossa was observing *Drosophila* larval chromosomes with light microscopy and noted an unexpected puffing pattern (Ritossa 1962). He also noted that the larvae had been accidentally placed in a 37°C incubator. He described that the numerous puffs that represent active genes (i.e. DNA regions where transcription is actively taking place) had disappeared, and new large puffs had appeared in places where they did not exist prior to the heat shock (HS). At that time, it was considered to be a laboratory artifact. Almost 15 years later, two groups of researchers showed that in these puffs observed in response to HS, RNAs that encode a special group of proteins that were mistakenly called "heat shock proteins" (HSP) are synthesized. Given that they arise after HS and other harmful effects, it would have been more correct to call them "stress proteins", but the original name HSP remained.

After this discovery, a vast literature accumulated describing a wide variety of cellular events in response to a heat shock stress. Coincidently, studies had revealed that protein folding and assembly events *in vivo* require the presence of accessory components, and they were identified as members of the heat shock protein family. Their main role is to protect newly synthetized polypeptides from misfolding ("protein quality control"), prevent improper interactions with other molecules facilitate transport across membranes and mediate conformational changes and formation of multimeric complexes. These new functions gave rise to the concept of "molecular chaperones" (Bascos and Landry 2019). Hsp70 plays an essential role in protein homeostasis.

3.3.7. *Hormone-like peptides: molecular mimicry*

The term "molecular mimicry" was coined in 1964 by R. Damian, who suggested that antigenic determinants of microorganisms may resemble antigenic determinants of their host (Damian 1964), which is thought to be one of the mechanisms that underlie the development of autoimmunity. The concept is based on a structural similarity between a pathogen or metabolite and self-structures. Recent studies have investigated molecular mimicry employed by viruses to generate molecules that resemble host growth factors, immunomodulatory cytokines and chemokines. Based on the diversity of viruses, it was hypothesized that viruses might encode proteins that can mimic human peptide hormones which in turn might stimulate or inhibit their cognate human hormone receptors. Investigators searched all available viral genomes in the viral genome database (National Center for Biotechnology Information) for sequences with significant similarity to sequences of human peptide hormones and metabolism-related cytokines. As a result, a great number of viral homologs or peptides were identified and had sequence similarity to different hormone families (Altindis et al. 2018).

Taken together, these findings introduce the concept of a system of viral hormones, and it reveals that viruses developed mechanisms that can directly target or mimic the host endocrine system.

3.3.8. *Telomerase and integrity of linear chromosomes*

In the first half of the 20th century, H. Müller (Nobel Prize 1945) working with the fruit fly and B. McClintock (Nobel Prize 1983) studying maize proposed the existence of a special structure at the chromosome ends (Chuaire 2006). This structure had the role of protecting chromosome ends from fusing together, thus enabling the correct segregation of the genetic material into daughter cells at completion of the cell cycle. Müller called these structures "telomeres" (from the Greek *telos* "end" and *meros* "part"), which allow the cell to distinguish between native chromosome ends and the DNA fragments resulting from double-strand breaks.

This protective function of telomeres is known as "telomere capping".

Telomeres are composed of repetitive sequences of DNA (tandem GGTTAG repeats) associated with a number of proteins which form a complex known collectively as the "shelterin" (Levis 1989). The shelterin complex serves to solve the end-protection problem, by preventing activation of mechanisms of DNA repair pathways in response to free DNA ends. This end-replication problem is solved by a ribonucleoprotein complex called telomerase which enables the synthesis and maintenance of these terminal structures (Greider and Blackburn 1985; Blackburn et al. 2006).

In adult somatic tissues, telomerase activity is insufficient and, during the S-phase of the cell cycle, telomeres shorten during each round of cell division (Blackburn 2001). Telomere shortening limits the proliferative capacity of somatic human cells to 50-80 cell divisions (Allsopp et al. 1992); hence, it has been suggested that it induces replicative senescence. Cellular senescence was first described in 1961 by L. Hayflick and P.S. Moorhead as the progressive and irreversible loss of proliferative potential of human somatic cells, and it is a stable phenotype (Hayflick and Moorhead 1961). In the 1990s, it was shown that telomere regions gradually shorten with cell division and that this correlates with the induction of cellular senescence (Harley et al. 1990). Then, a second observation indicated that ectopic expression of the enzyme telomerase, which is capable of maintaining elongation of telomeres, overcomes the senescence arrest (Bodnar et al. 1998). This finding clearly established that telomere length played a causal role in the process. Why do telomeres shorten? Telomere shortening occurs because the DNA polymerase that synthetizes DNA in the 5'→3' direction incompletely replicates the lagging strand, thereby causing progressive shortening of telomere sequences over time. It is currently explained by the intrinsic inability of DNA polymerases to fully replicate chromosome ends. It therefore exposes the telomere end which becomes recognized by the DNA repair system. However, several reports now suggest that telomere dysfunction inducing cell-cycle arrest and senescence can also occur irrespectively of the length of telomeres, supporting the concept that telomeres act as molecular sensors of stress. For instance, mild oxidative stress causes single-stranded breaks to accumulate at telomeres, leading to accelerated shortening and premature cell-cycle arrest (von Zglinicki 2000). However, it is possible that acute stresses induce telomeric double-stranded breaks which are not efficiently repaired. This results in a persistent DNA-damage response (DDR) signaling, preventing cells from undergoing further rounds of replication. It should be emphasized that the mechanism leading to senescence prevents uncontrolled cell division. However, if a cell escapes this non-replicative state and aberrantly reactivates telomerase expression to re-establish telomere length, it can reach "replicative immortality, which is a key feature of cancer cell" (Kim et al. 1994). The Nobel Prize in Physiology or Medicine 2009 was awarded jointly to E. Blackburn, C. Greider and J. Szostak for their work on telomeres and telomerase.

4

Integrative Aspects: From Cellular to Whole-Body Level

4.1. Homeostasis equilibrium: dynamic steady state

Homeostasis is a core concept necessary for understanding the many regulatory mechanisms in physiology. C. Bernard originally proposed the concept of the constancy of the "milieu intérieur", that would allow biological processes to proceed despite variations in the external environment (Bernard 1879). The concept was further explored by W. Cannon who introduced the term "homeostasis" in describing how key physiological variables are maintained within a predefined range by feedback mechanisms (Cannon 1929). In the 1960s, homeostatic regulatory mechanisms in physiology began to be described and focused on active mechanisms which keep a regulated variable in the internal environment within a range of values compatible with life. We should keep in mind that passive mechanisms such as water movement between capillaries are also involved in maintaining homeostasis, but these are not discussed here. There are additional concepts which are central to understand the function of a homeostatic mechanism. Firstly, it is important to distinguish two types of variables in homeostatic systems. Variables that are maintained at a stable level (near set point), by homeostatic circuit(s), such as blood glucose or core body temperature, are called "regulated variables". In contrast, variables that are manipulated in order to maintain the regulated variables within desired ranges are called "controlled variables", such as glycogenolysis, glycogenesis, glycolysis, gluconeogenesis and glucose transport from the blood into tissues. Multiple controlled variables typically contribute to the stability of a given regulated variable (Kotas and Medzhitov 2015). Thus, regulated variables refer to quantities, whereas controlled variables refer to processes. The concept of set point

---

For a color version of all the figures in this chapter, see www.iste.co.uk/gilbert/concepts.zip.

was introduced by J. Hardy, and it defines an optimal value of the regulated variable (Hardy 1953) such as approximately 5 mM for the normoglycemia. In a broad sense, the set point in a control system can be defined as the equilibrium between the factors that promote or suppress a cellular function or activity.

However, we must emphasize that there is never just a single set point that is maintained. Physiological variables oscillate within a range of values and are not completely fixed. Notably, the term homeostasis derives from Greek words *homoios* (similar) and *stasis* (state) rather than from *homos*, which means same, indicating that internal environment is always in a similar state, but is not completely fixed, which would be implied if the term "*homostasis*" were used. Regarding homeostasis, it can be broadly defined as a balance between the positive and negative extracellular and/or intracellular signals/regulators which govern all physiological responses.

Investigators who now address issues related to the regulation of homeostasis should consider it operating at three levels: (1) the whole organism (systemic homeostasis), (2) tissues/organ compartments and (3) within individual cells (Chovatiya and Medzhitov 2014) (Figure 2.1). At the level of the whole organism, the regulated variables include blood levels of glucose, ions, blood osmolarity and pH. Regulated variables of tissue homeostasis include cell–cell interactions, extracellular matrix abundance and integrity of structural components. Examples of regulated variables of cellular homeostasis include membrane potential, intracellular ion concentrations, nutrients and oxygen.

Homeostasis is a dynamically stable phenomenon, which can be illustrated when considering the regulation of systemic homeostasis and tissue homeostasis.

4.1.1. *Regulation of systemic glucose homeostasis*

When measuring the blood glucose level in the fasting state (basal state), it is quite stable, and we now know that this constancy is maintained by a dynamic process that is a combination of two fluxes: hepatic (and to a lesser degree kidney) glucose production and glucose utilization by the tissues. To demonstrate this, radioactive and nonradioactive tracers have been the principal tools and are briefly described below. Note that these experimental approaches also apply to the metabolism of different substances, such as amino acids, fatty acids, minerals, etc. These variables are maintained within an acceptable dynamic range by the endocrine and autonomic nervous systems, but this aspect is not discussed here.

Tracer kinetic studies

The initial idea of using radioactive "tracers" was conceived in the 1910s (Hevesy and Paneth 1913), when G. Hevesy published his work, "Research by a radiochemical method on the circulation of lead and bismuth in the organism", in Comptes Rendus 1924. In the 1950s, it was noted that despite a constant concentration of different metabolites in blood, this static situation represented a "dynamic steady state" where rates of appearance and disappearance of a metabolite are equal and constant (Wrenshall 1955). All metabolites in the body are in a state of steady flux or turnover, and quantitative information about the turnover of a metabolite can be obtained by labeling it with a radioactive isotope (^{14}C or ^{3}H) of one of its constituent atoms to form a radioactive molecule (=radioactive tracer). Radioactive isotopes have been used extensively to investigate the control of metabolism of different molecules such as glucose and amino acid, etc. These data are indicative of the glucose and amino acid homeostasis in both basal state and various diseases.

When considering radioactive tracers of glucose metabolic pathways, they fall into two categories: those that are radioisotopes of the parent compound (e.g. [U-^{14}C] glucose and [3-^{3}H] glucose) or those that are analogs of the parent compound (e.g. [^{14}C]2-deoxyglucose and [2-^{18}F]-2-fluoro-2-deoxyglucose (FDG)). The quantitative evaluation of the whole-body glucose utilization and hepatic glucose production uses the former tracers because they follow the same metabolic fate of the parent compound, whereas the latter compounds are used for quantitative assessments of individual tissue glucose uptake (Sokoloff et al. 1977).

There are two main experimental approaches to carry out tracer kinetic studies of glucose or other metabolites. The first one consists of a single injection of tracer (radioactive glucose) in the blood and its disappearance is measured in blood samples over a period of time. Thereafter, the data are analyzed mathematically and the most important concept in modeling this kinetic is that the glucose is present in homogeneous compartment(s). These mathematical concepts do not necessarily correspond to physiological compartment(s). A second concept is that the rate of transport of a tracer X out of a compartment is proportional to its concentration C, and the rate of loss from the compartment is $dC/dt = k \cdot A$. This approach requires an experimental situation called steady state where both production and utilization of a substance are equal and constant (Umpleby and Sönksen 1987).

A second approach carried out in steady state consists of a constant infusion of tracer (radioactive glucose) in the blood over a period of one or two hours until a steady concentration of tracer is achieved. At this plateau, the infusion rate of tracer is equal to the output rate of tracer. The same approach is used in non-steady state, which occurs when rates of production and disappearance of glucose are unequal.

An equation was described by R. Steele (1959) for calculating the change in rates of production and disappearance of glucose following the perturbation of a steady state.

Disruption of glucose homeostasis is a key feature of diabetic state, and with this technical approach investigators can establish the cause of hyperglycemia. It can be due to increased hepatic glucose production and/or decreased glucose utilization or both. Note that impaired insulin secretion can also be involved in this dysregulation of glucose level.

4.1.2. *Tissue homeostasis*

Tissue homeostasis is a dynamic process that maintains the size of tissues/organs constant in all adult living organisms. It requires mechanisms that control cell number, and it is achieved through a tight regulated balance between cell division and cell death. Most differentiated cells are regularly eliminated and replaced by the division progeny of adult stem cells. This process is called cell turnover, and in humans this flux is astonishingly elevated; we eliminate and in parallel replace a mass of cells equal to almost our entire body weight each year. The cellular turnover times of various organs range from less than two days to more than 50 days, and it has been estimated that the epithelium in the intestine and lung is completely renewed within about five days and six months, respectively. Other tissues, like white adipose tissue, approximately 10% of fat cells are renewed annually at all adult ages and levels of body mass index (Spalding et al. 2008).

4.1.2.1. *Stem cells and self-renewal concept*

Historically, many key milestones have driven progress in the field of stem cell research. The term stem cell appeared in the scientific literature as early as 1868 in the works of the biologist E. Haeckel in the context of the evolution of organisms. He used the term "Stammzelle" (German for stem cell) to describe the ancestor unicellular organism from which he presumed all multicellular organisms evolved (Haeckel 1868). The concept of stem cell emerged in the 1950s, with the demonstration that intravenously injected bone marrow cells can rescue irradiated mice from lethality by reestablishing blood cell production (Jacobson et al. 1951; Ford et al. 1956). These observations established the presence in the bone marrow of adult mice of cells with long-term hematopoietic repopulating activity. Thereafter, a remarkable series of experiments demonstrated the presence in the bone marrow of adult mice of quiescent hematopoietic cells that have the ability to differentiate into a variety of blood cell types and self-renewal without senescence. These cells were then termed stem cells by Till and McCulloch (1961), and the concept of self-renewing multipotent hematopoietic stem cells (HCSs) was born. Then, in 1990, C.S. Potten and M. Loeffler (1990) suggested defining stem cells as undifferentiated cells capable of: 1) proliferation, 2) self-renewal, 3) production of a large number of differentiated functional progeny and 4) regenerating the tissue after injury.

After the initial breakthrough in stem cell research, there were two major advances in the field: in 1998, the first human embryonic stem cells (hESCs) were isolated and in 2006, induced pluripotent stem cells (iPSCs) were derived from reprogrammed adult somatic cells with just four basic transcription factors. These two stem cells have shown great application promises in regenerative and transplant medicine.

Precise control of somatic stem cell (SC) activity is essential to the maintenance of tissue homeostasis in multicellular organisms. To ensure efficient replacement of damaged cells while limiting the potential for cancer, the proliferation rate of stem and progenitor cells has to be closely linked to tissue demands at any given time.

A stem cell self-renewal and differentiation is also regulated by the specialized microenvironment or niche which is composed of ECM structures, 3D architecture, chemical and mechanical signals, and cell-to-cell interactions. The niche contains cells that provide a local source of signals that nourish stem cells to support tissue homeostasis, maintaining a crucial balance between sufficient turnover versus neoplastic overgrowth (Santos et al. 2018).

It should be emphasized that stem cells, which are used for research purposes, are of different potencies and are obtained from different sources: (1) embryonic tissues, (2) fetal tissues, (3) specific tissues in the adult organism (skin, bone marrow, fat, skeletal muscle, blood, etc.) and (4) differentiated somatic cells after they have been genetically reprogrammed (i.e. iPSCs). For instance, embryonic stem cells derived from morula are totipotent, i.e. capable of differentiating into all types of cells. But at a later stage of embryonic development, stem cells from blastocyst are pluripotent (Bacakova et al. 2018).

History of cancer stem cells started in 1937 when J. Furth and M. Kahn established that a single cell from a mouse tumor could initiate a new tumor in a recipient mouse (Furth et al. 1937). In the 1970s, *in vitro* experiments performed on cancer cells reported that a subpopulation of malignant cells displays stem cell-like characteristics (Welte et al. 2010). They were called “cancer stem cells”, and they have been proposed to originate either from malignant transformation of a normal somatic stem cell or a progenitor cell, confirming Makino’s theory (Lapidot et al. 1994). It created a great excitement in the research community, and the cancer stem cell concept postulated that the growth of tumors depends on limited numbers of dedicated stem cells that have the capacity to self-renew. It is not clearly established yet whether cancer stem cells play a role in the tumor formation and metastasis processes. Readers should refer to a review by Clevers (2011).

For centuries, a central dogma in neuroscience was that new neurons are not added to the adult mammalian brain. It was assumed that neurogenesis occurred only during development, stopping soon after birth. But in 1962, with the introduction of autoradiography technique, J. Altman reported the first evidence, that radioactive thymidine is incorporated in DNA of neurons, proving that new neurons are present in different areas of the postnatal brain (Altman 1962). However, for decades, the existence of adult human neurogenesis was controversial, and it was unconceivable that mature neurons could reenter the cell cycle to divide and give rise to new neurons. The concept of neuronal stem cells (NSC) was recently introduced, and it was demonstrated that NSCs are mainly restricted to specific brain regions of the hippocampus. Within the subventricular zone, thousands of cells are born each day and approximately 60% that reach the olfactory bulb die and 40% survive (Winner et al. 2002). It has been estimated that approximately 10% of the existing cellular population within the olfactory bulb is replaced. The existence of adult neurogenesis has important functional contributions to neural plasticity and cognition (see review Ribeiro and Xapelli (2021)).

4.1.2.2. *Maintenance of tissue homeostasis*

To study the mechanisms involved in the maintenance of tissue homeostasis, experiments were mainly carried out on the intestinal cells of the epithelium. They constitute the protective layers that line our internal organs and perform an array of functions such as absorption of nutrients, secretion of hormones and enzymes and formation of protective barriers.

It is widely accepted that a constant renewal of intestinal epithelium occurs approximately every four to five days. It requires a tightly regulated balance between intestinal stem cell (ISC) proliferation and differentiation to maintain proper lineage ratios and support its major functions (Bankaitis et al. 2018). These stem cells also contribute to tissue repair and proliferate according to tissue requirements, while avoiding aberrant proliferation.

The homeostatic constant regeneration of the intestinal epithelium is driven by ISCs at the crypt bases, which give rise to all the different epithelial cell types (Barker et al. 2007). Progressing from the crypts towards the villus tips the transit cells differentiate into secretory lineages (enteroendocrine cells), goblet cells or enterocytes.

Regarding the maintenance of the hematopoietic system, it is also driven by stem cells mentioned above and called HSCs. Like ISCs, their function must be finely tuned to protect their self-renewal capacity and prevent their exhaustion which is crucial for blood system homeostasis. Hematopoiesis is a continuous process which gives rise to various blood lineages (B and T lymphocytes, erythrocytes, etc.). As mentioned earlier, they have the capacity to reconstitute the entire hematopoietic system after transplantation into an irradiated recipient. This capacity provides the scientific basis for HSC transplantation, and it is now the only curative therapy for several hematological malignancies (leukemia and lymphoma). Detailed cellular mechanisms of hematopoiesis are reviewed in Boulais and Frenette (2015), Eaves (2015) and Wilkinson et al. (2020).

4.1.2.3. *Restoring tissue homeostasis: wound healing, liver regeneration*

The body of tissues that repair and regenerate throughout the life span of an organism argues in favor of the presence of long-lived stem cells in adult tissues that serve as a source of cells for renewal. Here are two examples:

Wound healing

The epidermis is a stratified epithelium composed of several layers of keratinocytes, and the basal layer continually renews to maintain a tissue homeostasis. They also have a key role in wound healing which necessitates effective mechanisms that respond to and repair injury, through a process termed epithelialization. These cells acquire new behaviors such as migration, proliferation and differentiation. For the process of migration to begin, the cell–cell interactions need to be loosened; then, keratinocytes start to proliferate, to supply cells for covering the wound. It is noteworthy that the regenerative capacity of the skin relies on the local populations of epidermal stem cells (ESCs) (Taylor et al. 2000). Recent work is beginning to shed light on how integrated cellular and molecular mechanisms facilitate an effective wound repair response. The most significant pressure for our tissues is to maintain function in the face of continuous turnover and insults, either from injury or the progressive accumulation of mutations. Most cellular and molecular aspects of wound healing have been recently reviewed in Rodrigues et al. (2019).

Liver regeneration

Before scientific reports, legends from the Greek mythology described the fate of Prometheus. As punishment for defying Zeus and revealing the secret of fire to human, Prometheus was chained to a rock and each day a part of his liver was ripped out by an eagle. This torture was repeated every day because his liver regenerated itself overnight (Smith 1867). In this legend, the speed of regeneration is somewhat greater than that observed in the laboratory, but the myth emphasizes the ability of the liver to repeatedly regenerate. Such a capacity to regenerate implies that it is a property of all liver cell types (Mangnall et al. 2003).

In the adult liver, most of the cells are in the quiescent G_0 state and only about one hepatocyte of 20,000 (0.005%) is dividing. It is worth noting that after regeneration, the size of the liver appears to be within a few percent of that of the original liver. So far, the nature of the mechanism, which regulates the liver size, remains elusive. The concept of liver regeneration was introduced in modern medicine in the early 19th century (Power and Rasko 2008) and it was confirmed in studies showing that liver mass increases a few hours after partial hepatectomy and reaches normal size by 72 h.

The existence of liver stem cells that act as a source of regeneration after acute liver injury has been controversial. Nevertheless, several lines of evidence suggest that cellular plasticity in the liver plays a role in regeneration (see review Tsuchiya and Lu (2019)).

4.1.2.4. *Tissue cell turnover rate*

Cellular turnover was studied in the 1950s (Van Putten 1958) and earlier. To determine the life span of a certain cell type, investigators measured the incorporation of ^{51}Cr, $^{3}HTdR$ (methyl-tritiated thymidine) into the cells of interest, over months or even years. It therefore raised concerns regarding the effect of label toxicity on cell life span. In response, new methods used nontoxic labels such as deuterium in $^{2}H_2O$ to address the cellular turnover rate issue (Macallan et al. 1998). Then, the environmental ^{14}C, produced by aboveground nuclear weapons between mid-1950s and early 1960s, was used for determining the turnover of large tissues before this was banned by international treaties.

Recently, an estimate of the number of cells in an adult human body has been published (Bianconi et al. 2013; Sender et al. 2016), followed by a study in which investigators characterize the daily human cell turnover for all cell types. It was carried out in the adult male, between 20 and 30 years of age, weighing 70 kg and 170 cm in height body; the number of cells has been estimated to be approximately 30×10^{12}, and ≈90% represent the hematopoietic lineage, mostly red blood cells (RBCs) (Sender and Milo 2021).

The life span of a cell varies greatly across cell type and tissue, from three to five days for gut epithelia, to years (and even a lifetime) for cardiomyocytes or neurons. The daily cellular turnover rate is defined as being the average daily number of cell deaths in a specific cell population. The cellular turnover time is defined based on cell death and not on cell differentiation. The total turnover rate of the human body is $\approx 0.3 \times 10^{12}$ (330 billion) cells/day (equal to about 4 million cells/second). About 86% of these cells are blood cells, mostly of bone marrow origin. Almost all of the remaining 14% are gut cells. The three major contributors to the cellular turnover of the human body are RBCs, neutrophils, intestinal and stomach epithelia and they represent about 96% of the total turnover.

The total cellular mass turnover rate in human is about 80 g/day. Despite their ubiquity, blood cells account for only ≈40% of the total cellular mass turnover, due to their relatively low cellular masses of ≈100–300 pg. This contrasts with their ≈80% dominance over total cellular turnover. The epithelial cells of the gastrointestinal tract also contribute about 40% of the cellular mass turnover because of their greater single cell unit mass (≈1,000 pg) and short life span. Note that due to their exceptionally long life spans, adipocytes and muscle cells contribute only approximately 5% to the total mass turnover, although they represent ≈75% of the total cellular mass. Overall, we find that a reference 70 kg person with 30 trillion host cells and 46 kg of cellular mass (the other 24 kg are due to extracellular fluids and solids) produces this number of cells every 80 days and this amount of cellular mass every year and a half.

4.1.3. *Muscle and bone mass homeostasis*

4.1.3.1. *Muscle mass homeostasis*

The maintenance of skeletal muscle mass depends on the overall balance between the rates of protein synthesis and degradation. Thus, age-related muscle atrophy and loss of function, commonly known as sarcopenia, may result from decreased protein synthesis, increased proteolysis or simultaneous changes in both processes governed by complex multifactorial mechanisms. In addition, aging is associated with loss of motor neurons, which leads to denervation and loss of muscle fibers.

The size of skeletal muscle organs can be regulated by different mechanisms, and the number of cells in some tissues is controlled by a systemic feedback mechanism (Raff 1996). This is best understood for skeletal muscle, in which growth and differentiation factor 8 (GDF8), also known as myostatin, is secreted from myocytes and negatively regulates the generation of new muscle cells and thereby sets the number of cells (Joulia-Ekaza and Cabello 2006). Loss-of-function mutations in GDF8 result in a large increase in the number (and size) of myocytes in animals and humans.

4.1.3.2. *Bone mass homeostasis*

Bone is a dynamic organ, and old bone is continuously replaced by new tissue during the whole life in all higher vertebrates. In 1990, Frost defined this phenomenon as bone remodeling (Frost 1990). The process involves the resorption of old or damaged bone by osteoclasts and its replacement by new bone formed by osteoblasts (Sims and Martin 2014). The process is tightly regulated to ensure that no net change in bone mass occurs after each remodeling event. In fact, it has been estimated that the turnover of bone is 10%/year and most of the skeleton in adult humans is approximately 10 years old (Parfitt 2002). The rate of remodeling is rapid during growth and much slower after puberty, and it is influenced by physical activity, local and systemic factors, and nutrition. Imbalance between osteoclast and osteoblast activity can arise from a variety of hormonal changes, enhanced production of inflammatory cytokines or growth factors, resulting in either decreased bone mass (osteoporosis) or increased bone mass. These pathological aspects are discussed in section 4.2.

Bone remodeling proceeds in an orderly fashion, and bone resorption and formation are linked in time and space; this spatial and temporal relationship is termed coupling. This physiological process is coordinated and tightly regulated by the PTH and the calcitonin which stimulate the osteoclast (indirectly) and the osteoblast, respectively. As mentioned above, they exert a dual hormonal control of plasma calcium, and this mechanism explains the very precise regulation of the calcium in body fluids. Notably, mechanical loading favors bone formation and suppress bone resorption (Robling et al. 2006).

4.1.4. *Whole-body energy homeostasis*

Energy is essential for the survival of living organisms, and energy metabolism generated by nutrients should meet the cellular energy demand. When considering obesity, the fundamental law of the energetics is that the weight gain takes place when energy intake is greater than energy expenditure. Despite large fluctuations in daily energy supply (frequency and amount) and energy demand, there is a tight control of the whole-body energy homeostasis, as the majority of the population keeps body weight remarkably constant. It suggests the existence of systems for monitoring the body's internal status and of systems for effectively compensating for perturbations of its stability. This is demonstrated by the fact that if we consider a cumulative error in the adjustment of food intake to energy expenditure of only 1%, or about 20 kcal/day, would lead to a body weight gain of about 1 kg/year (taking into account that fat tissue has a caloric value of ~7 kcal/g), or >50 kg over the adult life span. Consequently, weight loss and weight gain induced by under- and overfeeding are rapidly corrected by compensatory increases and decreases of energy intake, respectively. Such weight stability is also achieved through energy expenditure adjustments, which are divided into basal metabolic rate, physical activities and thermogenesis. The search for a thermogenic mechanism is briefly reported below.

Studies that dealt with the control of energy intake were based on the observation of Claude Bernard who recognized the essential role the nervous system must play in coordinating and integrating the various components of a regulatory system. The first demonstration was reported in 1940 (Hetherington and Ranson 1940) and 1951 (Anand and Brobeck 1951) when severe aphagia was observed after large lesions of lateral hypothalamus in rats, and this was viewed as a hypothalamic "feeding center".

Since then, an impressive amount of knowledge has accumulated concerning the regulation of energy homeostasis, which is a highly integrated and regulated process aimed at maintaining the stability of body energy stores over time. Neurons located in the hypothalamus are the most important components of the complex system that regulates food intake and energy expenditure. The arcuate nucleus, in particular, is a major hub for integrating nutritionally relevant information originating from all peripheral organs and mediated through circulating hormones and metabolites and/or neural pathways mainly from the brainstem. Finally, the resulting optimal adaptive responses chosen are executed through a complicated interplay that occurs between hormonal factors, energy intake and expenditure and substrate utilization.

With respect to the different components of energy expenditure, a great deal of work has focused on the thermogenesis. The concept of dissipating excess energy intake as a heat was first suggested in 1961, and it was hypothesized that the brown adipose tissue (BAT) was the thermogenic organ. Note that this hypothesis was put forward 400 years after its first description by C. Gessner in 1551, who named it "hibernating gland" which suggested a putative role in thermogenesis (Afzelius 1970). The search for the thermogenic mechanism in BAT became a central focus, when Nicholls et al. applied the chemiosmotic theory of Mitchell, who was awarded the Nobel prize in Chemistry in 1978 for the contribution to the understanding of biological transfer, through the formulation of this theory. This group showed that in the BAT, the heat is produced through an uncoupling of oxidative phosphorylation leading to proton leakage across the inner mitochondrial membrane (Nicholls and Locke 1984). Then, Ricquier's group identified the mitochondrial uncoupling protein (UCP) responsible for regulating the proton conductance (Ricquier and Kader 1976). In the late 1990s, several UCPs were identified, and a link between BAT and obesity was unequivocally established in rodents, but there was no evidence that it could play a role in the etiology of human obesity, as a component of energy expenditure. However, in 2009, a procedure – fluorodeoxyglucose positron emission tomography (FDG-PET) – which is employed in cancer investigation to track the metastasis of tumors revealed that BAT is definitively present and functional in adult humans (Virtanen et al. 2009). This approach also revealed that the BAT activity is inversely proportional to BMI. Thus, in the 1970s, there were uncertainties about the importance of BAT to energy expenditure, but 40 years later, there was a renewed interest in the activation of BAT as a therapeutic approach to control the body gain weight.

Details regarding signals regulating the energy intake and expenditure are discussed in section 4.3.3.

4.1.5. *Metabolism and cellular energy homeostasis*

Metabolism encompasses anabolic processes that require energy to build macromolecules, as well as catabolic processes that produce energy to sustain cellular integrity and function. Given that both nutrient availability and energy demand are constantly changing, it therefore results in marked oscillations in metabolism, including blood substrate levels and their utilization as well as energy expenditure. Despite these variations, cellular homeostasis is maintained, and it is achieved through the ability of the cells to sense immediate environment, such as hormones, cytokines and growth factors. Upon binding of these extracellular signals to a plasma membrane receptor, it initiates signaling cascades that transduce information regulating metabolism. But cells can also sense intracellular environments.

A tight regulation of the balance between anabolic and catabolic processes is exemplified by the metabolic adjustments that occur in the feeding and fasting states. After feeding, anabolism prevails and fatty acids and glucose in excess are channeled towards lipogenic and glycogenic pathways in adipose tissue and liver and skeletal muscle, respectively. In the fasting state, catabolism prevails, resulting in a lipolysis and glycogenolysis in adipose tissue and liver, respectively. This metabolic response makes it possible to continuously maintain a flux of energetic fuels to the tissues and, hence, maintaining a metabolic homeostasis. It should also be emphasized that metabolic regulation in any physiological situation (fed, starved, exercise, pregnancy) is complex and it is partly due to the fact that the metabolic status differs between organs, tissues and cells (see section 4.3.6).

We should remember that mitochondria display key cellular functions such as oxidative metabolism, lipid metabolism and respiration. Consequently, the control of mitochondrial biology is a crucial aspect for cellular homeostasis. Recently, it was suggested that mitochondrial fusion and fission of outer and inner membranes are key regulators of mitochondrial metabolism and therefore essential for energy homeostasis. Mitochondrial fusion is positively associated with increased ATP production, while the inhibition of fusion is associated with impaired oxidative phosphorylation and reactive oxygen species production. For a review, see Wai and Langer (2016). It underscores the concept that mitochondria are highly dynamic organelles to continuously adapt shape to functional requirements.

4.1.6. *The gut microbiome and glucose homeostasis*

The original definition of the microbiome term was suggested by Whipps et al. (1988), but recently it has been extended by two explanatory sentences differentiating the terms microbiome and microbiota. The reader is referred to the excellent review by Berg et al. (2020) for more detail of this field.

The microbiome is defined as:

> a characteristic microbial community occupying a reasonable well-defined habitat that has distinct physio-chemical properties. The microbiome not only refers to the microorganisms involved but also encompasses their ecosystem (a system formed by an ecological community and its environment that functions as a unit).

The microbiota consists of "the assembly of microorganisms belonging to different kingdoms (Prokaryotes (Bacteria, Archaea), Eukaryotes (e.g. Protozoa, Fungi, and Algae))".

Recent studies strongly suggest that the gut microbiome could also influence host glucose homeostasis. Thus, it was observed that type 2 diabetes is associated with unique shifts in the collection of microbes and their genes residing in the gut (Karlsson et al. 2013). In line with this observation, the disease is associated with the type of metabolites derived from microbial interactions with the diet (Sonnenburg and Bäckhed 2016). It was therefore hypothesized that dietary microbial breakdown products exert effects throughout the body that would impact host glucose homeostasis. A detailed discussion of this theme is beyond the scope of this section, but for a comprehensive recent overview of this field, the reader is referred to Howard et al. (2022). In this review, the authors highlight some mechanisms by which the gut microbiome can influence glucose regulation, including changes in gut permeability, gut–brain signaling and production of bacteria-derived metabolites like short-chain fatty acids and bile acids.

4.1.7. *Synaptic homeostasis*

The main excitatory (E) and inhibitory (I) neurotransmitters in the central nervous system (CNS) are glutamate and γ-aminobutyric acid (GABA), respectively. A coordinated and dynamically regulated balance between these two opposite inputs (E/I) is critical for the proper function of the CNS. A perturbation in E/I has been associated with neuronal damage in a myriad of neurological and neuropsychiatric disorders. It appears likely that glutamate and/or GABA synapses may be possible sites of homeostatic change. Because these two neurotransmitters released into the extracellular space are not enzymatically broken down in the synaptic cleft, a tight control of their clearance rate from the synapse is the main strategy for termination of glutamate and GABA signaling. It is achieved by a reuptake of these neurotransmitters through transporters. This process contributes to the maintenance of physiological balance between excitatory and inhibitory inputs. It therefore participates to the synaptic homeostasis (see the review by Sears and Hewett (2021)).

4.1.8. *Open issues: membrane lipid homeostasis and acid–base homeostasis*

Cellular membranes are composed of hundreds of distinct lipid species, and this lipid complexity is maintained within a narrow range in steady-state conditions. It strongly suggests the necessity of a homeostatic control mechanism to keep this complexity fairly constant under widely varying conditions. Furthermore, the balance between esterified saturated and unsaturated fatty acids should be tightly regulated to keep basic physical properties of the membrane, including water permeability and fluidity. For instance, an increase in saturated fatty acids causes a tighter lipid packing and increase the membrane viscosity.

The molecular mechanisms underlying membrane homeostasis are starting to be elucidated through combinations of lipidomics, cell biology, genetics and computational modeling. Recent studies have reported how the membrane sense and response proteins monitor membrane properties and adjust its composition. Most of them display a dual function. For instance, inositol-requiring enzyme 1 (IRE1) senses both membrane defects in the ER and misfolded proteins. It is interesting to note that most of the membrane defect sense-and-response proteins identified so far are activated by membrane rigidification. For a comprehensive review, see de Mendoza and Pilon (2019).

With respect to the acid–base homeostasis, its historical development started in 1909 when Sorensen introduced the "pH" and the "acid/base" concepts (Sörensen 1909). Then, in 1923, Bronsted highlighted the role of the hydrogen ion in the definition of acids and bases. The regulation of pH and acid–base homeostasis are critical for both normal physiology and cell metabolism and function. The kidneys and lungs have a pivotal role in the regulation processes. A detailed discussion of these processes is beyond the scope of this review. The reader is encouraged to consult the source in Hamm et al. (2015).

4.2. Homeostasis disruption

In section 4.1, we emphasized that at all levels (systemic, tissue and cellular), the regulated variables are maintained stable by homeostatic control systems (Buchman 2002; Goldstein and Kopin 2007). Although homeostasis suggests constancy of values for variables, ranges of acceptable values are now recognized to be inconstant. There are diurnal variations in body temperature, heart rate and blood pressure, etc.

Disruption of homeostasis occurs in response to agents or conditions, named "stressors", which can be biological, chemical, physical, etc. They can produce "stress" and endanger homeostasis. Regulation of disrupted homeostasis is best characterized at the level of the whole organism, and the regulated variables are maintained within an acceptable range by the endocrine system.

For instance, in response to hypoglycemia, there is a secretion of counter-regulatory hormones (glucagon, epinephrine) that restores a normoglycemia. If these hormones do not reestablish a normoglycemia, it leads to a hypoglycemia, which may impair brain function and may lead to loss of consciousness and death. When a physiological increase in glucose concentrations occurs after a meal, there is a secretion of insulin which also restores a normoglycemia. If the hormone does not restore a normoglycemia, it ensues a chronic hyperglycemia which is a key feature of early type 2 diabetes. Therefore, most diseases can be broadly defined as a lack of a tight control of mechanisms reestablishing the homeostasis.

4.2.1. *Endocrine disorders: excess or impaired hormone secretion*

There are two basic groups of pathologies due to either an excess or impaired hormone secretion. A detailed discussion of these hormonal disorders is beyond the scope of this review; only a few of them are reported.

Excess of hormone secretion

Hypercortisolism occurs in two clinical manifestations: (1) Cushing's syndrome and (2) in patients with major depression. The former case is for instance observed in patients who are treated with exogenous glucocorticoids, which is the most frequent cause of Cushing's syndrome, or in those with pituitary or adrenal cortex adenomas. Pituitary adenomas secrete ACTH in an uncontrolled manner, which leads to excessive cortisol secretion. This condition is referred to as Cushing's disease. Sometimes, ACTH secretion may occur outside pituitary (i.e. ectopic ACTH syndrome); for instance, some lung carcinomas may produce ACTH, which also causes hypercortisolism. In both cases, clinical signs and symptoms result from high concentrations of cortisol as well as ACTH. For instance, high levels of ACTH lead to increased skin pigmentation. Alternatively, cortisol may be secreted from adrenal adenomas independently of ACTH. In this case, pituitary ACTH secretion is suppressed.

Regarding hypercortisolism in subjects (increased concentrations of cortisol in plasma) with major depression, it results from the downregulation of the glucocorticoid receptor (GR), secondary to persistent hypercortisolism in key brain regions. Consequently, negative feedback through this receptor is less effective which leads to a hyperstimulation of the HPA axis. Numerous studies support the hypothesis that abnormalities in the GR contribute to the pathophysiology of major depression. Thus, treatment with tricyclic antidepressants such as desipramine and imipramine increases both GR protein expression and HPA activity associated with a marked improvement of the feedback inhibition and the depression syndrome. See a detailed review in Pariante and Miller (2001).

– Hyperthyroidism in Graves' disease is due to actions of stimulating autoantibodies against the TSH receptor (so-called thyroid stimulating immunoglobulin, TSI). These autoantibodies stimulate TSH receptor and thus cause hypersecretion of the thyroid hormones. Other causes of hyperthyroidism include a hormonally active thyroid adenoma that secretes thyroxine and a pituitary adenoma that secretes TSH, which subsequently stimulates over secretion of thyroxine from the thyroid gland.

– Hypersecretion of growth hormone (GH) results in gigantism or acromegaly. Gigantism occurs in childhood and is associated with GH-secreting pituitary tumors. Acromegaly occurs when hypersecretion occurs after the fusion (closure) of the epiphyseal growth plates, leading to the growth of bone and soft tissue in specific locations, such as hands, feet and face, including nose and lower jaw, which results in characteristic alterations of facial features (see review in Lodish et al. (2016)). Note that estrogens are important for the control of longitudinal growth during puberty. In boys, a tall stature was reported due to a lack of conversion of androgens into estrogen (Morishima et al. 1995).

Lack or impaired hormone secretion

– Lack of insulin secretion: type 1 diabetes; autoimmune disease. For a comprehensive recent overview of this field, the reader is referred to Ilonen et al. (2019). Over the last two decades, there were tremendous advances in elucidating the causes and treatment of the disease based on extensive research both in rodent models of spontaneous diabetes and in humans (Bluestone et al. 2010).

– Impaired insulin secretion: type 2 diabetes.

Lack of hormone action

It is one of the main causes of type 2 diabetes. It is characterized by a decreased tissue insulin sensitivity, and it is also named insulin resistance. It leads to a disruption of different molecular pathways involved in the regulation of carbohydrates, lipids and protein metabolism. A review of these mechanisms is reported in Yaribeygi et al. (2019). A decreased leptin sensitivity has also been reported in obese patients.

– Hypothyroidism. In 1912, H. Hashimoto described four patients with a chronic disorder of the thyroid that he termed "struma lymphomatosa". More than 40 years later, the presence of antithyroid antibodies was reported in patients with this disorder and is now recognized as an autoimmune thyroid disease. It is due to a destruction of the gland via mechanisms of humoral as well as cellular immunity (Li et al. 2011).

4.2.2. *Muscle energy wasting*

Muscle wasting may result from decreased protein synthesis, increased proteolysis or simultaneous changes in both processes. Major factors implicated in the etiology of sarcopenia include decreased physical activity, loss of α-motor neurons, malnutrition, increased cytokine activity, oxidative stress, mitochondrial dysfunction and abnormalities in growth hormones and sex steroid hormones (Argiles et al. 2005).

4.2.3. *Energy*

Energy homeostasis is dictated by a balance between energy intake and energy expenditure. Given that both nutrient availability and energy demand are constantly changing, these two components should be tightly regulated to prevent a body weight gain. In the presence of obesity, the question which is often raised is: how do we restore energy homeostasis? This issue is often addressed; see the review by De Lorenzo et al. (2020).

4.2.4. *Cell number and activity*

Obesity

Obese subjects have increased white fat depots, and several cellular and molecular events participate in this dysregulation. At the cellular level (adipocyte), there are two ways to augment the fat mass: hypertrophy (increase in size of existing adipocytes) or hyperplasia (formation of new adipocytes through differentiation of preadipocytes = adipogenesis). In normal-weight subjects, the number of adipocytes stays quite constant during adulthood and this is explained by the fact that death of adipocytes occurs at a similar rate as adipogenesis. In people suffering from obseity, there is an increased rate of death, but adipogenesis is also elevated, thus increasing adipocyte numbers (hyperplasia). Hyperplasia is mostly associated with early-onset and severe obesity (Hirsch and Batchelor 1976).

In addition, the adipocyte fails to respond properly to systemic signals (Lamy et al. 2014) to either store or mobilize fat, which contributes to the persistent increase in fat mass. This metabolic decline is caused by a defect of insulin action on adipocytes, which is a characteristic of the insulin resistance state (reviewed in Stern et al. (2016)). Furthermore, there are two main fat depots, subcutaneous and visceral, which differ in many physiological and molecular properties. For instance, visceral fat depots display a higher release of fatty acids, which are known to increase the development of insulin resistance state.

Recently, it was suggested that some lifestyle factors would also contribute to excessive weight gain. Thus, studies have clearly shown that shift work is associated with increased adiposity, and it would be the result of physiological maladaptation to altered diurnal activity/sleep and eating patterns, termed circadian rhythm or chronodisruption. The concept of circadian clocks was then coined, and numerous studies have shown that they are present in almost every tissue and cell, including adipose tissue and their role is to coordinate 24 h rhythms of gene expression within the cells (as discussed in more detail in section 5.2). This clock network is synchronized to the external light/dark cycle, and disruption of this cycle (chronodisruption) affects the regulation of gene transcription. In humans, clock gene rhythms have been reported in subcutaneous and visceral white adipocytes in vivo (Garaulet et al. 2011). Disruption of the clock leads to enhanced adipogenesis, and two clock proteins (BAML1 and REV-ERBα) promote the differentiation of preadipocytes into adipocytes (Fontaine et al. 2003).

Cancer

Cancer is one of the major examples of disruption of cellular homeostasis.

In the 1980s, it was postulated that the etiology and pathogenesis of cancer was the result of a combination of genetic predisposition and environmental factors. Mutations in oncogenes and in tumor suppressor genes were sufficient to explain the carcinogenesis process and cancer progression. However, this view did not provide a satisfactory explanation for the mechanisms underlying metastasis. In 1889, S. Paget pointed out that metastatic colonization is dependent on the properties of the organ and there is a relationship between cancer cells and their microenvironment, thereby affecting tumor evolution (Paget 1889). Studies performed in the 1990s now demonstrated that the tumor microenvironment (TME), such as infiltrating non-tumor cells, extracellular matrix components and soluble factors interact with tumor cells. Through different dynamic interactions, they reprogram and shape the malignant phenotype of tumor cells. It provides a strong support for the concept proposed by Paget. A timeline of TME research is presented in a review showing the key concepts and findings relating to this field (Maman and Witz 2018).

This disruption of cellular homeostasis is also associated with marked changes in metabolism, which enable cells to meet the increased demand of four major types of macromolecules: carbohydrates, proteins, lipids and nucleic acids (see section 4.3.6.1 on metabolic adaptations of tissues).

One key question in cancer biology still to be addressed is the identity of the cells that sustain the initiating mutations. According to the cancer stem cell concept, tumor cells are generated and maintained by a small subset of undifferentiated cells that are capable to self-renew and differentiate into the bulk tumor cells.

Bone

Bone is formed and removed by specialized bone cell types, osteoblasts and osteoclasts, which are responsible for linear growth during childhood and adolescence. Tremendous skeletal growth occurs during puberty, and approximately 99% of the total body bone mineral content is attained by 27 years of age. Bone mineral accretion is osteoblast activity > osteoclast activity (especially during childhood and puberty until the closure of growth plates). Disruption of systemic bone homeostasis leads to either osteoporosis or osteopetrosis, a rare genetic condition characterized by increased bone density and fragility.

During lactation, there is extensive evidence of a loss of maternal bone mass, but in most women, complete remineralization occurs after weaning.

Immune system

Autoimmunity is the best example that illustrates the disruption of immune homeostasis. Before being able to react against infectious non-self-antigens, the immune system has to be educated in recognition and tolerance of self-antigens. This educational process starts in the thymus, and the development of an autoimmune response results from a dysfunction in program tolerance to self-proteins. Establishment and maintenance of immunological tolerance encompasses both central (thymus) and peripheral (lymphoid tissue) compartments. The development of the autoimmune process results from the dysfunction of the immune system in these compartments:

1) Thymus

It is in the thymus that T-cell precursors "learn" to distinguish between "self" and "non-self" antigens (Mathis and Benoist 2004). Hence, the T-cell that binds to a self-antigen, and does not undergo apoptosis as it normally should, is "autoreactive", meaning that it is capable of recognizing and acting against a body's own antigen. Subsequently, when it leaves the thymus and if it is exposed to the same self-antigen of a cell, it acts as a T cytotoxic cell, enabling a destruction of this cell and a progression of the autoimmune process. Therefore, a defect/a dysfunction in the central tolerance contributes to the autoimmune process.

2) Periphery (lymphoid tissue)

A dysfunction of the peripheral tolerance is due to the inhibitory signals of the activation (CTLA4 and PD-1) of autoreactive T-cells (see the review by Parry et al. (2005)). Recently, another mechanism inhibiting autoreactive T-cells has been identified after the seminal work of Nishizuka and Sakakura (1969). They performed a neonatal thymectomy in mice and it led to organ-specific autoimmune pathology. These experiments demonstrated that a population of immune cells present in the thymus contributed to the self-tolerance process and were termed T regulators (Tregs). Subsequently, two populations of Tregs were identified, one was capable of inducing autoimmunity and the second one was responsible for the inhibition of autoimmunity, by blocking unsuitable immune reactions directed to self-antigens (Kim et al. 2007). Investigations during the past two decades have clearly demonstrated that most autoimmune diseases display defects in either the number and/or the function of Tregs originating from thymus (Miyara et al. 2011). It is noteworthy that our understanding of Treg cell biology gave rise to three new concepts: Treg cell instability, plasticity and tissue-specific functions. Current efforts focus on the molecular mechanisms responsible for the induction of these functional states. For a comprehensive review, see Dominguez-Villar and Hafler (2018).

Mineral homeostasis

The physiology of calcium and phosphate is discussed in section 4.3.3, and the disorders related to their metabolism are described in detail in the reviews cited below.

– Hypocalcemia (low plasma calcium level): it occurs in conjunction with multiple disorders and can be life-threatening if severe. The main cause results from impaired secretion of parathyroid hormone (PTH). Disorders that disrupt the metabolism of vitamin D can also lead to chronic hypocalcemia, as vitamin D regulates the gut absorption of dietary calcium (Bove-Fenderson and Mannstadt 2018).

– Hypophosphatemia (low plasma phosphate blood level): the cause is often multifactorial and results from decreased intestinal absorption, vitamin D deficiency and increased urinary loss due to elevated PTH (hyperparathyroidism) (Gaasbeek and Meinders 2005). Another cause of hypophosphatemia is an excessive secretion of FGF23, a phosphatonin, which induces kidney excretion of phosphate.

4.3. Crosstalk between organs, tissues and regulatory systems

The three regulatory systems (endocrine, nervous and immune) should cope with environmental changes/variations to keep a harmonious functioning of the body. However, it is now well established that other organs, such as skeletal muscles, liver, adipose tissue, etc., actively contribute to the regulation of various processes on the systemic level. The concept that key organs influence the regulation of physiologic functions, like metabolism of other organs, by releasing factors into the circulation that can reach several other metabolically active organs, has led to interest in the role of any organ in metabolic regulation. This is one of the major principles of integrative physiology, the mutual dependence between organs.

4.3.1. *Axis concept*

4.3.1.1. *Entero-insular and gut–brain axes*

A rise in arterial glucose perfusing the pancreatic β-cell leads to insulin release. In addition to this direct effect of glucose on the β-cell, intestinal factors were suspected as early as 1906 as having a major role in the regulation of pancreatic endocrine secretion (Moore 1906). The observation that oral rather than intravenous glucose administration resulted in greater insulin secretion (the incretin effect) confirmed the existence of a connection between the intestinal tract and the endocrine pancreas (Mcintyre et al. 1964). These studies established that factors released from the gut in response to glucose absorption sensitized the β-cell of the pancreas and reduced the threshold for release of insulin. The concept of the "entero-insular axis" was thus born (Ranganath 2008).

The concept of "gut–brain axis" was coined in the 1960s and 1970s, and it originated from the so-called APUD (amine precursor uptake and decarboxylation) hypothesis that states that the gut and other organs derive from a common origin in the neural crest (Pearse 1969). It was based on the observation that the gastrointestinal endocrine cells express markers for neuronal differentiation, including those involved in the biosynthesis of neurotransmitters. Despite these similarities shared with neurons, it does not constitute a proof that they originate from the neural crest, and this hypothesis had to be subjected to experimental testing. The subsequent experiments demonstrated that neurons and gut endocrine cells are only related by their common transcriptional control mechanisms operating during their differentiation, rather than sharing a common embryological origin (neural crest) (Andrew et al. 1998; Schonhoff et al. 2004). These specialized cells, which represent the enteroendocrine cells (EECs), secrete in the gastrointestinal (GI) tract many different types of signaling molecules which enable a crosstalk between the brain and the gut. The brain receives inputs from the GI tract and initiates an integrated response to target cells in the GI tract (Mayer 2011). Signals carried from the gut to the brain are predominantly hormones, immune mediators (cytokines) and vagal afferent neurons, while the communication from the brain to the gut is via neuroendocrine signals and autonomic efferent neurons. Notably, gut peptides/hormones (e.g. ghrelin and cholecystokinin) signaling to the brain can act through an endocrine route and/or a neural manner by activating receptors on the vagal afferent terminals. Vagal afferent pathways convey messages enabling the regulation of motility, secretion, inflammatory response and food intake. Regarding the regulation of energy intake, the interplay between the gut and the brain has been addressed in detail to decipher the molecular mechanisms of obesity. It is briefly presented in section 4.3.3.4.

4.3.1.2. *The neuroendocrine axes: reproductive, growth and stress*

Neuroendocrine systems control the most fundamental physiological processes, for example, metabolism, adaptations to stress, reproduction and growth. The key players of these systems as well as their regulation have been delineated from experiments carried out in rodents (rats and mice) (Oyola and Handa 2017).

The principal components of the three major neuroendocrine axes comprise hypothalamic regulatory centers, the anterior pituitary gland and peripheral hormone secretion. In the 1970s, it was demonstrated that the secretion of most hormones occurs in a pulsatile manner. Investigators recognized that it was a physiological mechanism to enhance hormone concentrations rapidly and to convey distinct signaling information to selected target tissues. Numerous studies assumed that the generation of pulsatile peripheral hormone secretion is a result of the neural pulse generator in the hypothalamus. However, there is no good evidence for this. It was also suggested that the pulsatility may arise from the complex network of excitatory and inhibitory interactions between signals originating from the hypothalamus, pituitary gland and endocrine glands. The regulation of pulsatility in these axes likely involves several factors, and experimental studies have so far been unsuccessful in identifying either the mechanistic or anatomical origin of the ultradian rhythm.

These neuroendocrine systems are controlled by mechanisms of negative and positive feedback signals, which collectively preserve stable hormone levels.

Hypothalamo–pituitary–adrenal (HPA) axis

This neuroendocrine axis uses three primary structures, allowing it to respond appropriately to stressful life events. These include the paraventricular nucleus of the hypothalamus (PVN), the anterior pituitary gland and the adrenal gland. The PVN integrates neuronal and humoral inputs to activate neurons that synthesize and secrete corticotropin-releasing hormone (CRH) and arginine vasopressin (AVP) into the median eminence. Then, these neuropeptides reach the anterior pituitary gland and stimulate the synthesis and release of adrenocorticotropic hormone (ACTH) into the blood circulation. In turn, ACTH, upon reaching the adrenal cortex, stimulates the synthesis of glucocorticoid hormones which are immediately released into the blood. The HPA axis keeps its own secretion in check, thereby allowing a return to the baseline homeostatic state. This occurs through inhibition of the entire axis by adrenal glucocorticoids (negative feedback) acting in the PVN and the anterior pituitary, inhibiting the synthesis and release of CRH and ACTH (see review in Papadimitriou and Priftis (2009)).

Under normal conditions, the secretion of CRH, ACTH and cortisol follows a pulsatile circadian rhythm with increased amplitude and frequency of the secretory pulses in the early morning. Then, there is a progressive decline in their levels throughout daytime, a minimal secretory activity around midnight and an abrupt elevation in late sleep resulting in an early-morning peak (Arble et al. 2012) (Figure 4.1). The circadian rhythm is generated in the suprachiasmatic nucleus of the hypothalamus by the circadian molecular CLOCK system (see details in section 5.2, Nicolaides et al. 2014).

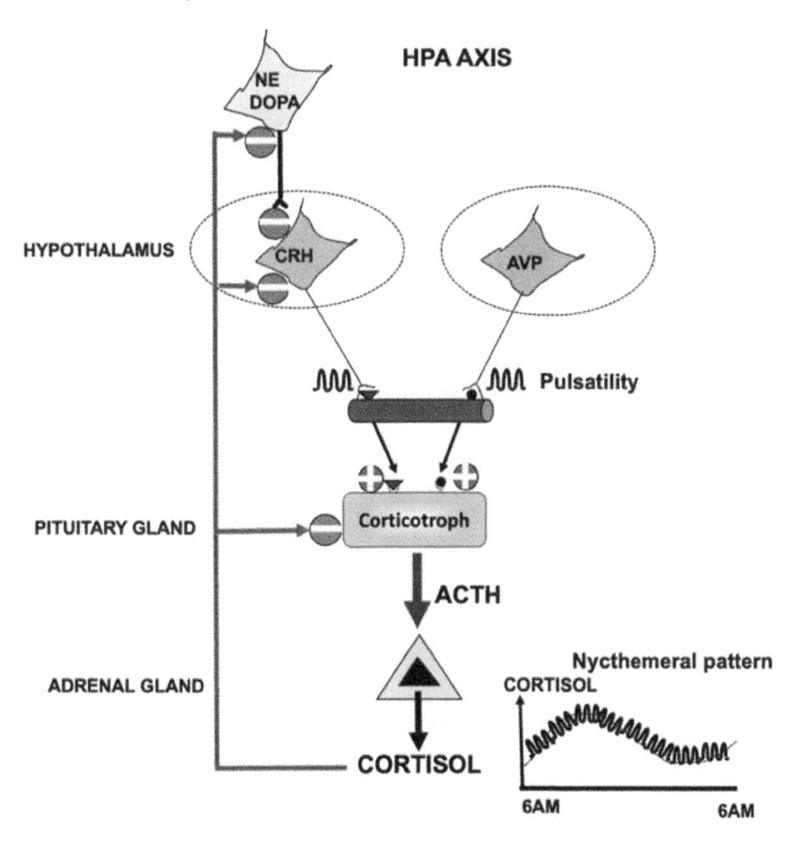

Figure 4.1. *Hypothalamo–pituitary–adrenal (HPA) axis*

Hypothalamo–pituitary–somatotropic (HPS) axis

The lack of a growth-stimulating factor in the humans who exhibited severe proportionate short stature has been recognized throughout history. However, the relationship between the pituitary gland and growth was not apparent until 1921, when H. Evans and J. Long treated rats with extracts from bovine anterior pituitary glands and showed increased growth (Evans and Long 1921). B. Houssay (1936) showed that hypophysectomized rats displayed an increased sensitivity to insulin, and that it was reversed when they were injected with pituitary extracts Thereafter, the gland was linked to metabolism. In 1965, Rabinowitz et al. (1965) demonstrated that it was attributed to a direct effect of GH; as an injection, the hormone blocked the action of insulin when both hormones were co-administered. Shortly thereafter, investigators discovered the key components of the GH axis as well as the central and peripheral components that regulate the axis (reviewed in Murray et al. (2015) (see Figure 4.2)).

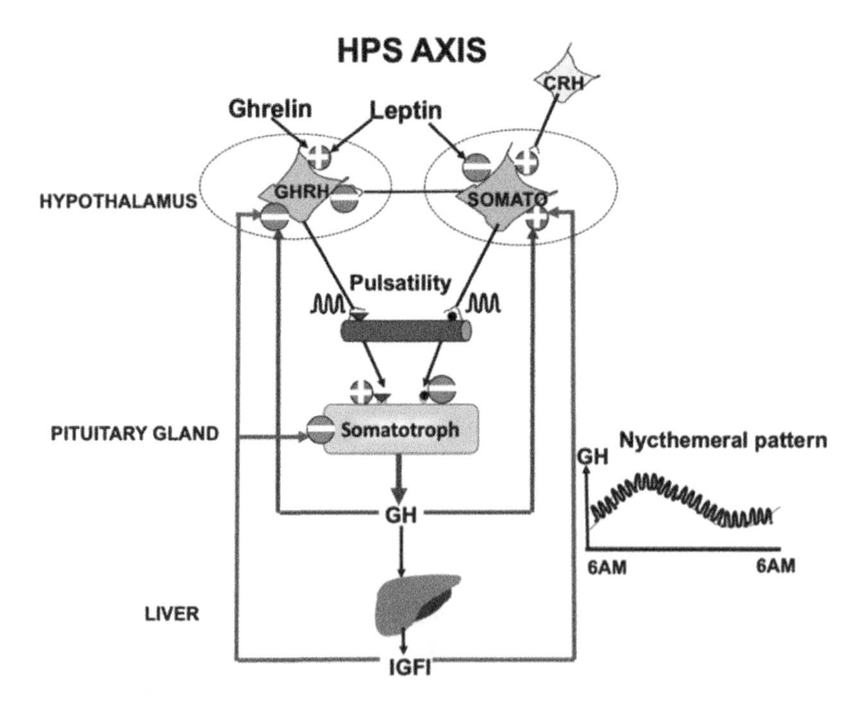

Figure 4.2. *Hypothalamo–pituitary–somatotropic (HPS) axis*

The GH (also known as somatotropin) release observed in the peripheral blood is generated by two hypothalamic hormones, a stimulatory GH-releasing hormone

(GHRH) and an inhibitory somatostatin (SRIF: somatotropin release inhibiting factor) regulating GH directly and indirectly through central regulation of GHRH-containing neurons. The net result of these interactions in the rodent is an ultradian rhythm of GH release with high amplitude GH secretory bursts occurring at very regular intervals (Tannenbaum and Epelbaum 1999).

Hypothalamo–pituitary–gonadal (HPG) axis

The axis is under the control of the gonadotropin releasing hormone (GnRH) neurons, mainly located in the anteroventral preoptic area of the hypothalamus. They project to the median eminence where they release GnRH in a pulsatile manner. It reaches the anterior pituitary which stimulates the secretion of luteinizing hormone (LH) and follicle stimulating hormone (FSH), which ultimately drive the growth and maturation of germ cells and the production of gonadal steroids and gametogenesis. The activity of GnRH neurons is also under the influence of steroid hormones, neurotransmitters and growth factors (Watanabe et al. 2014). However, no clear picture has emerged on their role in other mammalian species.

In males, FSH and LH are essential for spermatogenesis and steroidogenesis. Testosterone is involved in controlling the HPG axis through its negative feedback action on the GnRH neurons and anterior pituitary gonadotrophs. This regulation mechanism has been demonstrated numerous times by studies using a classical two-step experimental approach. Briefly, castration of male rats results in increased levels of LH and FSH and testosterone administration drives secretion back to normal.

In females, under basal conditions, estradiol acts through a negative feedback loop, similar to that described in males, to restrict GnRH secretion and maintain constant hormone levels. However, in rodents, estradiol is known to have a positive feedback on GnRH secretion, suggesting that another upstream player is involved in the secretion control. It has been found that in the anteroventral periventricular (AVPV) nucleus, a population of kisspeptin neurons can be activated by estradiol, causing a concomitant activation of GnRH neurons, which would drive the preovulatory surge of GnRH. A second population of kisspeptin neurons in the arcuate nucleus (ARC) would be involved in the negative feedback regulation by estradiol. The concept of the "switch" has been put forward, and it postulates that one cell population takes over from another depending upon estradiol levels. The existence of a negative feedback was established after a gonadectomy of rats was performed, resulting in increased mRNA levels of kisspeptin, and this effect was reversed by the administration of estradiol. The proposed model that describes the pattern of GnRH secretion results from neural integration of diverse afferent pathways mediating the positive and negative feedback signals at the GnRH neurons (see review Herbison (2020)). In other words, the neurons of the AVPV that target the GnRH cells synapse with incoming axons that express either excitatory (E) or inhibitory (I) neurotransmitters. Therefore, in women during the cycle, the pattern of GnRH secretion is under control of a negative feedback due to a low E/I ratio (Figure 4.3(A)) and a positive feedback due to a high E/I ratio (Figure 4.3(B)) (Zsarnovszky et al. 2001). The inhibitory and excitatory neurotransmitters are the GABA, the kisspeptin and the galanin. It is noteworthy that this biological response, like others already mentioned, illustrates the importance of the ratio concept (Naftolin et al. 2007). Positive feedback of estrogen on the hypothalamic–pituitary axis is essential for ovulation in the female. The ratio of LH to FSH secretion rises as the frequency of pulsatile GnRH release increases during the late follicular phase of the normal menstrual cycle. Increased LH secretion stimulates estrogen production from the ovary which through positive feedback leads to the midcycle LH surge that causes ovulation.

This feedback response is a sexually differentiated phenomenon; male rats do not respond to elevated estradiol with a surge of LH. Elegant studies carried out in rats showed that neonatal castration of males allows them to respond to estradiol in adulthood with an LH surge. Conversely, testosterone treatment of neonatal females defeminizes the brain such that they can no longer respond to estradiol treatment in adulthood with a surge of LH (Handa et al. 1985). Thus, sensitivity of the post-pubertal female brain to respond to elevations in estradiol with a surge of LH is primed by the absence of testosterone in neonatal females.

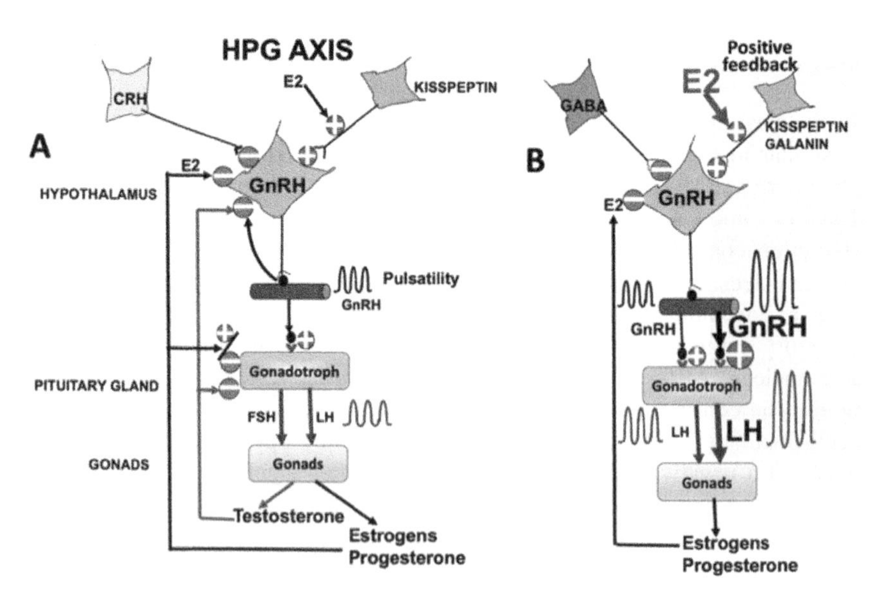

Figure 4.3. *Hypothalamo–pituitary–gonadal (HPG) axis*

Note that GnRH applied over time with high pulse frequency stimulated the transcription of β-LH but not β-FSH, whereas the reverse was true with lower GnRH frequencies.

With respect to the rhythms in the gonadotropic axis, they cover a wide range of frequencies, from episodic release in the ultradian range to diurnal rhythmicity and menstrual cycles (Arble et al. 2012). It is beyond the scope of this review to discuss these aspects.

Hypothalamo–prolactin axis

Prolactin (PRL) is named for its ability to stimulate lactation. In rats, it exerts its effects on about 300 biological events and the control of its secretion from the anterior pituitary is not fully defined. Early studies assumed that the regulation of PRL would be like all other pituitary hormones, being stimulated by a hypothalamic hormone (a prolactin releasing factor (PRF)) and inhibited by a specific hypothalamic hormone (prolactin inhibiting factor (PIF)). Several PRFs were tested (neuropeptides, steroids and growth factors) but their role is not clearly established, and the hunt continues for elusive factors involved in stimulating prolactin secretion. Regarding the inhibition of prolactin release, it was demonstrated that three populations of hypothalamic dopaminergic neurons are effective at suppressing PRL secretion (Demaria et al. 2000). Currently, it appears that the regulation of PRL secretion occurs primarily through an inhibition process, whereas secretion of all other pituitary hormones requires stimulation from hypothalamus. Thus, under most conditions, except during lactation and late pregnancy, PRL stimulates dopamine neurons which in turn inhibit prolactin release. This mechanism known as a "short loop" negative feedback maintains low levels of PRL. It should be emphasized that it is not the classical hormonal feedback system that operates with other anterior pituitary hormones. The latter stimulate the release of hormones from endocrine glands which in turn suppress, through the feedback mechanism, the secretion of anterior pituitary hormones. Regarding the PRL, its primary target is an exocrine gland (mammary gland) that cannot feedback through a hormonal message.

With respect to the regulation of PRL secretion during lactation, several mechanisms could contribute to the stimulation of its release. The predominant mechanism is the suckling stimulus elicited by the offspring, which conveys somatosensory information from the nipple to the hypothalamus, which leads to an increase in PRL levels. A second mechanism called "dopamine disinhibition" would lead to the removal of the negative influence of dopamine on PRL secretion (Guillou et al. 2015). The latter can be explained by a stimulation of a subpopulation of dopaminergic neurons by PRL, resulting in a release of enkephalin which then reaches the anterior pituitary and stimulates further prolactin secretion. This positive feedback loop produces a sufficient amount of prolactin secretion to induce a dopamine–prolactin negative feedback loop, resulting in two independent loops with indirect interactions (Yip et al. 2019). This phenomenon persists during lactation, and this positive feedback overrides the negative feedback. In other words, PRL facilitates its own secretion, and the secretory response persists as long as the stimulus is applied (Figure 4.4).

In summary, the main regulation of PRL secretion is provided in the form of tonic inhibition by dopamine, which is counteracted by stimulatory actions of many neuropeptides, steroids and growth factors. Evidence suggests that this circuitry may extend from rats to humans and other mammals (Kennett and McKee 2012). Readers should refer to a review by Phillipps et al. (2020).

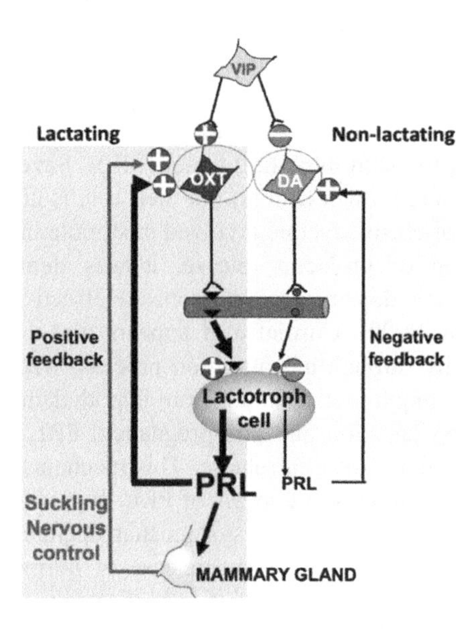

Figure 4.4. *Hypothalamo–prolactin axis*

Schematic illustration of the key mechanisms regulating the neuroendocrine axes is presented in Figure 4.5. These mechanisms aim to prevent marked changes in circulating hormone levels, both in amplitude and duration. Hormonal changes are initiated by excitatory hypothalamic neurons which alter a specific pituitary hormone release, resulting in the increased secretion of hormone release of endocrine target tissue. To prevent exaggerated cellular response both in amplitude and duration, the hormone level should go back to the basal value. Hence, the circulating hormone negatively feedbacks the activity of both stimulatory hypothalamic neurons and pituitary hormone release. It is termed direct "long feedback loop". In addition, there are neurons in the CNS that negatively regulate these hypothalamic neurons and they are also stimulated by the hormone which reinforces the negative back. It is termed indirect "long feedback loop". Besides this regulation, a population of hypothalamic neurons inhibits both the activity of excitatory hypothalamic neurons (crosstalk), mentioned above, and the release of the pituitary hormone. This dynamic interplay enables a tight control of circulating hormone levels.

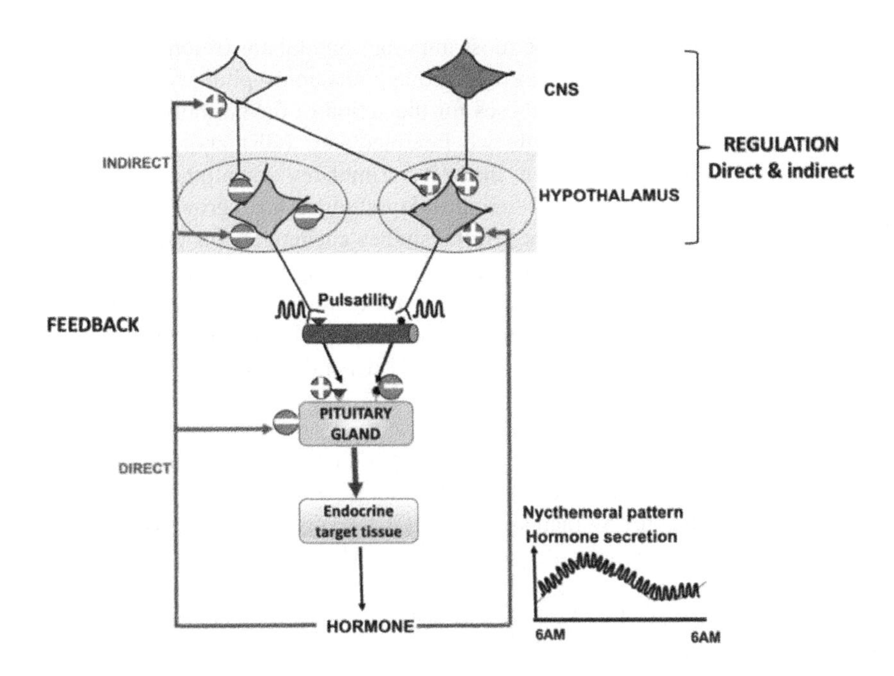

Figure 4.5. *Schematic illustration of the key mechanisms regulating the neuroendocrine axes*

4.3.2. *Crosstalk between neuroendocrine axes*

Crosstalk between HPA and HPG axes

It is schematically presented in Figure 4.6. It shows the reciprocal interaction between these two axes HPA and HPG. Hormones (testosterone, estrogen and progesterone) inhibit the HPA axis at the CRH neuron and the anterior pituitary. Cortisol inhibits the HPG axis at the GnRH neuron, the anterior pituitary and the gonads. HPG and HPA axes are influenced by upstream regulatory centers that secrete kisspeptin and GABA, respectively.

Notably, the HPA axis interacts also with the HPS axis through CRH that stimulates the somatostatin neurons, leading to an inhibition of GH release (Katakami et al. 1985).

Crosstalk between HPS and HPG axes

These axes, through GH/IGF1 and sex steroids, are involved in the control of bone growth. After birth, GH is the most important modulator of longitudinal bone growth and appears to be a key player of the hypothalamus–pituitary–somatotropic axis. However, there are still hypotheses for the action of the hormone, as it acts on its targets either directly or through two intermediates: IGF1 and IGF2. In 1957, Salmon and Daughaday suggested that GH stimulates somatomedin C (IGF1) synthesis in the liver which in turn stimulates longitudinal bone growth (Salmon and Daughaday 1957). In the mid-1980s, several studies challenged this view, and it was demonstrated that both GH and IGF1 contribute to longitudinal growth. Recently, elegant studies demonstrated that IGF1 effects on bone accretion during prepuberty are mediated predominantly via mechanisms independent of GH, whereas during puberty, they are mediated via both GH-dependent and GH-independent mechanisms (Mohan et al. 2003).

Sex steroids are critical contributors to longitudinal growth during puberty. At the start of puberty, the relatively low estradiol (E2) and testosterone (T) levels stimulate the growth spurt by increasing GH/IGF1 levels. This effect is attributable to T which modulates the pulsatile pattern of GH. Thereafter, during the late stages of puberty, the longitudinal growth is inhibited, and it was assumed that this phenomenon was achieved by E2 and T in girls and boys, respectively. Such a view has been challenged after the identification of an inactivating mutation in the estrogen receptor (ERα) in a male patient who displayed estrogen resistance, associated with the absence of growth plate closure, resulting in tall stature (Smith et al. 1994). A similar phenotype was reported in two male patients with an aromatase P450 deficiency whose role is to catalyze the conversion of T into E2 (Rochira et al. 2010). These findings highlight that in male, intracrinology may occur in the growth plate, resulting in increased intracellular E2. Consequently, at the end of puberty, epiphyseal fusion which shuts off the longitudinal growth, depends on E2. In boys, this process delays the cessation of growth, which explains why boys are taller than girls. More detailed information is presented in the review Tenuta et al. (2021), and Figure 4.6 summarizes the crosstalk between axes and feedbacks.

Steroid hormones can also shape the HPS axis. Testosterone facilitates GH (Roelfsema et al. 2018) secretion and E2 lowers liver sensitivity to GH on the release of IGF1. Likewise, ghrelin links the somatic growth with metabolism and reproduction, as it inhibits LH production and stimulates GH release (Tena-Sempere 2005). A number of other factors participate in the regulation of these axes, which are discussed in an excellent review by Veldhuis et al. (2012). A schematic representation of the interactions between the HPS and HPG axes and negative and positive feedback actions upon these axes are shown in Figure 4.6.

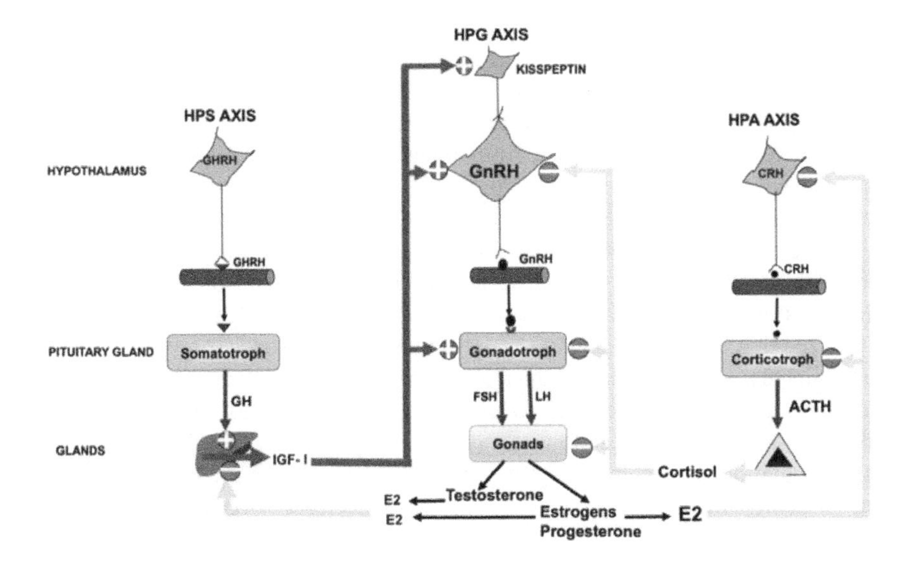

Figure 4.6. *Crosstalk between HPG axis and the two other major axes, HPA and HPS*

It should be emphasized that these interactions between axes are more complex than initially thought. Given that these axes are under the influence of external and/or internal factors throughout life, key questions can be raised, such as: which axis is dominant and which one acts first? Do reciprocal influences change over time, namely during development?

4.3.3. *Crosstalks between organs and brain*

These communications include essentially intestine, muscle, adipose tissue, liver, kidney, bone, pancreas and brain. A schematic illustration of these communications is briefly summarized in Figure 4.7.

One of the first pieces of evidence of the existence of interorgan crosstalk was provided by C.F. Cori and G. Cori, in which lactate produced by anaerobic glycolysis in muscles can be recycled by the liver and converted to glucose. The latter is returned to muscle where it is metabolized to lactate (Cori and Cori 1946). This example describes an efficient communication system between two organs which facilitates the metabolic adaptation to energy demands.

The views on organ crosstalk changed in 1987 when B. Spiegelman and J. Flier defined adipose tissue as a secretory organ/tissue after the identification of secretory protein named adipsin (Cook et al. 1987). Then, in 1994, Friedman's laboratory identified another protein secreted by the adipocyte and was named leptin (Zhang et al. 1994), which established adipose tissue as an endocrine tissue. After the identification of these signaling molecules, it was tempting to speculate that other humoral factors had to be released from other tissues, namely during metabolic adjustments which occur in widely varying states, including the fed state, fasting and exercise, so as to maintain energy homeostasis. For instance, during a transition from the resting state to exercise a marked and fast shift in whole-body fuel metabolism must occur in response to the rapid increase in the metabolic requirements of skeletal muscle (reviewed in Wasserman (1995). It was then hypothesized that enhanced energy demands in muscle at the onset of exercise likely generate signals which inform the rest of the body of these changes. Before such signals were identified, they were named "work factor". In 2000, the group of B. Pedersen showed that skeletal muscle produced and released interleukin-6 (IL-6) into the circulation (Pedersen and Febbraio 2012) and has multiple effects on other organs. It demonstrated that skeletal muscle was a secretory organ with endocrine functions.

Omics approaches have allowed the identification of novel secreted molecules that can be produced by organs that were not suspected to perform endocrine functions. It is worth noting that these discoveries revealed additional physiological processes, which then reshaped the way we think about physiology and endocrinology. Thus, these disciplines do not focus solely on the molecular events in different cell types; rather, they redefine the role of these novel molecules and integrate them in new circuits of inter-organ communication. Failure in the coordination of communication between organs potentially leads to a rise in pathophysiological situations.

The following section provides an overview of the diversity and the complexity of interorgan crosstalks, with a special emphasis on the role of the central nervous system in orchestrating different functions.

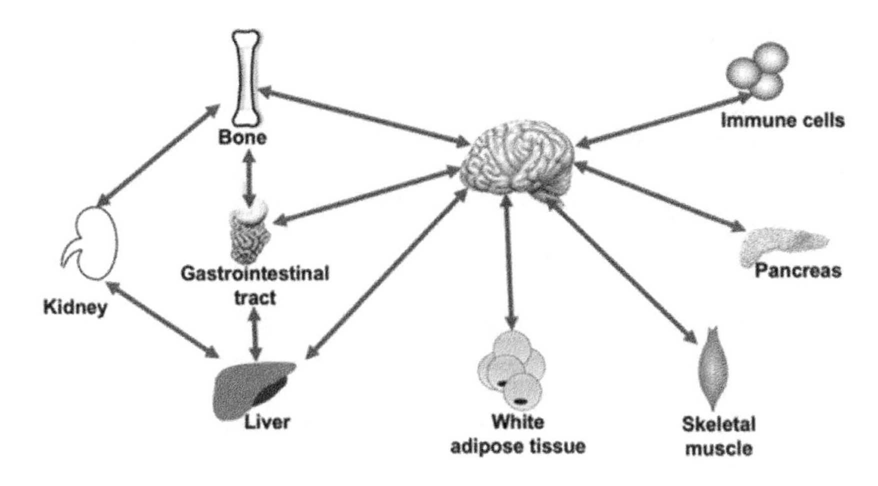

Figure 4.7. *Crosstalks between organs and brain, and crosstalks between bone, kidney, liver and gastrointestinal tract. See details below*

4.3.3.1. *Intestine–liver–kidney–bone*

Calcium (Ca^{2+}) and phosphate ($H_2PO_4^-$ and HPO_4^{2-}) have a key role in numerous biological processes, including intracellular signaling, enzymatic reactions, energy metabolism and bone structure. It is therefore important for the body to tightly regulate their blood levels through mechanisms which fine-tuned the balance between their dietary intake and cellular utilization. This regulation is a clear example of inter-organ communication, since it involves bones, kidneys, intestines, parathyroid gland and thyroid C cells. Calcium and phosphate homeostasis is mainly regulated by the PTH–vitamin D–FGF23 axis and the calcitonin.

PTH role

PTH release occurs in response to a lowered blood calcium level, which is detected by the calcium sensing receptor of the parathyroid gland. PTH role will be to restore the calcium level through different mechanisms: 1) stimulation of calcium and phosphate releases from the bones; 2) increase in calcium reabsorption and a concomitant decrease in phosphate reabsorption in the kidney; and 3) increased synthesis of $1,25(OH)_2D3$, which increases the dietary absorption of calcium and phosphorus in the gut (Gruson et al. 2014).

1) Release of calcium and phosphate from the bones.

PTH acts on osteoblast lineage and upregulates RANK-L gene expression and therefore stimulates RANK-L release. Binding of RANK-L to RANK, a membrane receptor expressed by osteoclast progenitors stimulates their proliferation and differentiation into osteoclast (Wein 2018). Within this context, the osteoclast activity leads to a net increase in bone resorption, accompanied by a concomitant release of calcium and phosphate in the bloodstream. Simultaneously, PTH suppresses the expression of osteoprotegerin (OPG) which is a decoy receptor of RANK-L and therefore prevents the binding of RANK-L to its receptor RANK (Martin and Sims 2015; Lee and Lorenzo 1999). In the presence of PTH, the RANK-L/OPG ratio is elevated which favors the RANK-L effect on the osteoclastogenesis. The role of this ratio was discussed in section 3.3.5.

2) Handling of calcium and phosphate by the kidneys (Figure 4.8).

PTH directly and indirectly induces phosphaturia (i.e. increased urinary phosphate excretion). It directly binds to its receptor on the renal tubule and induces a decreased expression of phosphate transporters which then promotes phosphaturia. Its indirect effect is more complex, and it occurs through FGF23 which is released by the osteocytes/osteoblasts in response to PTH (Fukumoto and Shimizu 2011). FGF23 binding to its receptor and co-receptor Klotho result in phosphaturia through the same PTH mechanism. It also decreases intestinal phosphate absorption. Taken together, these direct and indirect effects of PTH cause hypophosphatemia.

Another indirect action of PTH involves vitamin D3 (cholecalciferol), which is a natural form of vitamin D, and it is mainly generated in the skin after UV exposure (10–15% from the diet). In its native form, vitamin D3 is not biologically active; it is transported to the liver and hydroxylated by a 25-hydroxylase enzyme to produce 25(OH)D3, which is also a biologically inactive form. It is transported to the kidney and hydroxylated by a 1α-hydroxylase enzyme (Shinki et al. 1997) that converts the 25(OH)D3 in $1,25(OH)_2D3$. At this last step, PTH comes into play, as it stimulates this enzyme and therefore increases the amount of $1,25(OH)_2D3$ (Zierold et al. 2007), which in turn stimulates phosphate and calcium reabsorption in the kidney. Note that $1,25(OH)_2D3$ decreases the transcription of the PTH gene and therefore lowered PTH production in bovine parathyroid cells in primary culture, which represents a negative feedback loop (Silver et al. 1985). Given that FGF23 suppresses the production of $1,25(OH)_2D3$ by inhibiting the 1α-hydroxylase, it acts as a counter-regulatory hormone for vitamin and therefore contributes to the phosphaturia (Liu et al. 2006).

3) $1,25(OH)_2D3$ acts on the gut to increase the absorption of both dietary calcium and phosphate.

4) Signal attenuation in response to changes in circulating Ca^{2+} levels.

The restored blood calcium level provides a negative feedback signal to the parathyroid gland, lowering the release of PTH. Furthermore, $1,25(OH)_2D3$ has a direct action on the parathyroid gland to regulate PTH gene expression, thus completing an endocrinological feedback loop.

In summary, studies carried out on the calcium and phosphate balance have revealed a fairly complex inter-organ communications and highlighted the endocrine nature of bone. They also identified the PTH–vitamin D–FGF23 axis by which mineral homeostasis can be tightly regulated, and the net effect of the PTH on bone and kidneys is an increase in serum calcium concentration and a decrease in serum phosphorus concentration.

Calcitonin role

The major role of this hormone is to maintain calcium homeostasis. In contrast to PTH, it protects against hypercalcemia and the first evidence was obtained in the 1960s. Thyroidectomized patients were challenged with an intravenous calcium load, and their blood calcium levels remained elevated compared to controls with normal circulating calcitonin levels (Hirsch and Munson 1969). Lowered calcium level in controls is attributed to the presence of calcitonin which exerts an inhibitory effect on the osteoclast activity, resulting in decreased bone resorption, accompanied by a lowered efflux of calcium from the bone (Cornish et al. 2001). To reverse severe hypercalcemia, calcitonin inhibits renal calcium reabsorption. It is worth noting that the ability of the hormone to reduce the bone resorption depends on the rate of bone turnover. Thus, in childhood, which is characterized by a high bone turnover, calcitonin rapidly decreases the calcium level compared to adult patients.

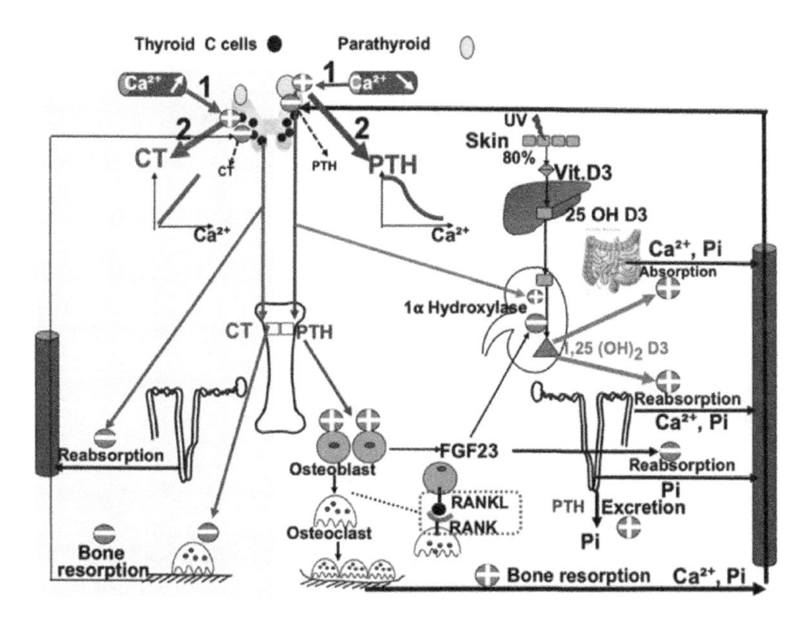

Figure 4.8. *Interorgan crosstalks between the bone, the liver, the kidney and the GI tract in the regulation of blood calcium (Ca^{2+}) and phosphate ($H_2PO_4^-$ and HPO_4^{2-}) levels*

4.3.3.1.1. Central nervous system–bone

The process of bone remodeling involves bone resorption by osteoclasts and bone formation by osteoblasts. The mechanisms regulating this process are not fully understood, but studies carried out over the last decades have reported that it is locally and centrally regulated. The first insight into the central regulation of bone mass was initiated in 2000, and it led to the seminal observation that mice lacking leptin (ob/ob mice) or its receptor (db/db mice) displayed low sympathetic activity and a high bone mass due to an enhanced bone formation (Ducy et al. 2000). Analyzing the site of action of leptin, it was shown that a knock-out of β-2 adrenergic receptor in mice produced a phenocopy of bone features observed in ob/ob mice. These findings strongly suggested that the sympathetic tone inhibits bone formation and favors bone resorption, thus resulting in a bone loss (Takeda et al. 2002). The demonstration of the role of the sympathetic nervous system (SNS) in the regulation of bone mass raised the question of whether the parasympathetic nervous system (PNS) would counterbalance bone mass regulation of the SNS. It was shown that cholinergic signaling favors bone accretion by suppressing the SNS signaling and directly by promoting apoptosis in osteoclasts (Eimar et al. 2013). This simultaneous control of bone formation and resorption by the SNS illustrates the concept of duality of control, mentioned earlier (Figure 4.9).

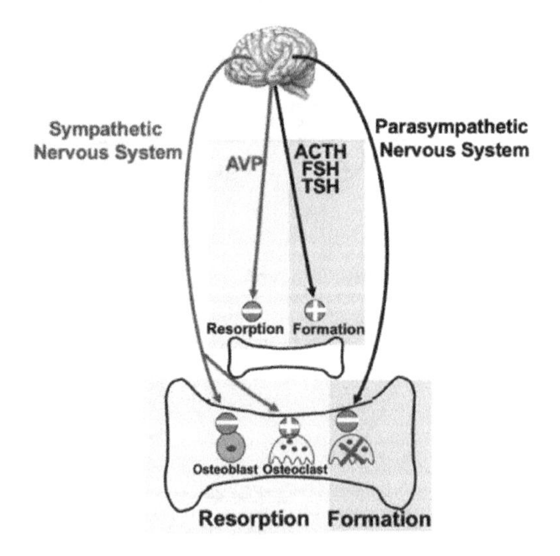

Figure 4.9. *Role of sympathetic nervous system and parasympathetic nervous system in bone remodeling*

It should be emphasized that the hypothalamic–pituitary–thyroid axis plays also a critical role in the process of bone remodeling (Bassett and Williams 2008).

Indeed, the deficiency of thyroid hormones in infants leads to marked growth retardation.

In this context, it is interesting to mention that the CNS influences bone remodeling through pituitary-derived hormones, such as ACTH, FSH and TSH, which increase bone mass while AVP decreases it. For a comprehensive review, see Chamouni et al. (2015).

4.3.3.2. *Central nervous system–metabolic organs*

One key strategy for regulating metabolic homeostasis is the communication between tissues and organs via secreted signaling molecules. Depending on the origin of the metabolic organs, they have been termed adipokines, myokines, hepatokines and batokines, when they are produced by the white adipose tissue, muscle, liver and brown adipose tissue, respectively. They exert either autocrine, paracrine or endocrine effects, and their secretion varies according to the metabolic status of the body, such as fasting and feeding cycles, exercise, cold exposure, thus maintaining energy balance and ensuring metabolic flexibility to meet specific energy demands.

4.3.3.3. *Gut–adipose tissue–pancreas–central nervous system: regulation of the energy intake*

In this section, a special emphasis is put on the historical concepts, which progressively shaped our understanding of cellular and molecular mechanisms controlling the energy intake.

Theories

During the 1950s and 1960s, it was hypothesized that the brain had a major role in regulating energy homeostasis. Two theories were proposed:

* *The lipostatic theory*

In 1953, Kennedy suggested that adipose tissue produces a signal, in proportion to its mass, that is sensed by the brain to regulate changes in intake or expenditure, and this keeps body fat within a predefined set point (Kennedy 1953). This negative-feedback system has been termed the “lipostatic” hypothesis.

* *The glucostatic theory*

In the 1950s, Mayer proposed that hypothalamic nuclei control the utilization and storage of carbohydrates. On the one hand, these nuclei would regulate the supply of energy substrates such as glucose, fatty acids and amino acids (Mayer

1953). On the other hand, these substrates could also be signal molecules that would convey the body's energy status to hypothalamic neuronal circuits.

Besides these theories, it should be emphasized that in the 1960s, indirect evidence suggested a crosstalk between the CNS and adipose tissue (WAT) through its sympathetic innervation. It was showed that electrical nerve stimulation stimulates lipolysis, strongly suggesting that the catecholamines observed in WAT are released from sympathetic nerves, as opposed to adrenal gland via the general circulation (Fredholm and Rosell 1968).

Hypothalamic nuclei lesion experiments

In 1940, Hetherington first demonstrated that lesions of the ventromedial hypothalamus (VMH) in rats caused hyperphagia (Hetherington and Ranson 1940). Given that destruction of these nuclei results in hyperphagia, their role before lesions is to restrain food intake. They are called "satiety centers".

When performing a lesion of the lateral hypothalamus (LH), it induces hypophagia in a mouse (Anand and Brobeck 1951). Given that the destruction of this nucleus leads to hypophagia, it implies that its role before lesion is to stimulate food intake. This nucleus is then called "feeding center". These experiments show that some hypothalamic nuclei regulate food intake.

Crosstalk between the gastrointestinal tract and the brain

The first hormone secreted by endocrine cells in the intestine was discovered in 1902 and was named secretin. It is released by the duodenal mucosa. A second hormone released from the enteroendocrine I cells was discovered in 1928 and named cholecystokinin (CCK). The most significant features of this hormone are briefly presented:

– it inhibits the food intake when injected into rats (Gibbs et al. 1973);

– it is present in the brain as well as its receptors (Saito et al. 1981);

– its effect is conveyed by vagal afferents, since vagotomy abolishes its inhibitory effect (Smith et al. 1981). It also acts directly, as hormone, in the central nervous system (CNS).

Taking account of these observations, the concept of "dialogue between the gastrointestinal tract and the brain" was then put forward. Physiologically, this means that a peripheral signal informs the CNS directly (circulating factor) or indirectly (via neural signal).

Control of the food intake by a circulating factor:

Kennedy's theory, mentioned above, predicts that genetic or acquired defects in the detection of signals by neurons should be interpreted by the brain as reflecting an energy deficit stored in fat tissue. In response, there would be hyperphagia and excessive weight gain.

It was then suggested that there was a circulating signal/factor that controls food intake. To explain an inappropriate brain response, the potential circulating factor would be either absent or inactive. To address this issue, parabiosis experiments, which consists of sharing blood between two mice, were carried out on obese mice (ob/ob and db/db strains) and nonobese mice (Coleman and Hummel 1969). This simple experimental approach clearly demonstrated that db/db mice were unable to respond to a circulating factor, whereas in ob/ob the factor was absent.

The next challenge was to specify its chemical nature and cellular origin. In 1994, this circulating factor (peripheral signal) was identified in J. Friedman's laboratory and named "leptin" (from *leptos* which means thin or lean in Greek) and it was secreted by the adipocyte (Zhang et al. 1994). Shortly after this discovery, its neural targets and their respective responses have been characterized. Then, this field regained a new dynamic.

Studies were aimed to identify the peripheral signals that are routed to the CNS and which carry information about the body's energy status. These signals originate from the adipose tissue, the endocrine pancreas and the enteroendocrine cells (EECs). As previously reviewed, EECs represent the first level of integration of the information present in the gut lumen (Latorre et al. 2016; Blanco et al. 2021). The sensing of nutrients by EECs activates an intracellular signaling pathway, leading to the release of signaling peptides, that convey information to the CNS via either the vagus nerve or the hormonal mode. These signals as well as the metabolic signals are briefly discussed.

Hormonal and neuronal signals

In response to a meal ingestion, there is a secretion of great diversity of signals from different organs/tissues (Figure 4.10).

– Leptin and insulin are hormones secreted by the white adipose tissue and β-cells of the endocrine pancreas, respectively.

– Peptide YY (PYY) and glucagon-like peptide-1 (GLP-1) (Spreckley and Murphy 2015).

Secreted by intestinal L cells and act directly on hypothalamic neurons and indirectly via the receptors located along the vagal afferent fibers that innervate the wall of the GI tract.

* *Cholecystokinin (CCK) (Rehfeld 2017)*

Secreted by I cells, CCK exerts mainly an indirect action on the CNS via the vagus nerve.

* *Ghrelin (Klok et al. 2007)*

Secreted by the A cells of the gastric fundus, ghrelin also acts directly and indirectly on the CNS.

All these gastrointestinal hormones that interact with the CNS constitute the gut–brain axis (Figure 4.10). They act in a coordinated way to modulate food intake and energy expenditure.

Metabolic signals

Many metabolic substrates, such as glucose, fatty acids and amino acids, can also serve as signal molecules that transmit information about the body's energy status to hypothalamic neuronal circuits. The reader is referred to excellent reviews (Spreckley and Murphy 2015; Latorre et al. 2016).

Three hypothalamic structures have a key role in the integration of peripheral signals:

– the arcuate nucleus (ARC);

– the paraventricular nucleus (PVN);

– the lateral nucleus (LH).

In the ARC of the mediobasal hypothalamus, two functionally opposing neuronal populations have been identified: the agouti-related peptide/neuropeptide Y (AgRP/NPY) expressing neurons and the proopiomelanocortin/cocaine and amphetamine-regulated transcript peptide (POMC/CART) expressing neurons. AgRP/NPY and POMC/CART are termed orexigenic and anorexigenic neurons, respectively. Both neuronal populations are targeted by insulin and leptin, and a detailed model of regulation of these neurons has been proposed (Cowley et al. 2001). Thus, in this model, leptin and insulin directly activate the anorexigenic POMC neurons, while simultaneously diminishing the release of NPY which is orexigenic. These dual mechanisms of leptin and insulin action make it possible to eliminate the orexigenic signal that would otherwise counterbalance the anorexigenic response to leptin. These mechanisms have been described in the regulation of opposite metabolic pathways by insulin (lipolysis vs. lipogenesis, proteolysis vs. protein synthesis, glycogenolysis vs. glycogenesis). The yield of an anabolic flux is maximum only if the catabolic flux is turned off; otherwise, catabolic processes partly or totally counterbalance anabolic processes.

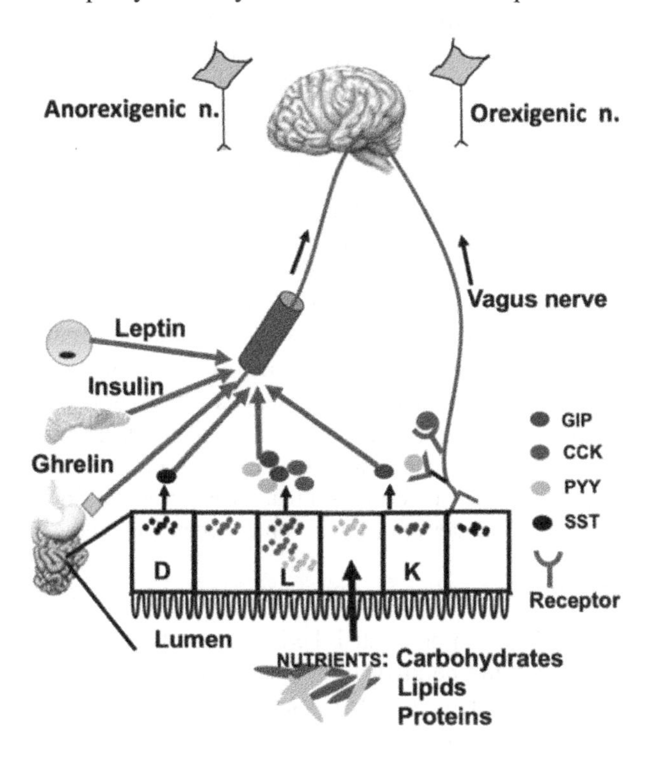

Figure 4.10. *Hormonal and neuronal signals regulating energy balance, in response to a meal ingestion. PYY and CCK convey both a hormonal and a neuronal signal through vagus nerve*

The role of gut microbiota in controlling food intake

A recently emerging concept is the ability of the gut microbiota to control energy homeostasis by regulating feeding and digestive processes (Butt and Volkoff 2019). The most important microbial metabolites are the short-chain fatty acids (SCFAs) produced by the anaerobic fermentation of dietary fibers. Their role was clearly established from a very simple observation made in germ-free mice displaying poor levels of SCFAs, due to the absence of microbiota; they display increased food intake to compensate for reduced energy harvest (Tremaroli and Bäckhed 2012). SCFAs activate EECs, triggering the release of peptides which in turn modulate the food intake (Huang et al. 2018) through a mechanism of satiety (see the meaning of this concept below).

Studies on the microbiota profile of the gastrointestinal tract revealed that bacterial diversity was significantly greater in obese subjects compared with nonobese subjects. Furthermore, it was observed that certain bacterial species, Firmicutes and Bacteroidetes, were significantly associated with each group and the ratio F/B was higher in obese subjects and in overweight subjects (Kasai et al. 2015).

There are also other mechanisms that make it possible to adjust energy intake to energy expenditure. In other words, even if humans or laboratory animals have free access to food, there are signals that inhibit intake, including sensory, cognitive, environmental, physiological and social factors. In 1971, pioneering studies carried out on rats postulated the existence of two mechanisms that contribute to the inhibition of the food intake: satiation and satiety (Larue and Le Magnen 1972; Magnen 2012). The concept of satiation represents a set of complex processes that takes place at the beginning of a meal. As food is ingested, inhibitory influences bring ingestion to an end and determine the size of the meal. With respect to the concept of satiety, it represents a mechanism that takes place after the eating episode and prevents the return of hunger for a certain time, depending on the subject.

4.3.3.4. *Muscle and adipose tissue–pancreas–liver–gut–brain*

In the 1960s, M. Goldstein postulated the existence of a humoral component in skeletal muscle, which could be involved in the regulation of glucose homeostasis (Goldstein 1961). Thereafter, it was demonstrated that the cytokine interleukin-6 (IL-6) is produced and released from skeletal muscle during exercise. The term myokine, denoting a cytokine released by the muscle, was then coined and the concept that skeletal muscle is an endocrine organ was then proposed.

Following the identification of the first myokine IL-6, it was demonstrated that it has a pivotal role in the interactions between muscle and different organs, including brain, bone, liver, adipose tissue, gut and pancreas. Thereafter, numerous studies revealed that a great number of muscle peptides are released, thereby showing that skeletal muscle does much more than produce power output for locomotion. Readers should refer to a review by Severinsen and Pedersen (2020). The major crosstalks reported in Figure 4.7 are discussed.

Muscle–bone–adipose tissue crosstalk

Historically, the coupling between skeletal muscle and bone was analyzed in terms of mechanotransduction, highlighting that the physical forces applied to muscle are transmitted to the bone to initiate its formation. It is now recognized that muscle and bone can receive, as well as secrete, molecular signals in a bidirectional manner, thus leading to the concept of the "muscle–bone" crosstalk. These signals, secreted by myocytes and osteocytes, are named myokines and osteokines, respectively, and they affect the metabolism of both tissues as well as the whole body (Hamrick 2011; Brotto and Bonewald 2015; Lombardi 2019; Gomarasca et al. 2020). It is worth noting that skeletal muscle traction applied to the skeleton is the main physiological stimulus for muscle anabolism. The reader is encouraged to consult some concepts of exercise physiology (Gomarasca et al. 2020). A third player in this crosstalk is the adipose tissue which secretes adipokines, like myokines secreted by skeletal muscle. Recent studies revealed that there is an overlap between myokines and adipokines, such as IL-6 and TNF-α which are secreted from both skeletal muscle and adipocytes (Raschke et al. 2013). These three organs communicate through the endocrine system as well as through autocrine/paracrine modes, and their respective signals interact in concert to regulate muscle and bone metabolism.

Muscle–brain crosstalk

Many studies have reported that physical activity can improve cognition via circulating exercise factors, termed "exerkines" (Tsai et al. 2019). Given that myokines and exerkines are simultaneously released during exercise, it has been challenging to establish their respective role.

Muscle–liver crosstalk

As mentioned earlier (Cori cycle: see section 2.1.1), lactate was described as the first molecule that could mediate the crosstalk between these two tissues. Shortly thereafter, it was discovered that the liver secretes other molecules (hepatokines) that regulate metabolic organs including skeletal muscle (see a review by Seo et al. (2021)). The mechanism underlying the effect of hepatokines on metabolism remains unclear.

Muscle–pancreas crosstalk

The concept that muscle-secreted molecules mediate interorgan crosstalk likely explains the beneficial effects of exercise in type 2 diabetes (Barlow and Solomon 2018). Exercise improves β-cell function in patients with prediabetes and diabetes (Eckardt et al. 2014). Despite unequivocal evidence that IL-6 is an exercise-regulated myokine, its specific effect on β-cell function is unclear.

Muscle–immune crosstalk

During exercise, skeletal muscle functions as an immunoregulatory organ via the release of IL-6 that exerts anti-inflammatory effects through two mechanisms. On the one hand, it inhibits the TNF production by the macrophage and, on the other hand, it stimulates cortisol release which in turn induces a lymphopenia (Pedersen and Hoffman-Goetz 2000).

4.3.3.5. *Gut–adipose tissue–reproductive axis: puberty*

The association of weight loss with infertility was first reported in 1694 when Morton observed that amenorrhea was linked to a condition that we now term anorexia nervosa (Silverman 1985). Given that pregnancy is a metabolic burden for the mother, it is therefore not surprising that under conditions of poor nutrition an adaptive mechanism takes place to restrain the reproductive function, thus preventing intrauterine growth retardation.

Several lines of evidence indicate that gut and adipose tissue hormones control the reproductive function through direct or indirect effects on the reproductive axis (HPG). Their respective role is briefly discussed; for a comprehensive review, see Comninos et al. (2014).

Gut hormones

* *Ghrelin*

The gastric mucosa is the predominant site of its synthesis. Its main action is to stimulate food intake. Given the high energy demands of reproduction imposed by the fetal growth, its role as a signal of energy status in reproductive physiology is expected. Here are some effects of this hormone on the reproductive axis (HPG) on feeding and metabolism. Administration of ghrelin inhibits both expression and release of kisspeptin whose role is to stimulate GnRH expression. It leads to a decreased LH surge required for ovulation. Clues to the effects of endogenous ghrelin levels come from several observations. Ghrelin is markedly increased both in anorexia nervosa and a decreased body weight, which occurs during a diet and intense physical exercise. A recent study showed that amenorrheic athletes have a higher ghrelin level and it causes a decreased GnRH and LH pulsatility. These observations suggest that ghrelin serves as a link between metabolism and reproduction by informing the brain of changes in peripheral energy status. As mentioned earlier, ghrelin is also a critical hormonal signal of nutritional status to the GH neuroendocrine axis, serving to integrate energy balance and the growth process.

* *Insulin*

The precise cell type(s) targeted by insulin in the reproductive axis are not clearly established. In animal models, it has been suggested that insulin inhibits NPY neurons which normally inhibit GnRH production. In response to this disinhibition, GnRH neurons are activated. It should be emphasized that this mechanism of indirect activation, which consists of relieving an inhibitory input on a target, is often involved in the regulation of biological processes.

* *Glucagon-like peptide I (GLP-I)*

It has stimulatory actions on the HPA axis in animals, but these effects have not been observed in humans.

Adipose hormones

* *Leptin*

As mentioned before, leptin plays a role in the regulation of energy homeostasis, but it also influences reproduction. Leptin-deficient ob/ob mice are obese and infertile (Pelletier et al. 2020), and mutations of leptin or its receptor in humans leads to infertility (Farooqi and O'Rahilly 2006). This phenotype is reversed after administration of recombinant leptin (Farooqi et al. 1999). Given that GnRH neurons do not express the leptin receptor, its action is likely mediated via an indirect intermediate. There is evidence that kisspeptin is likely a mediator, since its administration to women with amenorrhea stimulates LH and FSH secretion (Jayasena et al. 2009). However, the effects of leptin on kisspeptin have been challenged recently and it was suggested that glutamate signaling within the ventral pre-mammillary nucleus would be a mediator of leptin effects on reproduction (Donato et al. 2011).

* *Adiponectin*

Studies investigating its role suggest that it has an inhibitory role on the hypothalamus, pituitary and male gonads, but a stimulatory role on female gonads.

4.3.4. *Crosstalk between immune, endocrine and nervous systems*

Neural–immune interactions

These interactions were suggested when the presence of nerve fibers was demonstrated in bone marrow, lymph node and thymus. Subsequently, it was reported that neurotransmitters released by sympathetic and parasympathetic nerve endings bind to their respective receptors located on the surface of immune cells and initiate immune-modulatory responses. For instance, the parasympathetic neurotransmitter acetylcholine suppresses, via the vagus nerve, the systemic shock-like response to endotoxin (Tracey 2002). Likewise, activation of the acetylcholine receptor expressed in macrophages inhibits the secretion of IL-1 and TNF-α. The sympathetic nerve stimulation enhances the production of cytokines from Th2 lymphocytes, while inhibiting the production of cytokines from Th1 lymphocytes (Sanders et al. 1997).

* *Immune–neural interactions*

The immune system also targets various levels of the nervous system, which is another example of bidirectional interactions. For instance, the mechanism of the myasthenia gravis, which was described by T. Willis in 1672, was recognized in the early 1970s as a classic example of an antibody-mediated autoimmune disease. It is due to specific autoantibodies against the acetylcholine receptor in the neuromuscular junction (Patrick and Lindstrom 1973). In Rasmussen encephalitis, first described in 1958, the role of T cytotoxic lymphocytes (CD8+) seems predominant in the induction of brain damage, and the appearance of autoantibodies is likely a secondary pathological process.

A crosstalk between the immune and central nervous systems was definitively recognized in the 1970s and 1980s when Besedovsky and coworkers reported that an immune response induced an increase in the blood glucocorticoid level. The authors suggested that the immune system can produce polypeptide hormones which interact with the brain (Blalock and Smith 1980; Smith and Blalock 1981). Then, it was demonstrated that the release of IL-1 by the immune system results in activation of the hypothalamic–pituitary–adrenal (HPA) axis (Besedovsky et al. 1986). This interleukin reaches the brain through the OVLT (*organum vasculosum of lamina terminalis*) and stimulates astrocytes which release prostaglandins which in turn activate CRH neurons in the PVN. Therefore, the immune system detects microbial invasion and produces molecules that relay this information to the brain. This humoral route for communication between the immune system and the nervous system is implicated in the development of fever, anorexia and the hypothalamic–pituitary responses to infection. This latter response is detailed below. In recent years, this ability of the immune system to impact the central nervous system has become the most extensively studied aspects of neuroimmunology, because it has an enormous therapeutic potential for treating brain disorders through immune modulation. Over the past decade, numerous studies have revealed that both systems produced a common set of ligands (peptides, neurotransmitters and cytokines) for intra- and inter-system communications. It therefore provides a common repertoire of receptors and their respective ligands between these two systems (Schiller et al. 2020). Consequently, they are linked functionally, and they likely evolved in an intricate manner to cope with fever, stress, etc. Even though they share some features, such as the amount of information that can be stored, it remains to be known whether their respective memory has anything in common at the molecular level.

Recent studies report that immune cells can also sense environmental cues in the absence of infection. Notably, discrete immune cell subsets have been shown to integrate dietary, metabolic and unrelated tissue cues to promote intra-organ and inter-organ communications (Rankin and Artis 2018).

Another example of communication between the brain and immune system is given in the section "immune–nervous–endocrine".

Interactions between immune and endocrine systems

It has been reported that a variety of cytokines, originally known as being produced by immune cells, are also released by different endocrine glands. Although they both display the same cytokine receptors, it does not necessarily imply a transfer of information between immune and endocrine systems.

Accumulating data provide evidence that these two systems continuously interact to regulate diverse biological processes in a bidirectional manner (O'Connor et al. 2008). Several lines of evidence indicate that the immune system modulates the endocrine system, which illustrates the concept of pro-inflammatory cytokine-induced hormone resistance. The most studied example is the TNFα-induced insulin resistance, which was described in the early 1990s (Hotamisligil et al. 1993). A state of resistance in other important hormone systems has also been reported, such as glucocorticoids (Avitsur et al. 2005), GH (Lang et al. 2005) and IGF1 (Shen et al. 2002). With respect to IGF1, the pro-inflammatory cytokines TNF-α or IL-1β induce resistance to this growth factor, by diminishing downstream signaling cascade and some of its components interact with those of IGF1-R to attenuate the proliferation induced by IGF-1 in myoblasts (Broussard et al. 2004). Another well-studied example is activation of HPA axis by pro-inflammatory cytokines, which leads to increased secretion of cortisol.

Conversely, the endocrine system modulates the immune system, and glucocorticoids play a central role in these interactions when considering the inflammation process that occurs after acute or chronic infections. In response to bacteria, there is an increase in blood glucocorticoid levels whose role is to simultaneously inhibit the production of pro-inflammatory cytokines and to stimulate the production of anti-inflammatory cytokines from the lymphocytes Th1 and Th2, respectively (Elenkov 2004).

Glucocorticoids act on almost all types of immune cells and perform salient immunosuppressive and anti-inflammatory functions through various genomic and non-genomic mechanisms, thereby restoring a homeostatic state. Likewise, GH, IGF1 and prolactin also restrain immune processes (Arkins et al. 2010).

Interactions between immune–nervous and endocrine systems

Inflammation illustrates the crosstalk between these systems. It is initiated by an infection with bacteria, viruses, fungi, tissue damage and inappropriate responses to autoantigens. Given that inflammation is an unstable state, it either resolves or persists and is then termed acute and chronic inflammation, respectively. The question raised was how do we turn off inflammation? Similarly, to the initiation of inflammation, the resolution of inflammation is an intricate and active process and if left unchecked, could have deleterious effects on the body.

The mechanisms underlying chronic inflammation were investigated in autoimmune disorders such as rheumatoid arthritis (RA). In the 1940s, Hench et al. (1949) discovered that patients with this disease produced a substance that had anti-inflammatory properties and hence, improved the clinical symptoms. In parallel, Kendall demonstrated that gland extracts prolong the life of adrenalectomized animals. In 1941 he synthesized the compounds of these extracts and their chemical structures were recognized as steroids. Among them was cortisone, which has become, along with other steroids, one of the most powerful molecules in the treatment of inflammatory diseases. It was later realized that cortisone is an inactive precursor and is converted to cortisol, which is the most potent glucocorticoid in humans. In 1950, both researchers shared the Nobel Prize in Physiology or Medicine for their discovery.

As mentioned earlier, H. Besedovsky and A. del Rey focused on the interactions between immune and nervous systems and demonstrated that after endotoxin (lipopolysaccharide) administration, inflammatory stimuli can activate anti-inflammatory signals from the central nervous system. Thereafter, the same group provided evidence that inflammation in peripheral tissues alters neuronal signaling in the hypothalamus. Activated immune system regulates the release of glucocorticoids, which in turn controls immune responses (Besedovsky and del Rey 1996). These findings demonstrated a bidirectional communication between the neuroendocrine and immune systems. Detailed mechanisms underlying the interactions between these systems were described in the model of experimental inflammation disease (Sternberg et al. 1989). In this model, the administration of streptococcal cell wall (SCW) induces an RA in the Lewis strain rats, whereas the Fisher strain rats are protected. Increased susceptibility to the induction of the disease in the Lewis strain is due to a defective activation of the HPA axis and the biological response is a defect in the release of corticotropin releasing hormone (CRH), which is still sometimes called corticotropin releasing factor (CRF), and a diminished adrenal release of corticosterone, the major glucocorticoid in rodents. To unequivocally prove it, the disease can be induced after SCW administration to adrenalectomized Fisher strain rats and they recover from the arthritis after administration of glucocorticoids. Taken together, these findings indicate that chronic inflammation in Lewis rats elicits a blunted hypothalamo–pituitary–adrenal axis response (ACTH and corticosterone) (Figure 4.11).

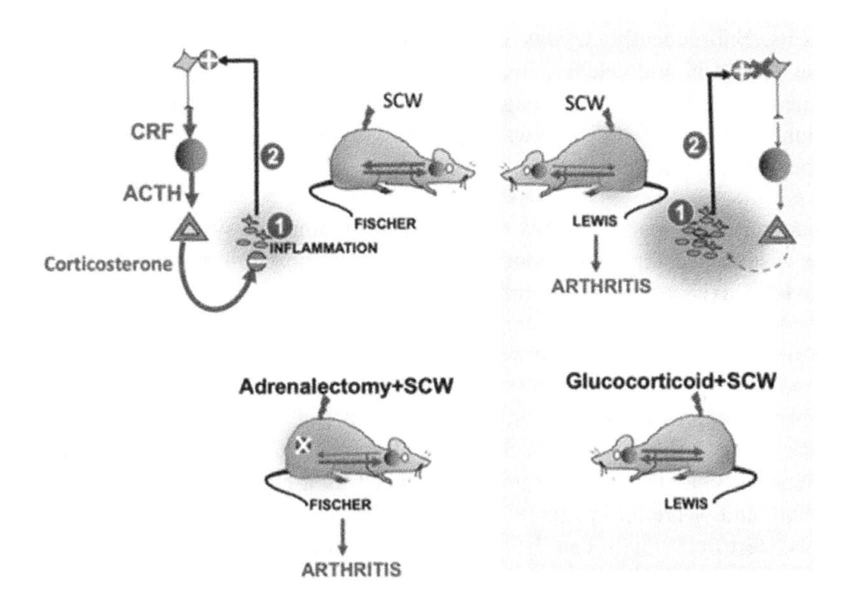

Figure 4.11. *Experiments showing a bidirectional communication between the neuroendocrine and immune systems. See details above*

In normal conditions, when an infection or injury occurs, the sequence of events is the following: (1) immune cells (macrophages, mast cells) are recruited to the inflammatory site and released pro-inflammatory cytokines (IL-1, IL-6 and TNF-α); (2) these cytokines bind to receptors on endothelial cells and induce, within the brain a release of other cytokines and prostaglandins, which target the hypothalamus; (3) they act on CRH neurons and stimulate the HPA axis abovementioned; and (4) the adrenal glands release glucocorticoids that suppress the generation of pro-inflammatory cytokines by cells of the immune system. It represents the bidirectional loop of neuroendocrine interactions, namely the role of HPA axis activity during bacterial and viral infections (Kapcala et al. 1995; Silverman and Sternberg 2012) (see Figure 4.12, immune HPA axis, inflammatory reflex and HPG axis).

Another mechanism involved in the regulation of the immune system in response to inflammation was explored by K.J. Tracey (2002) who was inspired by Molière (1622–1673) when quoting: "*The mind has great influence over the body, and maladies often have their origin there*". This group discovered that cholinergic neurons inhibit acute inflammation, strongly suggesting that the nervous system, via an inflammatory reflex of the vagus nerve, can inhibit cytokine release and thereby prevent tissue injury. In line of these observations, stimulation of the vagus nerve was shown to inhibit TNF release in animals receiving lethal injection of endotoxins. Subsequently, it was established that vagus nerve signaling inhibits cytokine synthesis and release, and was attributable to acetylcholine (ACh), the major neurotransmitter of the vagus nerve (Tracey 2007). In brief, the reflex activation of vagal efferent pathways dampens cytokine production, enabling the restoration of immune homeostasis.

In summary, two pathways link the brain and the immune system: the autonomic nervous system and the neuroendocrine outflow via the pituitary (see Figure 4.12, immune HPA axis, inflammatory reflex and HPG axis).

It is interesting that, in parallel with the activation of the HPA axis in response to acute and chronic inflammation (see above), the hypothalamic–pituitary–gonadal axis is partly suppressed. For instance, bacterial lipopolysaccharide (LPS) injection, to mimic acute inflammation, decreases GnRH mRNA levels (Lainez and Coss 2019). Similarly, a chronic inflammation induced by a high fat diet reduces GnRH expression and secretion (Lainez and Coss 2019). In both cases, it causes a decreased fertility which can be explained by the fact that pro-inflammatory cytokines target the GnRH neurons, leading to a lowered transcription and release of the neuropeptide (see review McIlwraith and Belsham (2020)). See Figure 4.12, immune HPA axis, inflammatory reflex and HPG axis.

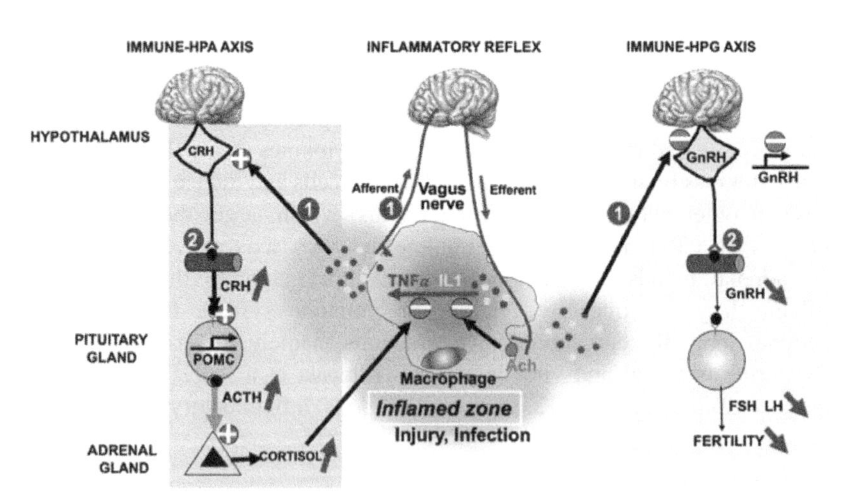

Figure 4.12. *Regulation of the immune system in response to inflammation: activation of HPA axis and inhibition of HPG axis associated with inflammatory reflex*

To summarize, historically, these systems have been studied independently of one another, and immunologists, endocrinologists and neurologists attended meetings of their respective discipline, talking only to expert of their field. We have now moved well beyond the narrow view of three systems that independently orchestrate the whole physiology. Multiple lines of evidence support that these systems influence each other and interact with each other at cellular and molecular levels to coordinate biological responses in terms of magnitude and duration. It is noteworthy to point out that a great number of molecular signals – hormones, neurotransmitters, and mediators of immunity – are at the crossroads of these systems. Additional studies will require novel interdisciplinary approaches for a further understanding of the dynamic and complex interactions between these systems.

4.3.5. *Immune system and cancer cell interactions*

In the early 1900s, Paul Ehrlich was perhaps the first to state that "cancer would be quite common in long-lived organisms if not for the protective effects of immunity" (Ehrlich 1909). Nearly 50 years later, a better understanding of the immune system combined with the demonstration of the presence of tumor-specific antigens favored the idea of the immune control of cancer (Old and Boyse 1964). Then, findings from different laboratories documented that the immune system can function as an extrinsic tumor suppressor, reviewed in Vesely et al. (2011). These emerging discoveries resulted in the development of the immunosurveillance concept which was defined by Burnet and who stated in 1957 that: "it is by no means inconceivable that small accumulation of tumor cells may develop and because of their possession of new antigenic potentialities provoke an effective immunological reaction with regression of the tumor" (Burnet 1957). This was rapidly followed by many experiments aimed at testing this concept, and all these studies failed either to prove or disprove this hypothesis. By 1978, the concept was abandoned, and the field of tumor immunology lay dormant for almost two decades awaiting advances in molecular biology. Between 1994 and 1998, a series of experiments reactivated interest in the concept of immunosurveillance. They validated the concept, demonstrating unequivocally that the immune system can protect mice from outgrowth of many different types of primary and transplantable tumors (Kaplan et al. 1998). It was clearly established that cancer cells express antigens that differentiate them from their non-transformed counterparts and these antigens could be recognized by the immune system (Shankaran et al. 2001).

In 2001, a key finding revealed that tumors formed in mice that lacked an intact immune system were more immunogenic than similar tumors derived from immunocompetent mice and were termed "unedited" and "edited", respectively. It was concluded that the immune system not only destroys the nascent tumor cells but also shapes the immunogenic phenotype of a developing tumor. This notion is the basis of the cancer immunoediting concept which points out the dual host-protective and tumor-promoting actions of immunity on developing tumors. The process of immunoediting proceeds sequentially through three phases, termed, "elimination", "equilibrium" and "escape", also called the three Es of cancer immunoediting (Shankaran et al. 2001) (Figure 4.13).

A brief outline of these phases is discussed. For a comprehensive review, see Dunn et al. (2002), Smyth et al. (2006), Swann and Smyth (2007), Vesely et al. (2011).

First phase: elimination

This process encompasses the original concept of cancer immunosurveillance. During this first phase, both innate and adaptative immunity are involved, leading to the elimination of nascent tumor cells. The process is initiated when the tumor reaches a certain size which disrupts the surrounding tissues, thereby inducing inflammatory signals. It leads to a recruitment of cells of the innate immune system (macrophages, dendritic cells) that recognize transformed cells and in response produce IFNγ, whose role is to transactivate these immune cells. Then, the newly immigrated dendritic cells induce $CD4^+$ that in turn stimulate the proliferation of $CD8^+$ which destroy tumor cells. However, most of the time, a tumor cell variant survives the elimination process and enters into a dynamic equilibrium phase during which new mutations occur.

Second phase: equilibrium

The immune system maintains residual tumor cells in a functional state of dormancy, which is characterized by a balance between elimination and proliferation of new tumor cell variants. The latter carry other mutations, a key feature of any cancer cells, that confer an increased resistance to immune attack. Many of the original variants which escape the elimination phase are destroyed. It should be emphasized that the equilibrium phase is the longest of the three processes and can last many years.

Third phase: escape

Some of the tumor cell variants that emerge from the second phase have acquired insensitivity to immunological detection and/or elimination, which can be due to a loss of antigens and/or expression of new antigens by the tumor cells. Hence, these variants have acquired the ability to circumvent immune recognition. They become "invisible" to the immune system and begin to expand in an uncontrolled manner.

Numerous studies have addressed the critical question: how do tumors evade immune surveillance? Here are some mechanisms that confer the ability of the tumor to escape immune control, leading to the appearance of overt cancer. They have been discussed in recent review articles (Vesely et al. 2011).

– Defective antigen presentation: impaired capacity to process and present antigens and peptide epitope and to load it onto MHC I (Algarra et al. 2000).

– Secretion of immune suppressive mediators: tumor cells release VEGF which prevents endogenous dendritic cell function.

– Expression of immune-inhibitory ligands: tumor cells express PD-L1, B7 H1 which interact with receptor on the cell surface of T-cells to dampen their cytotoxic actions.

– Reduced sensitivity to apoptosis: tumor cells express the ligand Fas-L that binds to the receptor Fas expressed on T-cells, resulting in the apoptosis of the latter.

In summary, throughout the last 100 years, there has been a vigorous debate as to whether the immune system protects the host from cancer development of nonviral origin (immunosurveillance function) as well as promotes tumor growth. This dual action has been termed "cancer immunoediting", and experimental and clinical data support the validity of this concept.

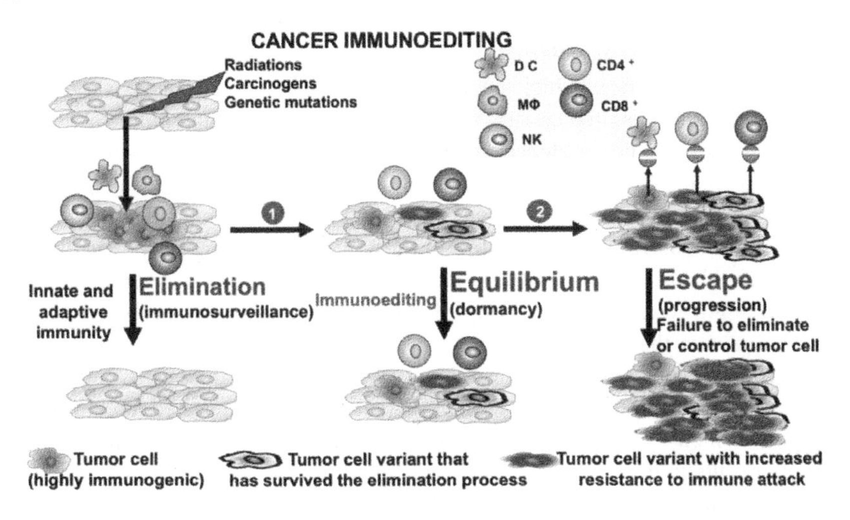

Figure 4.13. *Cancer immunoediting. It describes how the immune system can facilitate tumor progression, at least in part, by shaping the immunogenic phenotype of tumors as they develop. This process is comprised of three phases that are collectively denoted the three Es of cancer immunoediting: elimination, equilibrium and escape*

4.3.6. *Adjustments of intermediary metabolism: brain, skeletal muscle, cancer cells*

4.3.6.1. *Metabolic adaptations to starvation: brain, skeletal muscle*

Brain

Numerous studies have described metabolic adaptations of tissues or organs that occur during starvation in humans. In the 1960s, the knowledge about whole-body metabolism during human starvation was not clarified, partly due to a lack of methods for accurately measuring fuels. For instance, during early stages of starvation (two to three days), glucose was clearly the key fuel metabolized by the brain. The possibility that other fuels were also metabolized by this organ was completely ignored. Furthermore, there was a widely held misconception that ketone bodies (β-hydroxybutyrate and acetoacetate) were "unhealthy", because of their association with diabetes, reflecting a disease state. Until 1965, there were classic studies which reported that glucose was the only source of fuel for the brain. In 1967, when G. Cahill studied the energy requirements of organs during starvation, the above concept was not valid anymore (Owen et al. 1967). He demonstrated that during 41 days of starvation, obese subjects survived by metabolizing ketone bodies and markedly diminishing glucose utilization. In other words, the brain has the ability to switch from glucose oxidation to ketone bodies oxidation. The second finding clearly demonstrated that the brain obtains most of its energy requirements from ketone bodies, a fuel derived from fatty acids. These adaptations make it possible to spare glucose, which enables hepatic gluconeogenesis to slow down. Under these conditions, fewer substrates, including glucogenic amino acids, are needed for gluconeogenesis. Hence, the rate of proteolysis in skeletal muscle, which serves as the main source of glucogenic amino acids, is lower during prolonged starvation compared to the early stages of starvation, when blood levels of ketone bodies are still insufficient to fuel the brain and gluconeogenesis proceeds at its maximal rate. Thus, a switch to ketone body oxidation spares muscle proteins, which is thought to markedly prolong the ability of humans to survive starvation. Estimates indicate that humans cannot survive the loss of more than half of total muscle protein. Notably, death due to starvation is rarely due to hypoglycemia. Indeed, usually it is the muscle weakness that leads to impaired breathing and pneumonia, which – aggravated by suppressed function of the immune system – ultimately results in death. For a comprehensive review of metabolic switching from glucose to ketones as a fuel, see Owen (2005) and Cahill (2006).

Skeletal muscle

In 1963, Randle et al. introduced a new dimension of fuel control by adding a nutrient-mediated fine tuning on top of the hormone control. These investigators demonstrated that in the presence of elevated extracellular concentration of fatty acids, there is a reduction in cell glucose uptake, utilization and oxidation without hormonal mediation, in insulin-sensitive tissues. This phenomenon was first described in the heart and then in muscles, and it clearly demonstrated that these tissues possess mechanisms that allow them to shift readily back and forth between carbohydrate and fat as oxidative energy sources, depending on the availability of fatty acids. It was coined the "glucose–fatty acid" cycle (Randle et al. 1963; Hue and Taegtmeyer 2009).

Biochemically, the reduction of glucose utilization by fatty acids is the consequence of inhibition of key enzymes in the glycolysis pathway (Figure 4.14). Briefly, fatty acids fuel the β-oxidation pathway, which generates acetyl-CoA and NADH. These metabolites stimulate PDK (pyruvate dehydrogenase kinase), which then phosphorylates PDH (pyruvate dehydrogenase) and inhibits its activity. This reduces the conversion of pyruvate from glucose (glycolysis pathway) to acetyl-CoA. The increase in intramitochondrial concentration of acetyl-CoA from β-oxidation increases citrate production which migrates from the mitochondria to the cytosol and inhibits 6-phosphofructose-1-kinase (PFK). Glucose-6-P (G6P) accumulates in the cytosol and it inhibits the hexokinase, which consequently leads to an increase in intracellular glucose (reviewed in Randle et al. (1994)). The molecular mechanism of the inhibition of glucose uptake is less clear.

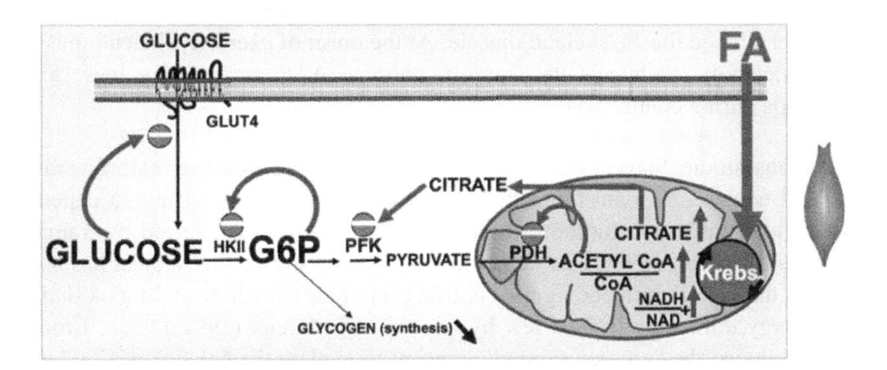

Figure 4.14. *Randle cycle also named the "glucose–fatty acid" cycle. It explains the reciprocal relationship between fat and carbohydrate oxidation in skeletal muscle*

For many years, the concept of a reciprocal relationship between fat and carbohydrate oxidation in muscle has been invoked to explain insulin resistance in the muscle of patients with type 2 diabetes or obesity. However, in contrast to Randle's hypothesis, other studies involving elevated fatty acid levels, associated with NMR spectroscopy, showed that the reduction in glucose uptake is not preceded by an accumulation of intracellular glucose and glucose-6-P, but rather a diminution of their concentrations (Roden et al. 1996; Cline et al. 1997). It implies that other mechanisms are also involved in the fatty acid-induced insulin resistance. These data strongly suggested that defects in insulin signaling were likely involved in decreased glucose transport.

4.3.6.2. *Fuel selection during exercise*

In the 19th century, the fuel used by skeletal muscle for work was suggested to be protein. However, in 1907, it was observed that there is a rapid accumulation of lactic acid in muscle exercised to fatigue (Fletcher 1907). O. Meyeroff published the concept that lactic acid formation was an indispensable energy source for muscular contraction. This eventually led to his awarding of the Nobel Prize in 1922. In 1930, E. Lundsgaard disproved this concept when he reported that muscle contraction occurred when the lactic acid formation is suppressed in response to glycolysis inhibition (Lipmann 1969). The remainder of the 20th century was devoted largely to understanding the fuel sources supplying ATP for muscle contractions.

Given that the muscle stores of ATP are almost negligible, exercise represents a metabolic challenge for the skeletal muscle. At the onset of exercise, skeletal muscle activates metabolic pathways that rapidly produce ATP to precisely meet high energy needs during contraction.

Numerous studies have clearly established that during exercise, carbohydrates (CHO) and fats are the main fuel substrates and their relative contribution depends on two factors, intensity, duration of exercise, but it is also influenced by training status (Holloszy et al. 1998). The large amount of literature on this subject has been focused on the factors influencing the contribution of carbohydrate and fat oxidation to total energy utilization (see review by Brooks and Mercier (1994)). G.A. Brooks (1998) introduced the "crossover point" concept to explain the balance of CHO and lipid metabolism. Thus, during rest and mild-to-moderate-intensity exercise, fatty acids are the major energy sources and their oxidation spares CHO. But as exercise intensity increases, there is a shift in substrate utilization towards CHO. However, the mechanism by which fatty acid oxidation spares carbohydrate is currently not clear. It was thought that the classical glucose–fatty acid cycle (mentioned above) was operating, but studies carried out on people exercising do not support this view. Inversely, increased availability and oxidation of carbohydrate decrease the contribution of free fatty acid oxidation to energy supply during exercise (Sidossis et al. 1996; Coyle et al. 1997). Despite many years of investigation, a number of fundamental questions remain, and they have been brought up by J. Holloszy, such as: (1) how is hepatic glucose production controlled so accurately to glucose utilization to maintain blood glucose level constant during prolonged mild to moderate exercise?; and (2) what are the factors responsible for the lower rate of muscle glucose uptake in the trained state compared to in the untrained state?

4.3.6.3. *Impaired metabolic switching*

In lean subjects, a key metabolic feature of skeletal muscle, in the transition from fasting to insulin stimulation, is its capacity to switch from predominately lipid oxidation to glucose oxidation. This metabolic flexibility allows an organism to adapt fuel oxidation to fuel availability. This ability to switch from fat to carbohydrate oxidation is lost in obese subjects and those with type 2 diabetes. They manifest a higher lipid oxidation during insulin stimulation, despite a lower rate of lipid oxidation in the fasting state, compared to lean subjects. This metabolic response was termed "metabolic inflexibility" and is mostly the consequence of impaired cellular glucose uptake (Storlien et al. 2004; Kelley 2005).

4.3.6.4. *Metabolic reprogramming in cancer*

Most tissues are comprised of differentiated and nondividing cells, implying that their metabolism is wired towards catabolic processes that provide energy to maintain cellular homeostasis. In contrast, cancer cells rewire intermediate metabolism towards anabolism, to promote the synthesis of macromolecules and energy for cell growth and proliferation. These new metabolic properties acquired by the cancer cell are termed "metabolic reprogramming" and are caused by oncogenic gain-of-function and loss-of-function of tumor suppressor genes that impact signaling pathways and transcriptional machinery.

Metabolic reprogramming is one of the strategies for cancer cells to obtain nutrients, metabolize them through pathways that produce ATP and generate all of the components necessary to duplicate the mass of the cell before cell division (see the review by Pavlova and Thompson (2016)). The link between cancer and altered metabolism was identified by O. Warburg in the early 1920s. He observed that cancer cells avidly take up glucose and excrete excess glucose-derived carbons as lactate even under conditions of adequate oxygen supply, i.e. without exposure to hypoxic conditions (Warburg et al. 1924). This metabolic trait was termed "aerobic glycolysis", later referred to as "Warburg effect", and he attributed to a mitochondrial defect (Warburg 1956) (Figure 4.15). This interpretation was questioned in the early 1950s, but during the past decades there has been clear evidence that in most tumor cells, mitochondria are not defective in their ability to carry out oxidative phosphorylation reviewed in Vaupel et al. (2019). Instead, their mitochondrial metabolism is reprogrammed to meet the challenges of macromolecular synthesis which requires mitochondrial enzymes (Ward and Thompson 2012).

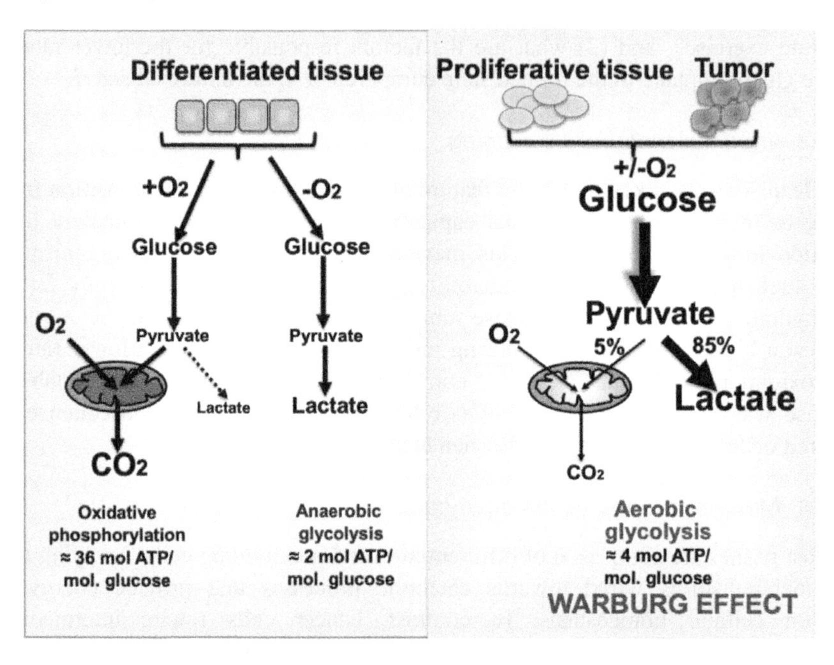

Figure 4.15. *Tumor cells or proliferative cells shift from oxidative phosphorylation to glycolysis. It occurs even in the presence of oxygen. This process is now called the "Warburg effect", and it can be interpreted as a lack of "Pasteur effect". The latter effect consists of increased glycolysis only under anaerobic conditions*

In 1955, H. Eagle observed that cancer cells also take up large quantities of glutamine (Eagle 1955). Given that glucose and glutamine constitute major fuels for central carbon metabolism, the critical question to address is: what is their metabolic fate? Glucose-derived pyruvate enters the tricarboxylic acid (TCA) cycle and participates in the mitochondrial citrate which is then transported to the cytoplasm and fuels the *de novo* fatty acid synthesis pathway. This pathway results in a loss of citrate from the TCA cycle, and it was termed cataplerosis. The concept was introduced by H. Kornberg (1965), which defines a pathway that removes carbon from the TCA cycle. Given that there is no net loss or gain of TCA cycle carbon mass, there are pathways that balance the consumption of TCA cycle intermediates, and they were termed anaplerosis.

In addition to glucose, glutamine is also an essential nutrient, and excessive amounts are taken up by cancer cells. A mitochondrial-associated enzyme converts it to glutamate which in turn is converted to α-ketoglutarate and enter the TCA cycle, which represents an anaplerotic pathway (see the concept above).

The concept of metabolic reprogramming as a cancer hallmark suggests that the mutations responsible for cancer initiation enable cells in nascent tumors to acquire metabolic properties that support cell survival, evasion of immune surveillance and unregulated growth. For instance, among the two major pathways involved in cell growth and metabolism, the phosphatidylinositol-3-kinase (PI3K/Akt) and the extracellular signal-regulated kinase-mitogen-activated protein kinase (Ras/ERK); the former is perhaps the most common lesion in spontaneous human cancers, and its activation leads to enhanced glycolysis (Buzzai et al. 2005). In this chapter, three transcription factors, c-Myc, HIF-1α and p53, which coordinately regulate some aspects of metabolism in proliferative cells (normal and cancer cells), are briefly discussed (Yeung et al. 2008; Gomes et al. 2018). The following review is suggested for further reading (Ward and Thompson 2012).

* *Role of Myc*

In normoxia, Myc stimulates genes that increase the transport of glucose, its catabolism to pyruvate and ultimately to lactate (Shim et al. 1997; Elstrom et al. 2004; Stine et al. 2015). This observation suggests that Myc could contribute to the Warburg effect under adequate oxygen tension. In hypoxic conditions, HIF-1 (see below the regulation of its activation) also regulates the same set of genes than Myc which highlights the interplay between these two transcription factors (Figure 4.16).

Myc-transformed cells exhibit a stimulation of glutaminolysis, which represents the secretion of glutamine-derived lactate (Wise et al. 2008).

* *Role of HIF*

HIF-1α seems to play an essential role in increasing glycolytic flux by stimulating the transcription of several genes of glycolysis for promoting both energy source and building blocks of nucleic acids (Semenza 2011, 2013, see also Figure 4.16). It is worth noting that HIF-1α induces nucleotide biosynthesis by promoting the flux of glucose carbon through a non-oxidative arm of pentose phosphate pathway (Zhao et al. 2010). It also decreases oxidative phosphorylation by upregulating PDK, which leads to the suppression of PDH.

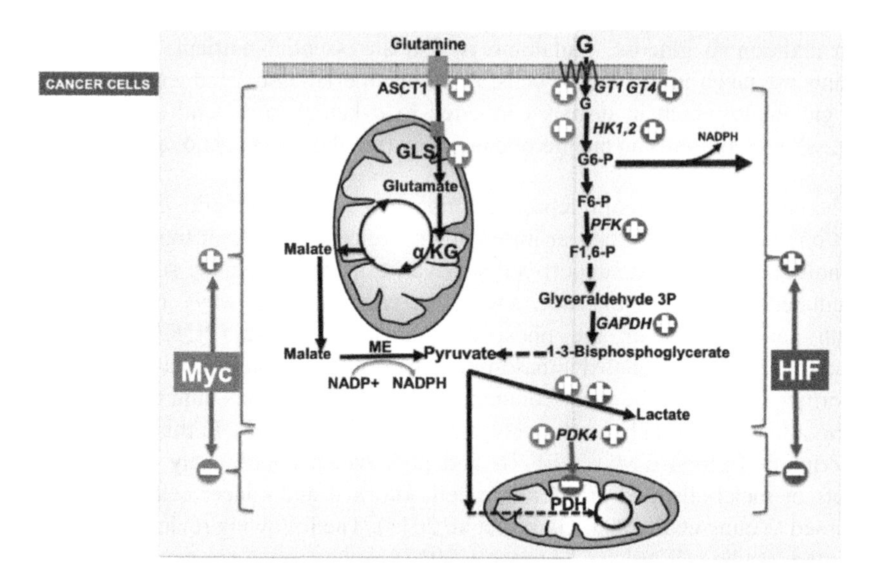

Figure 4.16. *Role of Myc and HIF-1 in the control of glycolysis and mitochondrial oxidative phosphorylation, in proliferative cells (normal, and cancer cells). Myc also contributes to increased lactate production through a stimulation of glutamine utilization*

* *Role of p53*

The p53 protein has largely been defined as tumor suppressor, but recently other roles have been reported, including the control of cellular metabolism during nutrient and oxygen stress. For instance, perturbations in metabolite availability are rapidly relayed to p53 through its tight interaction with key nutrient sensing pathways of the cell, such as AMPK and mTOR-signaling network. Thus, during starvation, these sensors transiently engage p53 activity, which increases cell survival and promotes efficient nutrient utilization. Similarly, during transient hypoxia, a signaling crosstalk between p53 and HIF-1 facilitates redirection of cellular metabolism towards energy generation through non-oxidative pathways. However, in deleterious situations, p53 can engage the death of damaged cells.

Unlike Myc and HIF-1, mentioned above, p53 represses glycolytic flux through the transcription of genes regulating this flux leading to a reduction of pyruvate production (Figure 4.17) (Levine and Puzio-Kuter 2010). It has been shown that p53 is mutated in over half of all human cancer p53 mutations, leading to a loss of function of the protein. It strongly suggests that the lack of functional p53 contributes to the increased glycolysis, i.e. the Warburg effect. For a comprehensive recent overview of this field, the reader is referred to Humpton and Vousden (2016).

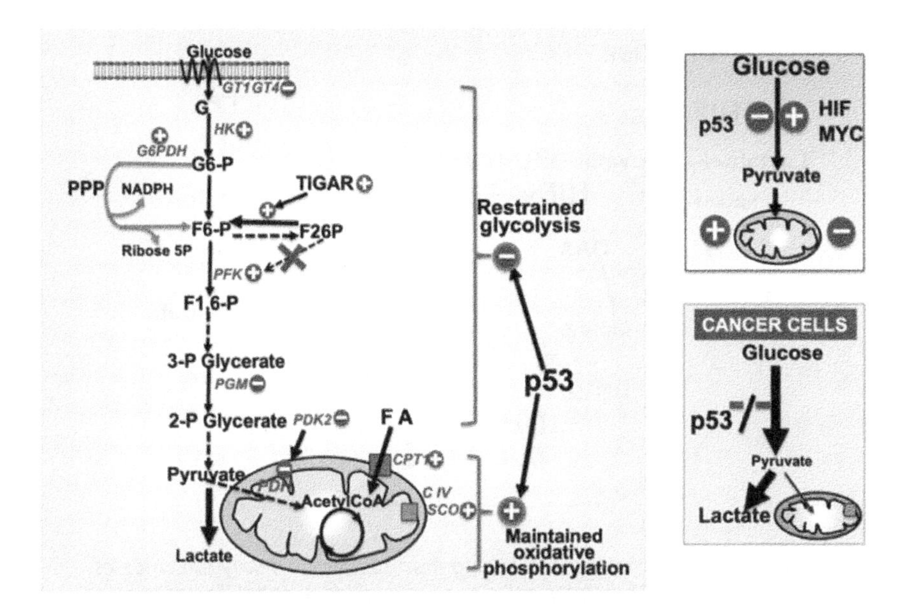

Figure 4.17. *Role of p53: it restrains the glycolytic flux and maintains oxidative phosphorylation. The gene encoding p53 is mutated in about 50% of cancer cells and it likely participates in the Warburg effect*

* *Role of mTOR*

It exists in two functionally distinct complexes 1 and 2, but mTORC1 seems to have the most direct influence on the energy balance. Both signaling pathways, mentioned above, are potent activators of mTORC1, which is considered as the master regulator of anabolic reprogramming. mTORC1 is also a major negative regulator of autophagy (role of autophagy in the next section). In the presence of nutrient excess, mTORC1 inhibits autophagy and the opposite response prevails in the presence of nutrient deprivation (Saxton and Sabatini 2017). In other words, autophagy feedbacks to regulate mTORC1.

Recent evidence suggests that metabolites themselves can be oncogenic by altering cell signaling and blocking cellular differentiation. For instance, somatic mutations in $NADP^+$-dependent isocitrate dehydrogenase I gene (IDHI) confers to the enzyme a new reductive activity, allowing a conversion of α-ketoglutarate to 2-hydroxyglutatrate (2HG). This end-product is found in trace amounts in normal cells, but it is synthetized at a high rate in acute myeloid leukemia. Whether or not 2HG plays a role in tumorigenesis remains uncertain (Ward and Thompson 2012, see also Figure 4.18).

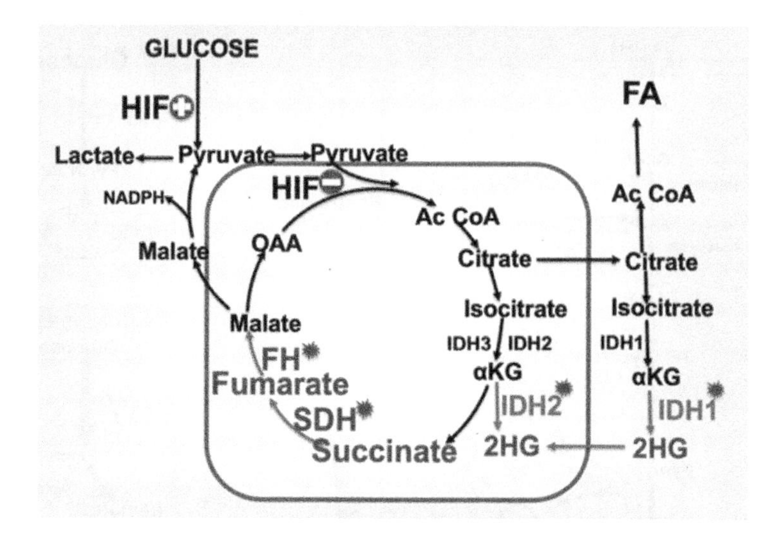

Figure 4.18. *Oncometabolites arising from mutations of key enzymes of the Krebs cycle. They alter cell signaling and block cell differentiation*

4.3.6.5. *Autophagy*

The term "autophagy" (from the Greek, "auto" oneself, "phagein" to eat) was already in use in the 1860s, which referred to a concept that the human body has a self-nourishment system that enables the individual to survive by feeding off itself in response to nutrient deprivation (Ktistakis 2017). In 1955, Christian de Duve discovered the lysosomes and in 1963 used the term autophagy to describe the cellular process through which cytoplasmic materials are delivered to the lysosome for degradation (Klionsky 2008). Substrates of autophagy include soluble factors, membrane-bound organelles, extracellular material and portions of the plasma membrane. Three major types of autophagy are used: macroautophagy, microautophagy and chaperone-mediated autophagy and refer to any cellular degradative pathway that involves the delivery of cytoplasmic cargo to the lysosome (reviewed in Mizushima (2018)). This mechanism helps cells to recycle their endogenous materials and builds macromolecules to maintain cellular homeostasis. In the absence of stress, basal autophagy serves a housekeeping function; it eliminates damaged components that could otherwise become toxic. This mechanism is important in quiescent differentiated cells, where damaged components are not diluted by cell replication. The presence of stressors induces autophagy which then degrades protein aggregates, oxidized lipids, damaged organelles and intracellular pathogens. Defects in autophagy are linked to liver disease, neurodegeneration, aging and cancer.

During the 1960s–1990s, detailed descriptions of the ultrastructural morphology of autophagy-related structures were published. In the 1990s, Y. Ohsumi identified a great number of autophagy-related genes termed APGs, in yeast (Takeshige et al. 1992). In 2003, the nomenclature of these genes was unified under the name "autophagy-related genes" (ATGs) and they are conserved from yeast to mammals and regulate the autophagy of intracellular cytoplasm proteins, and organelles. In 2016, Ohsumi was awarded the Nobel Prize in Physiology or Medicine for his work in this field.

It is worth noting that the effects of autophagy in cancer cells are paradoxical; they can prevent the initiation of some tumors, but they may also support tumor cell survival in hypoxic tumor regions associated with lowered nutrient supplies (Degenhardt et al. 2006).

It should be emphasized that many open questions remain to be addressed such as: (1) what are the molecular functions of the ATG proteins?; (2) how can autophagy be selective?; and (3) what are the functional and pathophysiological effects of autophagy at the tissue level?

5

Epigenetics and Circadian Rhythms: Role of Environmental Factors

5.1. Epigenetics: general overview

The purpose of this chapter is to present an overview of the roles played by epigenetics in many physiological and pathophysiological situations. In the 19th century, one of the pioneers of genetics, G. Mendel, demonstrated that inheritance could be studied on its own without including development. As mentioned earlier, in Chapter 1, in the history of endocrinology, neurology and immunology, biologists considered that there were no links between these disciplines, and the same reasoning prevailed between genetics and developmental biology disciplines. Towards the middle of the 20th century, biologists realized that these disciplines were indeed related and should come together in a common discipline.

Before discussing epigenetics, it is noteworthy to recall a brief history of the concept of the gene, which has evolved continuously since 1909, when the word was first used (Johannsen 1909). At first, the gene was indeed just a concept because it was unclear whether it existed in a physical form, and if it did, there was uncertainty as to its chemical nature. In the 1930s, the gene was regarded as the indivisible unit of inheritance, subject to mutations and genetic recombination. In this period, the prevailing view was that only proteins, unlike nucleic acids for instance, were sufficiently complex to be able to contain genetic information. In the 1940s, experiments carried out in yeast enabled them to link the concept of the gene to the synthesis of a given enzyme (Beadle and Tatum 1941), followed by a key discovery by Avery and colleagues that DNA carries the transmissible information that determines the virulence of pneumococcus (Avery et al. 1944). In 1953, it was a

---

For a color version of all the figures in this chapter, see www.iste.co.uk/gilbert/concepts.zip.

turning point when Watson and Crick solved the structure of DNA. It became widely accepted that inheritance resides in DNA. Then, until the middle of the 1970s, the co-linearity involving gene, RNA and protein was the standard model. The introduction of new technologies led to discoveries and the concept of the gene evolved. For instance, the discoveries of (1) the absence of co-linearity and the existence of a surplus of genetic material repeated genes that do not code for proteins (Chow et al. 1977), (2) DNA sequences able to move within a genome (see McClintock's discovery below) and (3) the overlapping genes, etc. In parallel to these novel discoveries, studies focused on the tridimensional structure of DNA, suggesting that in terms of hereditability, information can also be associated with other elements beyond a simple linear sequence of nucleotides. In other words, it is necessary to consider the structural-phenotypic aspect of DNA; this is discussed below.

In 1942, C. Waddington coined the term "epigenetics" as the crosstalk between genetic information and the environment. In other words, the term refers to the ways in which the developmental environment can influence the mature phenotype. He suggested the existence of mechanisms of inheritance in addition to (over and above) standard genetics (Waddington 1940; Bard 2008). Until the 1950s, the word epigenetics was used to define the development events leading from the fertilized zygote to the mature organism, but the original question was: How can a single fertilized egg give rise to a complex organism with cells of different phenotypes? After the discovery of DNA structure in 1953, D. Nanney suggested that the phenotype was driven by "genetic systems" and "epigenetic systems" (Nanney 1958). The existence of phenotypic differences between cells with the same genotype underlines that the expressed specificities are not determined entirely by "genetic systems".

During the mid-1970s, R. Holliday and J. Pugh proposed that DNA methylation was the molecular mechanism behind Waddington's hypothesis (Holliday and Pugh 1975). Subsequently, it was suggested that the chemical or structural modifications of chromatin, i.e. complexes of DNA and proteins, mainly histones, might influence gene expression. It is worth noting that the concept of chromatin was developed by cytologists and biochemists and the term "chromatin" was introduced by W. Flemming in 1879. Interestingly, its molecular composition was resolved much later and research on covalent modifications of DNA and histone modifications developed from one another for about two decades. In the 1990s, increasing evidence showed that genes are regulated not only by classical signaling pathways but also by mechanisms that are independent of changes in a DNA sequence. This new regulation led to the concept of epigenetics, which literally means "above genome". The epigenetic marks or molecular factors (described below) represent the epigenome that can be modified by environmental factors and, in response, genes are either turned on or off (Wolffe and Matzke 1999; Tammen et al. 2013). In addition, these gene expression patterns influenced by environmental factors can lead to phenotypical changes throughout the lifetime leading to the potential risk of some diseases without any change in the nucleotide sequence of DNA. Recent studies have also revealed that environmental exposures modify epigenetics marks that are mitotically stable (Gluckman et al. 2009). Furthermore, many observations demonstrate meiotic stability in these markers, which allows these epigenetic modifications to be transmitted to offspring via germline epigenetic inheritance (Franklin et al. 2010). It should be emphasized that the idea that particular characteristics acquired in response to environmental exposure can be transferred from parents to progeny has its roots in the doctrines of Hippocrates (Adams 1891) which were later propagated by Jean-Baptiste Lamarck (Lamarck 1963) and others. The reader is encouraged to consult a review reporting a detailed discussion of the history of epigenetics (Choudhuri 2011, see Figure 5.1).

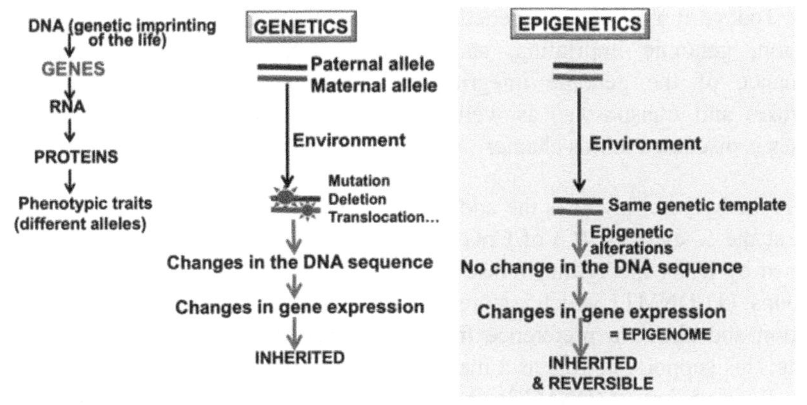

Figure 5.1. *Schematic illustration of the differences between genetics and epigenetics*

It should be emphasized that initial observations of how the environment can influence epigenetics and phenotype were shown in plants. In this field, Mendelian genetics dominated, despite exceptions which were reported, and as it is often the case they were left aside and considered somewhat secondary aspects. The discovery by McClintock in the 1940s of genetic transposition in maize was not in line with the Mendelian inheritance, and other cases were classified under the name "paramutation" (Brink 1956). This observation was a peculiarity of the plant, but the development of molecular biology tools demonstrated that in mice, heredity also could not be completely explained by the Mendelian rules (Rassoulzadegan et al.

2006). In recent years, multiple lines of evidence indicate that paramutation is a heritable epigenetic change and the plant and the mouse share significant common features.

5.1.1. *Epigenetic modifications of DNA and regulation of biological processes*

Over the last 20 years, studies have aimed to investigate the "epigenetic effects" of DNA methylation, histone modifications and non-coding RNA (ncRNA) in mammals. They will be discussed briefly.

DNA methylation

DNA methylation was first discovered in thymus DNA by R.D. Hotchkiss (1948). Today, it is a key epigenetic process involved in the control of gene expression, genomic imprinting, stabilization of X chromosome inactivation, maintenance of the genome integrity through protection against endogenous retroviruses and transposons, as well as carcinogenesis. Greater detail of these processes is discussed in this chapter.

DNA methylation involves the addition of a methyl (-CH_3) group to both DNA strands at the 5'-cytosine (C) of CpG dinucleotides. This covalent modification is performed by DNA methyltransferase (DNMT) enzymes, which are classified into two groups: (1) DNMT1 which restores the parental methylation pattern after DNA replication and shows a preference for hemimethylated DNA over unmethylated substrate; this supports its role as a maintenance methyltransferase, and (2) *de novo* DNA methyltransferases (DNMT3a and 3b) which perform new methylations due to passive and active demethylations (Figure 5.3). Although numerous experiments have correlated DNA methylation with gene repression and demethylation with gene activation, it was necessary to test whether DNA methylation has a causal effect on gene expression. Using the DNA-mediated gene transfer technique, it was shown that unmethylated genes are actively transcribed when inserted into the genome, whereas the exact sequences are repressed if the inserted DNA had been premethylated *in vitro*. Because the only difference between these templates is the presence of methyl groups, these data provided convincing proof that DNA methylation itself is responsible for gene inhibition (Pollack et al. 1980). This concept has been reinforced by subsequent experiments carried out in a transgenic model (Siegfried et al. 1999). However, the DNA methylation maintenance model, mentioned above, does not explain all experimental observations. A new concept has been presented, in which DNA methylation can be described by a dynamic stochastic model in which DNA methylation at each site is determined by the local rates of methylation (DNMTs) and demethylation (DNA demethylases) (Jeltsch and Jurkowska 2014).

DNA methylation represses transcription directly by inhibiting the binding of specific transcription factors and indirectly by recruiting methyl-CpG-binding proteins and their associated repressive chromatin remodeling activities (Figure 5.2).

Localization of DNA methylation

The majority of CpGs are concentrated in genomic regions called CpG islands, and in humans a small percentage acquires methylation during development. They are mainly located in three regions of the genome: (1) imprinted genes, which might account for approximately 1% of the total genes or approximately 300 genes, (2) transposable elements (TEs), also referred to as "jumping genes", have accumulated throughout millions of years as evolutionary ancient DNA in the form of transposons and retrotransposons repetitive sequences, called "transposons", that can be defined as "junk" DNA of viral infections which are reverse-transcribed long-terminal repeat (LTR) retroviruses (Waddington 2016). Given that they represent approximately 40% of the human genome, the bulk of DNA methylation in mammalian genomes is found at these sequences. It is critical to maintain these sequences in a transcriptionally silent stage through the key process of DNA methylation. However, any genes positioned near them can be at risk for inappropriate methylation, and (3) promoters of genes transiently transcribed.

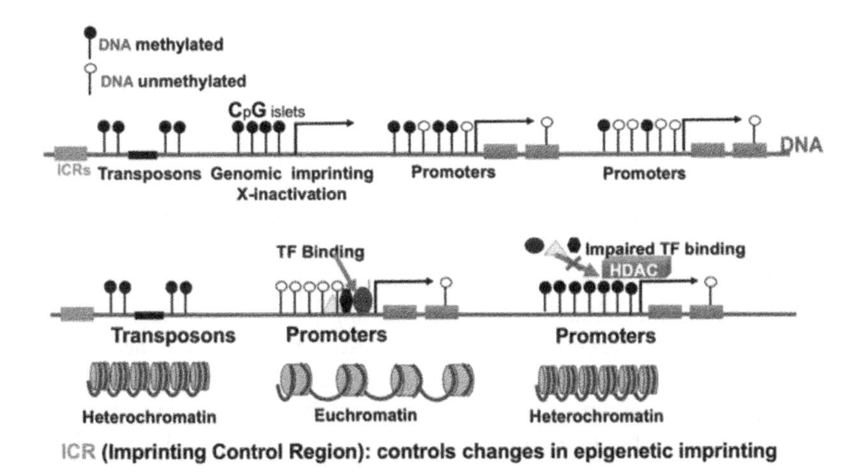

Figure 5.2. *Localization of DNA methylation: CpG islands are mainly located in three regions of the genome: (1) imprinted genes, (2) transposable elements and (3) promoters of genes transiently transcribed*

Notably, unmethylated CpG islands are located within genomic targets that are likely to be susceptible to gene expression changes, such as gene promoters of actively transcribed housekeeping genes, transposable elements that lie adjacent to genes and regulatory elements of imprinted genes. In contrast, CpG islands of silent genes are predominantly methylated (Figure 5.2).

There are many remaining questions such as: Do all CpG islands colocalize to sites of transcriptional initiation? Do tissue-specific methylation patterns have a mechanistic role in differentiated cells?

It should be emphasized that DNA methylation does not turn off genes, in the same way that the demethylation itself does not turn them on. "Basically, the transcription is first turned on, followed by the demethylation process" (Deichmann 2016).

Histone post-translational modifications

In the early 1960s, at a time corresponding to the end of the golden age of metabolism discoveries, V. Allfrey became interested in the potential role of chromatin in the regulation of gene expression. He showed that acetylation of histones lowered their ability to inhibit RNA synthesis and proposed "dynamic and reversible mechanisms for activation as well as repression of RNA synthesis" (Allfrey et al. 1964). Later, in the 1980s, it was reported that histone acetylation is not a consequence of transcription but rather a general enabling step for transcription (Hebbes et al. 1988). In 1996, the first histone acetyltransferase (HAT) was identified (Brownell et al. 1996), thereby validating Allfrey's hypothesis and providing a direct link between transcription. The question was then, if histone acetylation enables transcriptional activation, then the regulation of transcription requires the existence of an opposite mechanism, enabling histone deacetylation. In the same year, the first histone deacetylase (HDAC) was identified (Taunton et al. 1996). Acetylation of specific lysine residues in histone tails results in chromatin decompaction (i.e. euchromatin), thereby allowing access to transcription factors and other transcription co-activators. Inversely, deacetylation results in chromatin compaction (i.e. heterochromatin), thereby repressing gene transcription (Figure 5.2). Given that HATs add acetyl groups and HDACs remove these acetyl marks, resulting in chromatin remodeling, they act as "writers" and "erasers", respectively (Figure 5.3). Another group of proteins is also involved in chromatin remodeling and called "readers", because they contain a bromodomain (BRD) which binds acetyl marks in histone tails and recruits transcriptional machinery (Fujisawa and Filippakopoulos 2017) (Figure 5.3).

Non-coding RNA (ncRNA) molecules

Beyond these chemical modifications previously mentioned, emerging evidence for transcriptional gene silencing by the binding of small non-coding RNAs (ncRNAs) added a further dimension, which is exemplified with the X-chromosome inactivation (see section below). Although eukaryotic genomes transcribe up to

75%, approximately 3% of these transcripts encode for proteins; the majority are ncRNAs, which are classified according to size and function. They regulate gene expression at several levels: transcription, mRNA degradation, splicing and translation. They include microRNAs (miRNAs), small interfering RNAs (siRNAs) and long non-coding RNAs (lncRNAs). In addition to their gene silencing role, they participate in DNA methylation and histone modification, thereby altering chromatin conformation. For instance, miRNA in human blastocysts correlates with the maintenance of pluripotency in embryo development (see the review by Wei et al. (2017), Figures 5.3 and 5.6).

In summary, an epigenetic system should be heritable, self-perpetuating and reversible.

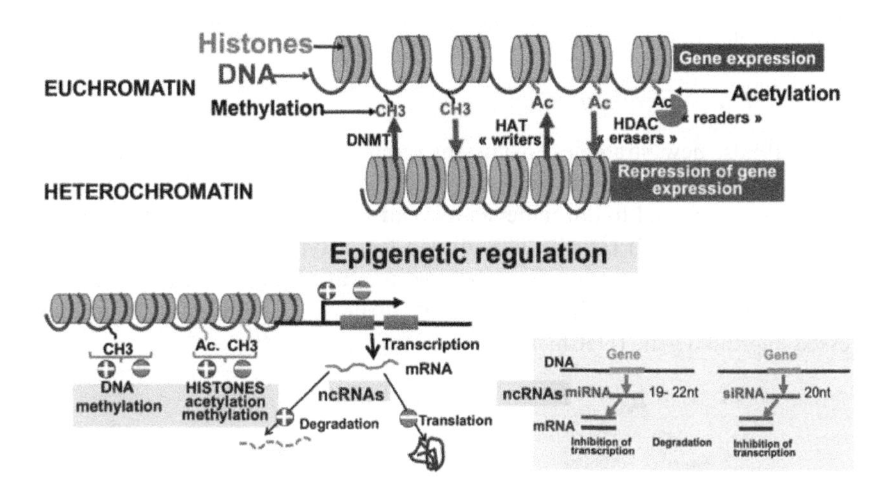

Figure 5.3. *Summary of epigenetic regulation: DNA methylation, histone modifications (acetylation and methylation) and non-coding RNAs*

Two of the most comprehensively studied, epigenetically regulated phenomena are genomic imprinting and X-chromosome inactivation, and they illustrate transcriptional silencing by DNA methylation. As discussed earlier, epigenetic changes can be inherited mitotically in somatic cells, indicating that environmental effects on the epigenome can have a long-term effect on gene expression, and epigenetic alterations can also be inherited transgenerationally.

In this context, a series of data, which are reported below, highlight that epigenetics changes provide an additional layer of gene regulation mechanisms in addition to genetics without changes in genotypes, but changes in phenotypes.

5.1.2. *Genomic imprinting*

In the 1980s, the concept of differential functioning of the maternal and paternal genomes was put forward. In 1984, a first study, using nuclear transfer technology, provided the first clue regarding the existence of genomic imprinting in mice. The experimental approach consists of a nuclear transplantation, which is achieved by using a fertilized egg that had been enucleated, followed by a transplantation of two maternal or two paternal pronuclei, from a fertilized egg, or a biparental pronuclei (wild type). Transplantation of two female pronuclei or two male pronuclei does not permit a complete normal embryogenesis. It was concluded that maternal and paternal genomic contributions are not functionally identical, and mouse embryogenesis requires both the maternal and paternal genomes (McGrath and Solter 1984; Surani et al. 1984). The difference between the parental genomes was believed to be due to gamete-specific differential modification (Figure 5.4(A)).

5.1.2.1. *Imprinted genes: Igf2r, Igf2*

In the 1990s, new strategies, using emerging technologies in mouse genetics, including positional cloning and gene knockout technology, led to the identification of genes that contributed to our understanding of the mechanism of imprinting. The first of these genes, Igf2r (insulin-like growth factor type 2 receptor), which binds Igf2, was identified as a maternally expressed imprinted gene (Barlow et al. 1991) (Figure 5.4(B)). A few months later, the Igf2 gene was identified as a paternally expressed imprinted gene (DeChiara et al. 1991) (Figure 5.2(B)).

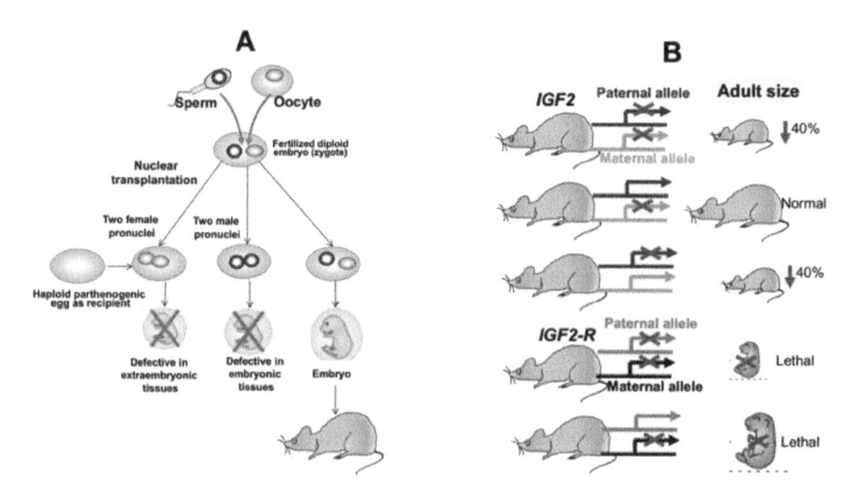

Figure 5.4. *Maternal and paternal genomic contributions are not functionally identical (A). Identification of imprinted genes, Igf2r and Igf2 (B)*

In the process of genomic imprinting, DNA methylation is one of the main mechanisms that marks either the paternal or maternal allele of certain genes for differential expression in various tissues in the offspring. Genomic imprinting is initiated in the germline and leads to the transcriptional silencing of one allele, usually called imprinted (Figure 5.5(A)).

One of the hallmarks of imprinted genes is that many are found in clusters throughout the genome, and they are regulated through an imprinting control region (ICR). To demonstrate the critical role of this region, a deletion of ICRs leads to a loss of imprinting of multiple genes within the cluster.

5.1.2.2. *X inactivation*

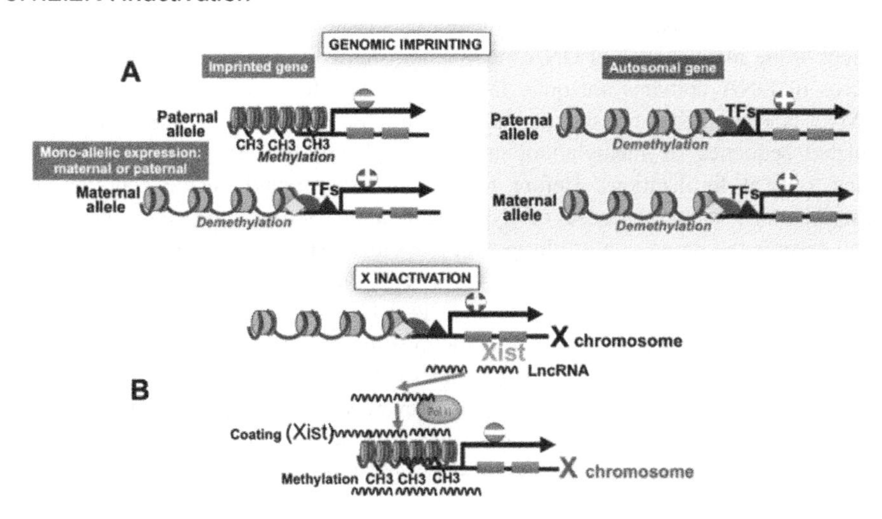

Figure 5.5. *Summary of genomic imprinting (A) and X inactivation (B)*

In 1949, M. Barr showed that the sex of cat cells could be deduced by a subnuclear structure, now called the "Barr body" (Barr and Bertram 1949). Then, S. Ohno demonstrated that the Barr body is a condensed X chromosome (Ohno et al. 1959), and it is the result of whole-chromosome silencing. We now know that female mammals must compensate for their double dosage of X-chromosome genes compared with XY males. To compensate, the majority of the genes on one of the female X chromosomes are quickly turned off, in early embryonic development, to equalize the expression of X-linked genes. Notably, the X chromosome is home to nearly 1,000 genes in humans. These genes are transcriptionally silenced in a process called X chromosome inactivation (XCI). The silencing of the X chromosome is controlled by the X chromosome inactivation center (XIC), which includes genes that produce long non-coding RNAs to silence genes in cis (review Starmer and

Magnuson (2009)). One of them, Xist (X inactive specific transcript), plays a critical role to regulate XCI, as its deletion abolishes X silencing (Penny et al. 1996). Xist coats the chromosome to inactivate it, and also induces a cascade of chromatin changes, including DNA methylation and histone modifications (see Figure 5.5(B) and the review by Gendrel and Heard (2014)). It should also be emphasized that this chromosome-wide silencing occurs during a developmentally restricted time-window, i.e. a critical time/period which is a concept often invoked.

5.1.3. *Setting and maintenance of DNA methylation*

DNA methylation patterns take place during gametogenesis, and they are different in the genomes of female and male gametes. This issue will be discussed below. The concept of the maintenance of DNA methylation stems from the fact that methylation patterns of DNA domains are quite faithfully propagated during development, and DNMT1 is responsible for this stability (Pfeifer et al. 1990). It implies a tightly regulated sequence of methylation and demethylation processes, as soon as life begins, (i.e. at fertilization). Before fertilization, sperm and oocyte genomes are transcriptionally silent, because of hypermethylation of their respective genomes, which ensures the repression of pluripotent markers. It should be emphasized that the gametes arise from primordial germ cells (PGCs). Upon fertilization, extensive epigenetic reprogramming takes place, and the two gametes reorganize their cellular and molecular signatures by an active demethylation, a process called erasure, and it suppresses almost all the methyl marks inherited from the parents and establishes totipotency of whole cells (Figure 5.6). However, some differences in methylation patterns remain at imprinted loci (see genomic imprinting, below), which are responsible for allele-specific gene expression and these differences between alleles remain throughout development. These events are reported in Figure 5.7 and represent a maternal-imprinted allele, which is maintained during embryonic development and throughout life. In the primordial germ cells of the embryo before sex determination, the process of erasure of methylation of the imprinted gene(s) takes place. Then, in the process of gametogenesis, there is a gametic imprint re-establishment of methylation according to the sex of the individual (Tycko 1997).

The global demethylation in the preimplantation embryo enables an activation of key early embryonic genes, such as Oct-3/4 and Nanog. About the time of implantation, the DNA methylation patterns are set up all over the genome and it is mediated by *de novo* methyltransferases (DNMT3a and 3b, mentioned earlier). Following implantation, the genome undergoes additional changes in methylation, and they are all of a tissue-specific or gene-specific nature (review Fraser and Lin (2016)). Shortly after implantation, the inactivation of one X chromosome occurs in female embryos (see the detailed process in Figure 5.5). It thus appears that the cycle of erasure followed by the reestablishment of a new methylation pattern is a basic aspect of epigenetic regulation

(Figure 5.7). However, the key question is whether methylation patterns can also be influenced by environmental cues, either during embryogenesis or even afterward in the newborn adult organism. These issues are addressed in the next sections.

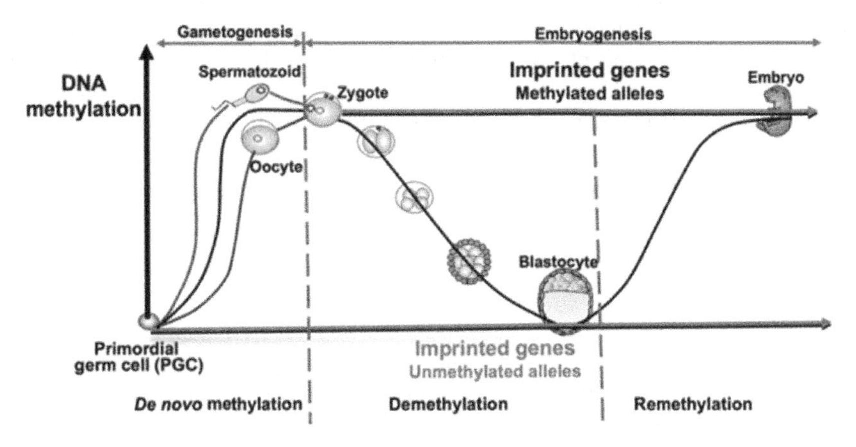

Figure 5.6. *Setting and maintenance of DNA methylation. Detailed description of the process is reported in section 5.1.3*

Figure 5.7. *Germline establishment of imprints: during oogenesis or spermatogenesis, pre-existing methylation needs to be erased in the primordial germ cells. Then, either a maternal or a paternal allele is methylated during the gametogenesis. One example of imprinting is depicted and represents maternally derived DNA methylation*

5.1.4. *Evidence for non-genomic inheritance: epigenetic mechanisms*

It is widely accepted that epigenetic modifications play a critical role in the regulation of gene expression and therefore contribute to the phenotype. These marks are erased in the germline and re-established at each generation. However, in the A^{vy} mouse model, recent reports indicate that: (1) the clearing of epigenetic modifications is incomplete and these modifications can be inherited across generations, and (2) epigenetic states can be modified by the environmental factors. These findings have renewed an interest from the scientific community in Lamarckism. Here is the best evidence for transgenerational epigenetic inheritance in the A^{vy} mouse.

A^{vy} *mice and agouti*

The *agouti* (*a*) gene derives its name from the South American mammals, *Agouti paca* and *Agouti taczanowskii*, which have the same grizzled coat pigmentation pattern as that conferred by the *agouti* gene in mice. Numerous phenotypically distinguishable alleles make the agouti locus unusual and fascinating to study at the genetic and molecular levels. The agouti viable yellow (A^{vy}) is one of the best-characterized alleles (reviewed in Wolff (1987)). These alleles are variably expressed, in the absence of genetic heterogeneity, and their activity is dependent upon their epigenetic state (methylation). Such alleles that can stably exist in more than one epigenetic state and that result in different phenotypes are termed "epiallele".

The *agouti* (a) gene encodes a paracrine signaling molecule that promotes follicular melanocytes to produce yellow pheomelanin rather than black eumelanin pigment. Transcription of the *agouti* (a) allele normally occurs in the skin, and there is a transient expression in hair follicles during hair growth which results in a yellow band on each black hair, causing the brown wild-type agouti fur (Wolff et al. 1986). Notably, if the agouti locus is null (A allele), it produces black mice (Figure 5.8(A)).

The *agouti* A^{vy} allele is the result of a spontaneous insertion of an intracisternal A-particle (IAP) retrotransposon upstream of agouti gene and promotes constitutive ectopic expression of agouti protein. A^{vy} mice have coats that vary in a continuous spectrum from full yellow, through mixed yellow/agouti to full agouti (termed pseudoagouti). The mechanism underlying the continuous phenotypic spectra is epigenetic in origin, due to the presence of IAP which renders the genomic region epigenetically labile. When IAP is silent (hypermethylation), agouti is expressed in its normal pattern. Inversely, hypomethylation at IAP results in pseudoagouti fur and partial methylation gives rise to mottling (Morgan et al. 1999; Rakyan et al. 2003) (Figure 5.8(B)).

Then, the question is whether or not these epigenetics modifications leading to different phenotypes are inherited because they are generally considered to be cleared on passage through the germline in mammals, so that only genetic traits are inherited. It was demonstrated that the distribution of phenotypes among offspring is related to the phenotype of the dam. In other words, when A^{vy} displays the agouti phenotype, her offspring displays the same phenotype. It results from incomplete erasure of an epigenetic modification. These observations demonstrate the inheritance of an epigenetic mark, i.e. the range of coat colors of the offspring correlates with the coat color of the dam (Morgan et al. 1999; Rakyan and Whitelaw 2003). Taken together, these findings demonstrate that the A^{vy} allele displays a labile epigenetic state in relation to differential activation of IAP. The resulting-agouti mice are a clear example of epigenetics modulating genetics. It challenges the dogma stating that heritable traits are only conferred by the sequence of DNA transmitted from parent to offspring.

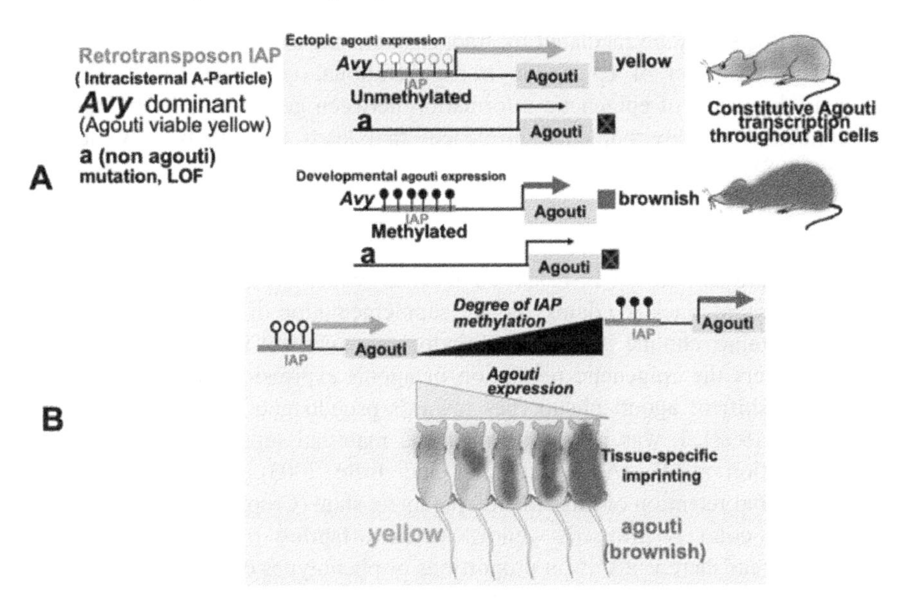

Figure 5.8. *The* A^{vy} *allele displays a labile epigenetic state in relation to differential activation of IAP.* A^{vy} *mice have coats that vary in a continuous spectrum from full yellow, through mixed yellow/agouti to full agouti (termed pseudoagouti). The mechanism underlying the continuous phenotypic spectra is epigenetic in origin, due to the presence of IAP*

Studies carried out in animal models and in humans support the notion that many adult-onset diseases are linked to intrauterine exposures, namely in the presence of both maternal undernutrition and overnutrition. The following sections describe how these adverse exposures that occur at the critical period of development are associated with metabolic abnormalities, including obesity, type 2 diabetes and cardiovascular disease, in later life. The investigators have therefore sought to understand potential programming mechanisms of these diseases by exploring epigenetic changes, throughout the life of offspring.

5.1.5. *Nutritional influences on developmental epigenetics*

Numerous studies show that early environment during development is a strong predictor of phenotype and disease in later life. These environmental influences can be dietary, physical, chemical or stochastic and can have effects beyond a single generation, which led to the concept of "maternal effects" (Mousseau and Fox 1998). Such heritability does not result from classical genetic inheritance, and numerous studies carried out in animals and humans have provided strong evidence that these effects may be mediated by transgenerational transmission of epigenetic marks. The definition of epigenetic transgenerational inheritance is "germline-mediated inheritance of epigenetic information between generations in the absence of continued direct environmental influences that leads to phenotypic variation" (Skinner 2008).

Animal and human studies

Studies were carried out in black a/a females, which were date-mated to A^{vy}/a males. In pregnant black *a/a* dams, dietary supplementation of methyl donors (folic acid, B12, betaine, choline isoflavones (phytoestrogens/soy)) from 5 to 15 days of pregnancy alters the epigenetic regulation of agouti expression in their offspring, leading to a shift of agouti phenotypes towards pseudoagouti in both F1 and F2 (Wolff et al. 1998). It was later confirmed that maternal supplementation induces hypermethylation at A^{vy} IAP (Waterland and Jirtle 2003), and it leads to an intragenerational retention of this altered epigenetic state (Cropley et al. 2006). Note that the coat color is primarily yellow in the offspring from unsupplemented mothers, whereas there is a shift in proportions of phenotypes of A^{vy} in the offspring from supplemented mothers (Figure 5.9). This is the best experimental model, making a molecular link between environmental influence (nutrition) during a critical period of pregnancy and a change of a phenotype over a long term. The question is whether this epigenetic modification is inheritable. Interestingly, the supplementation of dams with the soy phytoestrogen genistein before and during pregnancy induced offspring A^{vy} hypermethylation comparable to that caused by methyl donor supplementation (Dolinoy et al. 2006). It is worth noting that A^{vy} mice are extremely obese, due to ectopic *agouti* expression in the hypothalamus. The agouti protein binds to the melanocortin-4 receptor (MC4R) and antagonizes the endogenous ligand α-MSH (α-melanocyte-stimulating hormone) which negatively regulates the energy intake.

Similarly, in pregnant rats, nutritional manipulations induce transgenerational non-genomic inheritance and cause specific epigenetic changes in DNA methylation or histone acetylation in the offspring. For instance, dietary protein restriction in pregnant rats induces DNA hypomethylation, which correlates with the increased expression of glucocorticoid receptor and peroxisome proliferator-activated receptor genes in the liver of the F1 offspring at weaning (Lillycrop et al. 2005). All these epigenetic changes were prevented if the protein-restricted diet was supplemented with folic acid, suggesting the importance of methyl group provision.

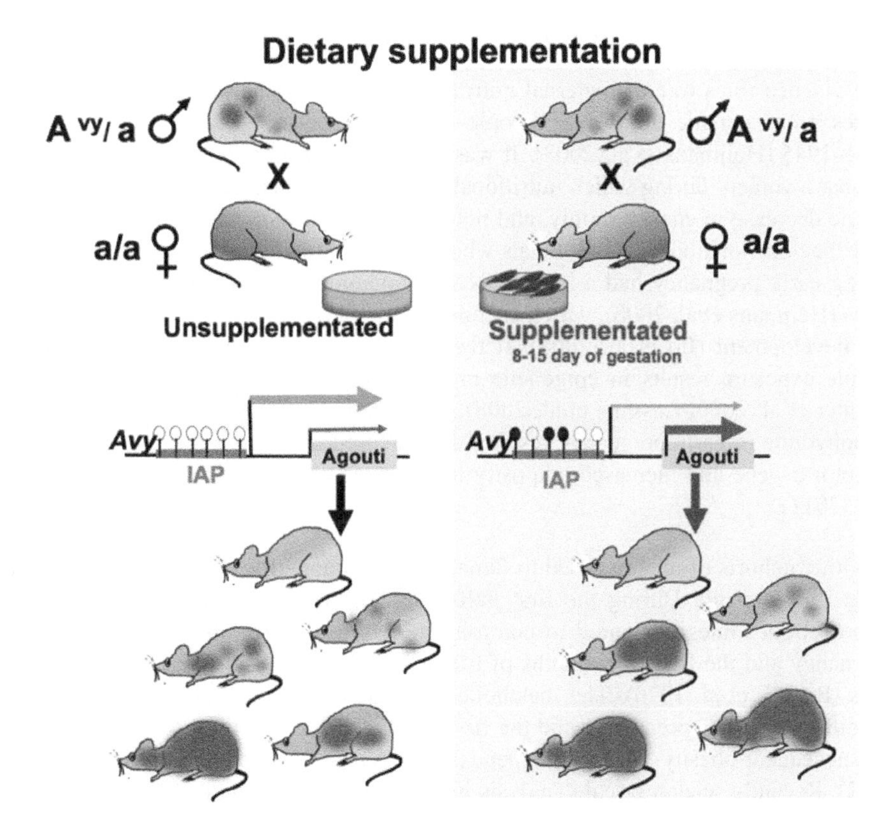

Figure 5.9. *Epigenetic regulation of the* A^{vy} *allele. The "pseudoagouti" phenotype of* A^{vy}*/a embryos is favored by feeding the mother methyl-supplemented diet during pregnancy*

Although studies performed in humans are not comparable to those previously discussed in mice, they also indicate that the effects of maternal diet persist throughout the life course. D. Barker hypothesized that imbalanced nutrient supply to the fetus can trigger changes in "fetal programming" associated with poor health outcomes in offspring (Barker et al. 1993). Since this hypothesis was proposed, epigenetics has emerged as an underlying mechanism explaining how environmental cues during pregnancy have the potential to influence epigenetics in utero and shape the offspring's propensity to chronic disease development later in life. To gain insight into how changes in fetal epigenetic programing influence long-term health trajectories, two opposite situations have been studied: maternal undernutrition and overnutrition.

5.1.5.1. *Maternal undernutrition*

Evidence for a role of maternal nutrition in regulating the offspring epigenome comes from genetic studies of people conceived during the Dutch famine of 1944–1945 (Heijmans et al. 2008). It was observed that there is a critical period in pregnant women, during which nutritional changes can cause effects far beyond the simple decrease in energy supply, and potentially over a long period of time (years) after the nutritional insult. Individuals who were 60 years old and exposed to famine during early pregnancy had a lower DNA methylation at the imprinted IGF2 gene locus (Heijmans et al. 2008), which encodes a growth factor involved in fetal growth and development (Ito et al. 2008). It therefore raises the possibility that prenatal famine exposure results in epigenetic programming that persist six decades later (Painter et al. 2006; Lussana et al. 2008). A recent study found that lower maternal carbohydrate in early pregnancy resulted in increased methylation of the retinoid X receptor α gene and increased adiposity in offspring at nine years of age (Godfrey et al. 2011).

Other cohorts of men exposed to famine (Dutch famine) were studied when they were 19 years old. During the first half of pregnancy, undernutrition resulted in higher obesity rates compared to controls. In contrast, during the last trimester of pregnancy and the first five months of life, undernutrition resulted in lower obesity rates (Ravelli et al. 1976). The metabolic response of the latter group supports the hypothesis that late pregnancy and the first months of life constitute a critical period for subsequent obesity. This rate depends on adipose-tissue cellularity (Brook et al. 1972). Recently, genome-scale analysis in this cohort revealed adult tissue-specific differentially methylated genes, and it is dependent on specific exposure to famine during pregnancy.

The Dutch famine also altered fetal growth, which of course led to a low birth weight (LBW) < 2.5 kg, and includes two groups of infants: small for gestational age (SGA) and intrauterine growth-restricted (IUGR). Regardless of the underlying cause, low birth weight is associated with poor health outcomes and genes that are related to metabolism and obesity are hypomethylated compared to controls (Diaz et al. 2020). The question is whether these epigenetic changes are the key mechanisms underlying the pathogenesis behind LBW. Recent studies suggest that the association between LBW and risk of type 2 diabetes could be explained by the combination of both genetic factors and fetal environment. Thus, the polymorphism of PPAR-γ2, a gene involved in the metabolic function of adipose tissue, is only associated with a higher risk of type 2 diabetes if birth weight is reduced (Eriksson et al. 2002). This finding points out a link between gene and environment. It should also be kept in mind that nutrient transport across the placenta directly influences fetal development. Hence, maternal undernutrition is known to lower transfer of substances between maternal and fetal circulations, and it might influence early developmental programming through epigenetic changes and secondarily predisposes to adult disease.

5.1.5.2. *Maternal overnutrition*

Intrauterine overnutrition occurs when pregnancy is associated with obesity or gestational diabetes, which can cause detrimental effects on offspring health.

Studies carried out in animal models and humans have clearly demonstrated that obesity cannot entirely be attributed to genetic component(s) (Rohde et al. 2019; Allum and Grundberg 2020). It was hypothesized that maternal obesity can alter fetal epigenome resulting in changes in fetal metabolic programming, which would extend to later life, thus leading to dysregulation of the metabolism and obesity (Barker et al. 1989; Hanson and Gluckman 2014).

It was therefore investigated whether a high-fat diet in pregnant rats resulted in epigenetic alterations in the liver and adipose tissue of the offspring (Masuyama et al. 2015). In adipose tissue, there was decreased acetylation and increased methylation of histone H3 in the promoter of adiponectin. These epigenetic modifications led to reduced adiponectin mRNA expression and plasma concentrations. These effects are reversed only after feeding a standard diet to three consecutive generations (Masuyama and Hiramatsu 2012), demonstrating that epigenetic modifications contributed to mRNA changes.

Besides histone modifications, differences in DNA methylation were observed in dams receiving a high-fat diet throughout pregnancy. Maternal obesity reduces DNA methylation in the promoter of a gene that encodes a protein regulating preadipocyte differentiation. Consequently, it increases its expression, leading to premature differentiation of preadipocyte into adipocyte. It reduces the pool of progenitor cells which therefore limits adipose tissue expandability. These epigenetic changes, DNA methylation and histone modifications in the adipose tissue of the offspring appear to be mechanisms influencing the offspring's phenotype in response to maternal obesity (Agarwal et al. 2018).

In addition to these epigenetic changes, the increased expression of miRNAs affects adipose tissue metabolism in the offspring of obese dams (Ferland-McCollough et al. 2012). It has been suggested that the mechanism might lead to insulin resistance in later life. Last, these studies were carried out in obese rodents, but similar clinical studies cannot be performed, mainly due to the impossibility to investigate samples of relevant organs. While gestational weight gain during whole pregnancy affects birthweight, only the first trimester of gestational weight gain affects child weight gain, suggesting that the first third of pregnancy duration has the biggest impact on childhood weight. Epigenetics is thought to mediate these effects, and epigenome-wide studies identified many differentially methylated sites in exposed offspring (Hjort et al. 2018).

As previously mentioned, the agouti mouse model shows a link between epigenetic changes in maternal diet and fetal outcomes. Thus, methyl supplementation of the mother's diet (from day 5 to day 15 of pregnancy) shifted the fur color from yellow to agouti and it is related to epigenetic modifications (Dolinoy 2008). Furthermore, there is a transgenerational obesity since the inhibition of the hypermethylation during development can prevent it, proving that the effects are mediated by epigenetics alterations that maternal obesity accumulates over several generations (Waterland et al. 2008). Another example showing that feeding soy-rich diets to pregnant *a/a* mice resulted in augmented DNA methylation during fetal development (epigenetic changes) that affected the fur color and also reduced obesity in the offspring (Dolinoy et al. 2006).

In summary, epidemiological studies and animal model experiments have demonstrated that obesity and diabetes during pregnancy are strongly associated with altered fetal growth and development as well as with adult disturbances in metabolic tissues. The mechanism behind these disorders is thought to be at least in part due to changes in the epigenome in response to these variations in nutritional conditions. Given the human life span, it is difficult to test the existence of transgenerational epigenetics effects.

Anecdotally, in bees (*Apis mellifera*), early life nutritionally induced changes are the underlying cause of queen and worker honeybee differentiation. Bee larva fed royal jelly, shifts development to the queen phenotype and DNA methylation is a primary mechanism by which it acts on the genome (Kucharski et al. 2008).

5.1.6. *Gut microbiome and epigenetic changes*

It is widely accepted that the gut microbiome can induce epigenetic changes in the host, which contributes to disease etiology (Wankhade et al. 2017). Depending on the composition of the gut microbiota, there is a great diversity of released metabolites from the bacteria. For instance, butyrate, which is mainly produced by *Bifidobacterium* and other strains, is a histone deacetylase that can influence histone modification processes (Hamer et al. 2008). In addition, butyrate can influence DNA methylation by regulating methyl-donor availability through modulation of folate production (Pompei et al. 2007).

In this context, it was interesting to investigate whether the maternal gut microbiota can shape microbiome profiles in the offspring gut (Koren et al. 2012). There is growing evidence that in human, maternal obesity or a maternal high-fat gestational diet alters the offspring gut microbiome through a maternal gut microbiota transfer *in utero* (Chu et al. 2016). Similarly, in pregnant rats, a high-fat diet induced changes in maternal gut microbiota (Paul et al. 2016). Thus, it is possible that the offspring can "inherit" special bacteria which would lead to epigenetic modifications through specific metabolite production. For instance, inhibition of histone deacetylase activity by butyrate influences chromatin structure and the epigenetic state of the cell (Davie 2003). These mechanisms suggest the existence of complex interactions between the microbiome, metabolism and epigenome (Cortese et al. 2016). Hence, maternal diets may impact the establishment of the fetal and neonatal microbiome leading to specific epigenetic signatures that may potentially predispose the development of late-life obesity.

5.1.7. *Metabolites and epigenetic changes*

The idea that gene transcription is influenced by intermediary metabolism products through epigenetic mechanisms was suggested several years ago. They have direct regulatory roles, linking environmental factors such as nutrition, to changes in cellular function such as epigenetic regulation of gene expression. Key cellular metabolites from diverse metabolic pathways are essential cofactors for the chromatin remodeling enzymes that regulate the epigenome. For instance, acetyl-CoA and NAD^+ influence gene expression by serving as cofactors for enzymes that mediate posttranslational modification of histones (Ladurner 2006). The activity of histone acetyltransferases (HATs) is dependent on nuclear acetyl-CoA levels (Wellen et al. 2009), and the activity of histone deacetylases (HDACs) also called sirtuins is dependent on NAD^+ levels (Imai et al. 2000). Other enzymes target histone residues and require metabolic cofactors to perform phosphorylation, methylation and glycosylation. These modifications are associated with changes in chromatin organization, gene activation and silencing.

Unexpectedly, the β-hydroxybutyrate (β-OHB), which is a major source of energy during prolonged starvation, inhibits HDAC activity. Sustained inhibition results in increased histone acetylation, since HDAC activity does not counteract the opposite process. It is associated with transcriptional activation and repression of a subcellular gene (Van Lint et al. 1996). Therefore, β-OHB can regulate cellular processes directly via binding to plasma membrane receptor (see section 3.2.1.2) and through inhibition of HDAC. It also indirectly alters the levels of regulatory metabolites including acetyl-CoA and NAD^+.

For a comprehensive recent overview of the mechanisms involved in the metabolic regulation of epigenetics, the reader is referred to Dai et al. (2020).

5.1.8. *Social environment and endocrine disruptor: epigenetic changes*

Social environment

In rats, it was found that the level of maternal care (licking, grooming their pups) during the suckling period impacts offspring physiology and behavior. It is attributed to the establishment of CpG methylation at the critical site in the GR promoter in the hippocampus. At birth, a key CpG site in the promoter is unmethylated and then becomes hypermethylated on postnatal day 1. Thereafter, from day 1 to day 6, there is a critical period/window (concept evoked earlier) during which the CpG methylation is either hypermethylated or unmethylated. Unmethylation occurs only in pups suckled by high-caregiving dams during the first six postnatal days. A high maternal care during this period activates GR transcription and expression, thereby enhancing corticosteroid feedback sensitivity (Weaver et al. 2004) (Figure 5.10). Conversely, hypermethylation leads to reduced GR expression, which reduces responsiveness of the HPA axis to secreted glucocorticoids. Under stressful conditions, which activate the HPA axis, negative feedback by glucocorticoids is less effective, thus leading to overactivity of the axis and excessive adrenal glucocorticoid secretion. Importantly, glucocorticoids act on the amygdala, an area of the brain associated with emotional processing, where they enhance the sense of fear, which affects the behavior of these animals. It can be concluded that early life experience, which alters epigenetic modifications, affects the ability to deal with stress, mental well-being and behavior later in life.

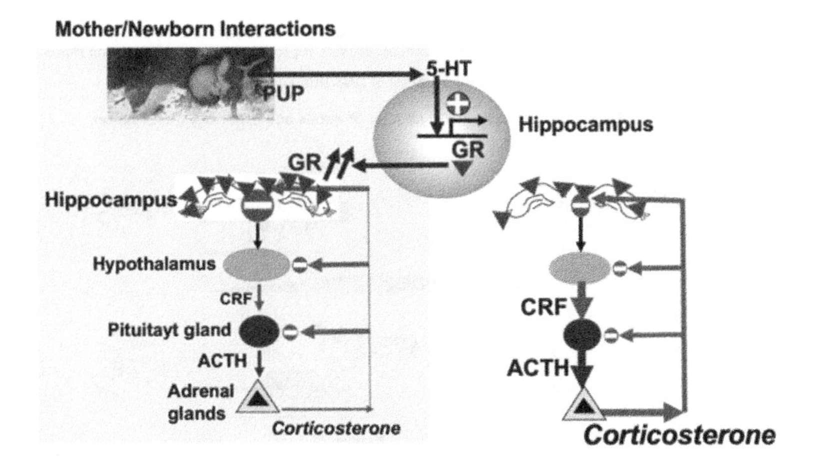

Figure 5.10. *Influence of maternal care on offspring neurobiology. Low maternal licking/grooming results in reduced hippocampal glucocorticoid receptor and elevated hypothalamic CRH (CRF) associated with increased ACTH levels in offspring exposed to stress. Following the HPA activation, the corticosterone negative feedback is dampened due to a lower GR expression*

Epigenetic action of endocrine disruptor

Bisphenol A (BPA), initially synthesized as a synthetic estrogen, is widely used in the manufacture of polycarbonated plastics and the epoxy lining of canned food. Given that it can bind to classic estrogen receptors, it was postulated that it could contribute to cancer initiation. To test this hypothesis, newborn rats were exposed to relevant doses of BPA. It induced hypermethylation of the promoter of a cAMP regulating gene PDE4 and it resulted in overexpression of the gene in the prostate. These aberrant changes are associated with an increase in the risk of developing prostate lesions in the adult rats (Ho et al. 2006). Another study has demonstrated that *in utero* exposure to BPA increased mammary cancer susceptibility in offspring (Betancourt et al. 2010) (Figure 5.11).

Vinclozolin, a fungicide commonly used in agriculture, is known to be an anti-androgenic endocrine disrupting compound, leading to a spermatogenic cell defect in the testis of adult male rats. Furthermore, additional transgenerational disease develops as animals age, including prostate cancer and kidney disease. This endocrine disruptor is capable of permanently altering the germline epigenome during the critical period of establishment of DNA methylation marks and reprogramming can persist transgenerationally (Guerrero-Bosagna et al. 2010).

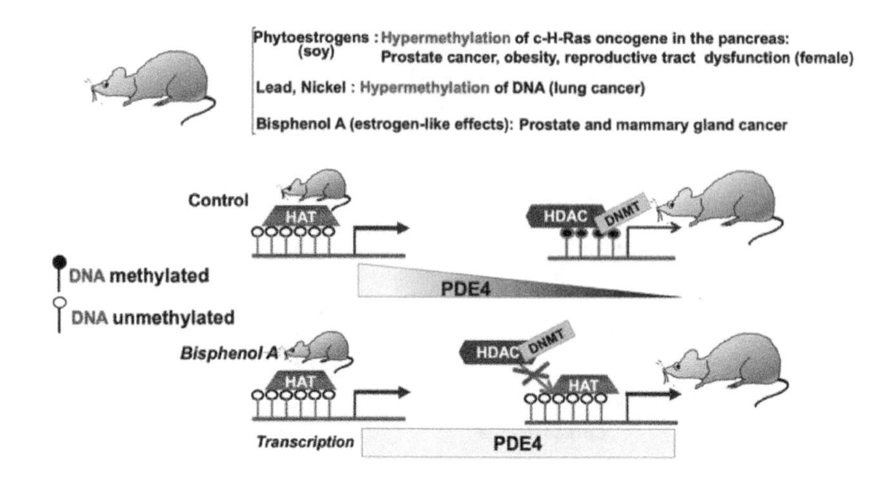

Figure 5.11. *Epigenetic action of endocrine disruptors. Bisphenol A prevents the progressive decrease in phosphodiesterase 4 expression, which normally occurs in adult mice. Other agents responsible for excessive DNA methylation are mentioned*

5.1.9. *Importance of epigenetics in the etiology of cancer*

Cancer is defined as a genetic and epigenetic disease, and genetic modifications (mutations, polymorphisms, deletions) occur in all cells, whereas epigenetic changes are cell and tissue specific (Legler 2010; Luteijn and Ketting 2013).

The concept that epigenetic dysregulation might underlie human tumors was suggested in 1983 by A. Feinberg and B. Vogelstein who showed that some genes were hypomethylated in tumors relative to adjacent normal tissues (Feinberg and Vogelstein 1983).. This initial finding pointed out that tumors were not strictly related to genetic mutations, but rather a contribution of both epigenetic and genetic alterations to cancer initiation and progression. Since epigenetic processes are mitotically heritable, they can have similar roles as genetic alterations in the progression of cancer. Darwin's hypotheses state that an evolving population results from mutations that give rise to a specific biological advantage, which increases reproductive success. These selective concepts might apply for epigenetic events which alter gene expression and give rise to a cellular growth advantage. This trait is therefore selected, resulting in an increase in the growth rate of the tumor.

Recent studies provide compelling evidence that cancer cells frequently display a global pattern of DNA hypomethylation and site-specific CpG island hypermethylation, and these alterations coexist in almost all human cancers (Gal-Yam et al. 2008; Plass et al. 2013). It is unlikely that all these numerous methylation changes play a causative role in tumor development but have a critical role for epigenetic gene regulation in carcinogenesis. Global hypomethylation is observed in some cancers and is known to be involved in carcinogenesis via microsatellite instability through the activation of repetitive elements or retrotransposons, loss of imprinting, and transcriptional activation and overexpression of oncogenes (Figure 5.12).

Aberrant DNA hypermethylation of CpG islands mainly occurs in the promoter regions of tumor suppressor genes, which causes a silencing of these genes (Herman et al. 1995, 1998; Yang et al. 2003) and other genes involved in the major cellular pathways. It therefore confers to tumor cells a higher growth advantage and an increase in genetic instability and malignant phenotype (Kanwal and Gupta 2012). The mechanism that triggers pathological *de novo* methylation is poorly understood. It should be emphasized that the silencing of tumor suppressor genes can also result from a combination of epigenetic and genetic defects. For instance, in multiple myeloma, the dysregulation of p53 (tumor suppressor gene) expression and activity are due to epigenetic silencing of the promoter of one allele, associated with mutations on the second allele (Hurt et al. 2006).

Unlike genetic changes, epigenetic alterations are reversible; it therefore raises the possibility of reverting these alterations via pharmacological therapies. Thus, inhibitors of DNA methylation and histone deacetylase would be a promising candidate for reactivating tumor suppressor gene transcription.

It should be emphasized that the role of p53 as a tumor suppressor is well established and its activity is altered in nearly half of all human cancers. In multiple myeloma, the dysregulation of p53 is due to mutations on one allele associated with epigenetic silencing of the promoter of the second allele.

Genome-wide hypomethylation has been observed in lung tumors, which is primarily caused by the loss of methylation from repetitive regions. It is often associated with a gain of function (GOF) of oncogenes, such as H-Ras, K-Ras and c-Myc (Cheung et al. 2009; Fearon 2011) (see c-Myc role in section 4.3.6.4). Hypomethylation can also induce loss of imprinting, such as the gene encoding for insulin-like growth factor 2 (IGF2). It is observed in a wide range of cancers, including breast, liver, lungs, ovaries, pancreas and colon (Plass and Soloway 2002). For a comprehensive review, see Jones and Baylin (2007) and Herceg and Vaissière (2011).

In addition to changes in DNA methylation, histone modification patterns are also altered in human cancers. Thus, HDACs responsible for deacetylation of histones have been found to be overexpressed or mutated in various cancer types.

Inactivating mutations in histone methyltransferases have also been reported (Jonasch et al. 2012; Liu et al. 2013).

Interestingly, altered expressions of various ncRNAs have been reported in several human cancer types (Gupta et al. 2010; Hassler and Egger 2012). Their new role is to act as decoys for miRNA when they contain specific miRNA binding sites (Deng and Sui 2013).

Recently, the concept of the generational threshold has been introduced when it was observed that the prostate risk is low in recently migrated populations, but within three generations, the risk of developing this cancer in this population is similar or even higher than the native host country population (Martin 2013). This threshold is partly due to epigenetic alterations in response to a more westernized lifestyle. It is primarily attributed to increased methylation of specific genes involved in prostate carcinogenesis (Kwabi-Addo et al. 2010).

Epigenetics alterations in the major types of cancer

Certain genetic changes play a key role in tumor initiation and progression, but epigenetic changes may also set the course of tumor development and be required for malignant transformation. As mentioned above, alterations in DNA methylation are the first epigenetic marker associated with cancer altering normal gene function. These alterations include hypermethylation, hypomethylation and loss of imprinting.

Global hypomethylation is recognized as a hallmark of all cancers (Feinberg and Vogelstein 1983), but CpG island hypermethylated genes in different cancers seem to be a key part of the carcinogenesis and progression of cancer.

Types of cancer

– Gastric cancer (GC)

It is a consequence of the gradual accumulation of various genetic and epigenetic alterations, leading to gain of function in oncogenes and loss of function in tumor suppressor genes (Wadhwa et al. 2013).

Overexpression of DNA methyltransferases (Fang et al. 2004) and inactivation of histone acetyltransferase activity (Koshiishi et al. 2004) play a key role in the onset and progression of gastric tumors. Furthermore, a large number of miRNAs and their targets are upregulated or downregulated in GC (Wang et al. 2013). Numerous studies have reported an increased activity of oncomiRNA, leading to cell proliferation and inhibition of apoptosis.

– Colorectal (CRC) and prostate cancer (PCa)

Epigenetic changes that have been investigated have been focused on genomic DNA methylation along with specific methylation at promoter CpG islands of tumor suppressor genes.

CRC: the most common genetic aberration in CRC is a mutation in adenomatous polyposis coli (APC), leading to uncontrolled cell growth and proliferation. This mutation is often followed by mutations of KRAS, TP53, PIK3CA and PTEN (Day et al. 2013). With respect to the epigenetic changes, they are predominantly characterized by a global decrease in DNA methylation along with specific hypermethylation at promoter CpG islands like MutL homolog 1 (MLH1). It therefore leads to a loss of expression of MLH1 resulting in mismatch repair and microsatellite instability. On the other hand, a dysregulation of histone modifications, primarily methylation and acetylation, have been reported (Fraga et al. 2005).

PCa: also marked by genetic and epigenetic events. The most common genetic aberrations include ABL, BRAF, EGFR, HRAS, KIT and KRAS (Sethi et al. 2013). In addition, these aberrations are associated with a global decrease in DNA methylation and a dysregulation of histone modifications, primarily methylation and acetylation, resulting in a more relaxed chromatin state, which may lead to genomic instability (Sethi et al. 2013).

– Lung cancer

There are two major types, small cell lung cancer (SCLC) and non-small cell lung cancer (NSCLC), which are characterized by a CpG island hypermethylation within the promoters of tumor suppressor genes and are usually associated with transcriptional downregulation of these genes.

– Pancreatic cancer

Pancreatic ductal adenocarcinoma is the most abundant form of exocrine pancreatic cancer. It is an aggressive tumor and the fourth common cause of cancer death. It is characterized by a global DNA hypermethylation.

– Breast cancer

In addition to genetic mutations in oncogenes and tumor suppressor genes, epigenetic alterations such as promoter methylation and histone modifications also lead to the initiation, progression and metastasis of breast cancer (Esteller 2008). Hypermethylation of promoter CpG islands occurs in breast cancer development, and more than 100 genes have been reported to be hypermethylated and many of them play critical roles in tumor suppression, cell-cycle regulation and apoptosis. In about 40% of invasive breast tumors, two key tumor suppressor genes, BRCA1 encoding a transcription factor (Turner et al. 2007) and PTEN, a phosphatase (García et al. 2004), display a promoter hypermethylation, resulting in a reduced expression of the proteins. Although breast tumors are also frequently hypomethylated on a genome-wide scale, it impacts a relatively small number of genes.

– Bladder cancer: urothelial carcinoma

This is the most frequent type of bladder cancer, and epigenetic aberrations play a key role in the development and progression of the carcinoma. Compared to other cancers, DNA hypomethylation, especially retrotransposons, and mutations in enzymes (genetic defects) establishing or removing histone acetylation or methylation are particularly prominent (Han et al. 2012). However, there is also a strong tendency for DNA hypermethylation to increase with tumor stage and grade. For instance, several genes involved in apoptosis regulation have been found to be hypermethylated, thereby promoting cancer cell survival (Sánchez-Carbayo 2012).

Cancer-associated infectious agents and epigenetic regulation

Seven major infectious agents contribute to cancer development: human papilloma virus in cervical cancer, hepatitis virus in hepatocellular carcinoma, herpes virus in Kaposi's sarcoma, Epstein–Barr virus in nasopharyngeal carcinoma, human T-cell lymphotropic virus type-1 (HTLV-1) in T-cell leukemia and lymphoma, *Helicobacter pylori* in gastric cancer. They often alter the major components of epigenetic machinery. Readers should refer to the review by Poreba et al. (2011) in which the epigenetic interactions are between oncogenic viral proteins and host epigenetic machinery. For instance, the small DNA tumor viruses including polyomaviruses, adenoviruses and papillomaviruses promote the cellular transformation events by a common strategy in which one or more virus-specific proteins bind to critical cellular regulatory factors and disrupt normal cellular control processes.

Epigenetic therapies for cancer

Epigenetics starts with two macromolecules – DNA and histone proteins – which intertwine in chromatin. In cancer cells, their functions are altered; hence, the epigenetic therapies seek to restore DNA methylation patterns and post-translational modifications on histones that maintain malignant phenotype (Figure 5.12).

– Therapies for aberrant DNA methylation and histone acetylation.

The aim of these therapies is to induce hypomethylation through the inhibition of DNMT activity. Two agents, azacitidine and decitabine, are currently the most advanced methyltransferase inhibitors. As aforementioned, tumor cells overexpress deacetylases (HDACs), resulting in reduced histone acetylation associated with decreased gene expression. Two molecules, vorinostat and romidepsin, have been proposed to inhibit HDACs and a marked clinical response was noted. More inhibitors of "writers", "erasers" and "readers" are in development.

Recently, epigenetic therapeutics have been combined with chemotherapies and immune checkpoint inhibitors to broaden response rates, namely in patients with hematologic cancers.

For a comprehensive recent overview of this field, the reader is referred to Bates (2020).

In summary, epigenetic changes are needed for normal development, but these modifications can be altered depending on genetic background and by environmental factors such as diet, physical activity and toxins, which lead to the development of cancer. Contrary to genetic marks that are static, epigenetics marks are dynamic, which makes this cancer research field very attractive.

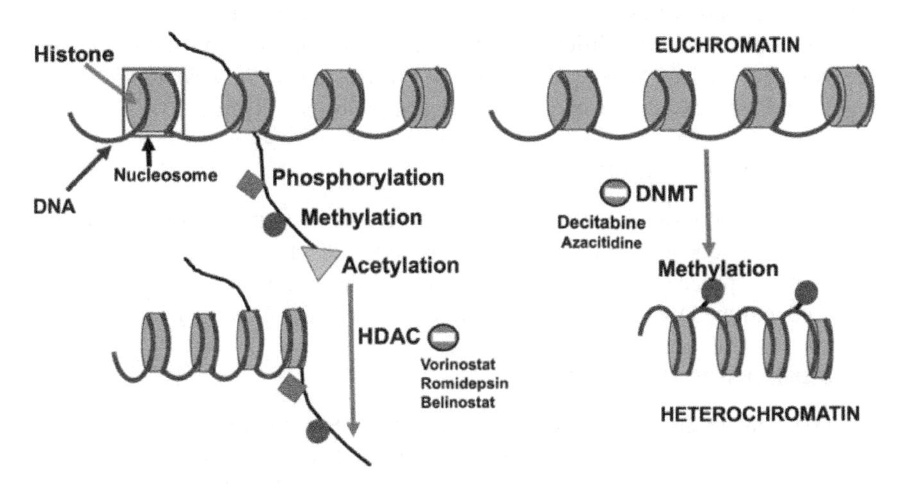

Figure 5.12. *Therapies for aberrant DNA methylation and histone acetylation: inhibitors of the DNMT and the HDAC*

Key questions for environmental epigenetics

– Which human genes result in enhanced disease susceptibility when they are epigenetically deregulated by environmental factors?

– What environmental factors deleteriously alter the epigenome, and at what doses?

– What role does the epigenome have in reproduction, development and disease etiology?

– Are there nutritional supplements that can reduce the harmful effects of chemical and physical factors on the epigenome?

– Can epigenetic biomarkers that will allow for the detection of early-stage diseases be identified?

– Can detection technologies that will allow for a quick and accurate genome-wide assessment of the epigenome be developed?

– Can epigenetics be integrated into systems biology as an important regulatory mechanism?

5.1.10. In vitro *reprogramming systems*

These systems restore pluripotency in somatic cells. It is worth noting that these reprogramming processes include the erasure of epigenetic marks (DNA methylation and histone modifications) that have been imposed on the chromatin, in the somatic nucleus during development. Interestingly, *in vivo* and *in vitro* reprogramming processes share common molecular pathways (Figure 5.13).

Somatic cell nuclear transfer

The ability of oocytes to erase epigenetic information from the somatic nucleus has been well documented elsewhere (Wilmut et al. 1997). The nucleus reprogramming proceeds through the erasure of the methylation pattern of the somatic cell and the DNA methylation, identical to that of embryonic pattern.

Induced-pluripotent stem cells

In 2007, Takahashi and Yamanaka's aim was to restore pluripotency in somatic cells. Therefore, differentiated cells were manipulated back to pluripotent stem cells. They identified and tested four factors: Oct4, Sox2, Klf4 and c-Myc, which in combination had the ability to convert somatic cells into pluripotent cells (Takahashi et al. 2007).

Cell fusion

A somatic cell can be reprogrammed to pluripotency by fusion with pluripotent embryonic stem cells. One of the major molecular mechanisms involved is the DNA methylation of the Oct4 promoter (Bhutani et al. 2010).

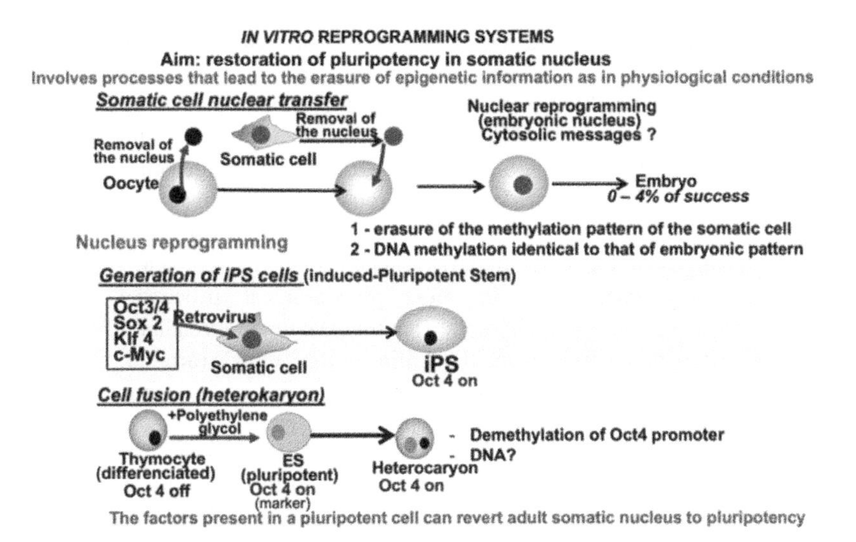

Figure 5.13. In vitro *reprogramming systems: somatic nuclear transfer, generation of iPS cells and cell fusion*

5.2. Circadian rhythms

This section covers the basic concepts of the circadian system and the underlying mechanisms causing phase adjustment of the circadian rhythms by light and environmental cues. A number of general reviews cover the different themes developed in this section, and they are suggested for further reading.

The rising and setting of the sun have captivated naturalists interested in the cause of daily rhythmic phenomenon, and for centuries, the prevailing view was that rhythmicity originated from an external electromagnetic force. The scientific literature on diurnal rhythms began in 1729 when French astronomer J. de Mairan observed that the daily opening of the leaves of the *Mimosa* plant placed in a dark box continued to open and close every 24 hours. His proposal was that flowering is intrinsically generated and driven by an internal clock (de Mairan 1729). Over two centuries later, scientists have described a wide range of processes in plants such as respiration, germination and photosynthesis, which are synchronized with the day/night cycle, and termed diurnal rhythm. We can hypothesize that the internal clock evolved to adapt the plant physiology to Earth's rotation. For instance, photosynthesis has a primordial role in the life cycle of a plant. They synthesize carbohydrates only in the daytime because the biochemical processes rely on sunlight, and they store them during nighttime. These stores are then used in daytime, which allows plants to grow in diurnal cycles.

5.2.1. *Circadian rhythms and the concept of a circadian clock*

In a broad sense, all living organisms, from cyanobacteria to mammals, are exposed to the rising and setting of the sun each day. As a result, evolution has given them a robust system for anticipating regular changes in light and other stimuli that fluctuate as a function of the Earth's rotation around its axis. This system is called a circadian rhythm, and it is used to synchronize the timing of organismal processes with cyclic changes in the external environment, such as light and food intake, which drive daily oscillations in most physiological processes, including metabolism, reproduction, immunity and behavior. The term "circadian rhythm" was coined by F. Halberg in 1959 from the Latin words "circa" (around, about) and "diem" (*dies* – day, *circa diem* – about (approximately) a day) and it defines a rhythm with a period of about 24–25 h.

To adapt physiological and behavioral processes to the appropriate time of the day, it was suggested that a system of endogenous timers, so-called circadian clocks, or molecular oscillators, keeps track of the duration of the daily light or daily dark period.

Here are some key points of circadian biology. In diverse organisms, biological processes that occur within cells of organs have the potential to oscillate within a wide range of periodicities, i.e. they display peak-to-peak intervals of activity. The molecules, cells and tissues involved are defined as oscillators. Oscillators known as circadian oscillators express periods of about 24 h and represent the circadian biological clock.

A circadian oscillator can drive a rhythmic output, but requires other oscillators (pacemakers) for its function. A pacemaker is an oscillator that drives an output and/or entrains another oscillator. A circadian pacemaker is a specialized oscillator that operates independently of other oscillators to drive rhythmic outputs.

Concept of a circadian clock

During the 1970s, the search for identifying circadian clocks was conducted in invertebrates and vertebrates. In *Aplysia*, the clock was described in the eye (Jacklet and Geronimo 1971), while in birds and rats, they were identified in the pineal gland and in the hypothalamus, respectively (Gaston and Menaker 1968; Moore and Eichler 1972). When these studies were published, there was a debate about the criteria that identify the cells of these clocks.

The fundamental question was also how circadian rhythms are maintained in synchrony with the environment. A conceptual model was developed, and it was organized around three major components: (1) an input pathway by which environmental cues (mainly light) are transmitted to a second component, (2) a central or "master" pacemaker, and (3) a set of output pathways by which the pacemaker modulates circadian rhythms in all cells in the body (Mohawk et al. 2012).

5.2.2. *Overview of the mammalian clock*

In the early 1980s, the structure of a biological oscillator was proposed from the analysis of biological and chemical oscillations. It was demonstrated that a system with a delayed negative feedback loop defines an oscillator. The delayed element of the system is necessary to set the system to oscillate, and the kinetics determines its periodicity. Tissues involved are known as oscillators, and some of them form the circadian biological clock, because they express a period of approximately 24 h. The amplitude of a rhythm is a measure of the level of expression and is measured from the midline (see Figure 5.14).

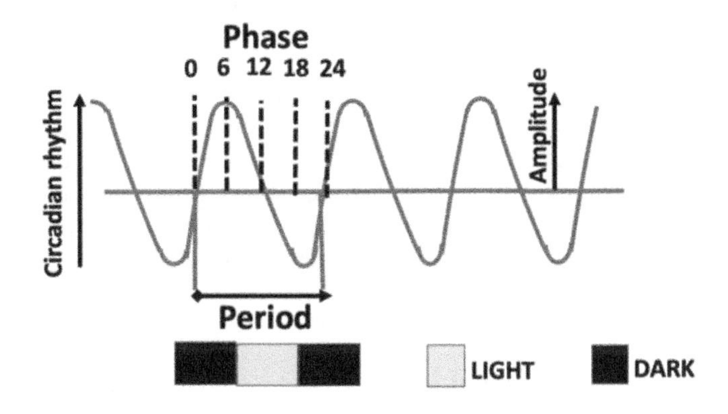

Figure 5.14. *Kinetics of an oscillator. The period is about 24 h, and the amplitude of the rhythm is the level of expression*

The next step was to identify the molecular components of the clock machinery, which generate circadian rhythms. In 1971, a seminal study by Konopka and Banzer described an arrhythmic mutation in *Drosophila melanogaster*, and they called the locus *per*, from "period". It clearly demonstrated the genetic nature of circadian clocks driving rhythms. Some years later, Hardin showed that the per protein feeds back to the *per* gene to regulate its own mRNA level. The contributions of other groups lay down the basis for the current knowledge on circadian clock genes. J. Hall, M. Rosbash and M. Young identified the first clock gene period, for which they shared the Nobel prize in Physiology or Medicine in 2017.

The core oscillatory mechanism responsible for the regulation of circadian rhythms is composed of a negative transcriptional feedback loop in which positive transcription factors stimulate the expression of their own inhibitors. It is also called an autoregulatory transcription–translation feedback loop (TTFL) model (Lowrey and Takahashi 2011) (Figure 5.15).

This model involves the following temporal sequences: the transcription factors circadian locomotor output cycles protein kaput (CLOCK) and brain and muscle ARNT-like 1 (BMAL1) heterodimerize (CLOCK/BMAL1) and bind to E-boxes in the promoters of the following genes, period (Per1, Per2) and cryptochrome (Cry1, Cry2), among others. Then, the proteins translated form heterodimers PER/CRY which enter back into the nucleus and interact with CLOCK/BMAL1 to inhibit the heterodimer function. This results in the suppression of PER/CRY transcription. Other genes have been identified and play additional roles in this network such as the feedback loop involving Rev-Erbα (a repressor) and the opposing RORα, which are also direct targets of CLOCK/BMAL1. They integrate circadian rhythms and metabolisms (Duez and Staels 2008).

In summary, circadian rhythmicity is supported by intracellular transcriptional and translational feedback loops that maintain oscillations in gene expression. Virtually all cells express these "clock genes" and have the capacity to generate autonomous circadian oscillations (Balsalobre et al. 1998). It therefore raises questions such as: (1) How does this give rise to rhythmic physiology in multicellular organisms? (2) How do environmental signals adjust clocks to geophysical time?

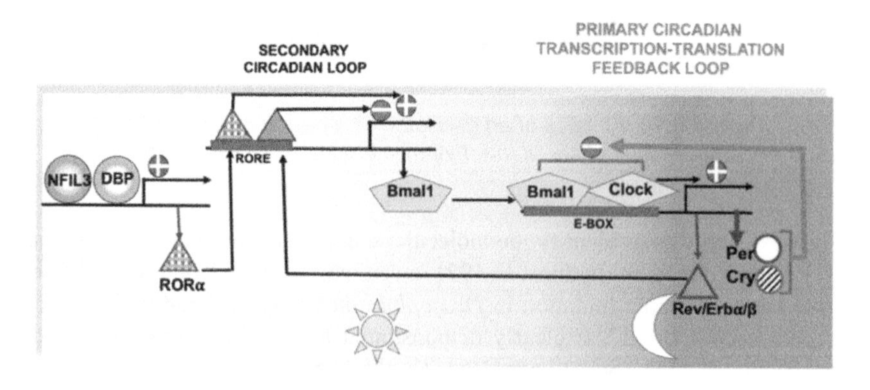

Figure 5.15. *Canonical transcriptional feedback loops in the core clock. See above for further details*

5.2.3. *Mechanisms by which circadian clocks govern biological processes*

In mammals, all physiological functions involved in the whole-body and individual cell homeostasis are coordinated by the circadian system. It is a highly specialized and hierarchical network of oscillators encompassing the central clock located in the suprachiasmatic nucleus (SCN) of the hypothalamus that aligns the phases of clocks in the peripheral tissues. The light–dark cycle sets the synchronization of the SCN on a daily basis, which raises the question of how does SCN receive the light information? The retina and SCN are linked directly via the retinohypothalamic tract. In a recent study, it was shown that SCN receives direct photic input from the retina from a photoreceptor cell type known as "intrinsically photoreceptive retinal ganglion cells" (ipRGCs). These cells are photosensitive and receive time-of-day information which are then transmitted to the SCN, which, in turn, sets the phase of oscillators in cells of the peripheral tissues. These oscillators confer circadian rhythmicity to tissue-specific functions. Hence, the SCN provides a crucial link between the outside world and the internal time-keeping mechanism. Notably, peripheral oscillations are phase-delayed by 4–12 h relative to the circadian patterns observed in the SCN, which represents the time taken for a signal from the SCN to reach the periphery and entrain peripheral oscillators.

The major challenge is to find out the mechanisms and components that enable the communication between oscillators, and their coordination by a pacemaker.

It has been demonstrated that there are marked differences between peripheral oscillators and the SCN pacemaker oscillator. Cultured peripheral tissues show a progressive decrease in amplitude of the rhythm, whereas explants of SCN can cycle for many weeks, showing that peripheral oscillators are less robust. In addition, co-culture models showed that neurons and fibroblasts in culture function as a cell-autonomous oscillator and have a wide range of periods, indicating that one oscillator is independent of the next. Using the same co-culture model, SCN cells but not fibroblasts impose metabolic and molecular rhythms on the co-cultured cells. It seems that the intercellular communication mechanisms are likely unique to the SCN for its function in synchronizing rhythmicity of peripheral oscillators. In mice, a disruption of the circadian clock in the SCN results in a desynchronization of peripheral circadian clocks (Yoo et al. 2004). Although the master pacemaker of the circadian clock had been assigned to the SCN, this view has been challenged with the development of experimental tools that described extra-SCN brain clocks and oscillators (Begemann et al. 2020).

It is noteworthy to mention that peripheral clocks receive additional environmental cues termed "zeitgebers" (from the German "time giver"), such as a light or feeding schedule. They have the capacity to permanently change the phase of a circadian oscillator, and the whole process is known as "entrainment". Some examples of rhythms with the potential to synchronize peripheral oscillators: feeding patterns, cycles of melatonin secretion, core body temperature and cyclic glucocorticoid production. In addition to controlling hormone secretion and body temperature directly, the SCN coordinates rhythms in behavioral processes: locomotor activity, feeding, which in turn can influence endocrine function and body temperature.

Notably, the intrinsic oscillation in cells can be entrained by specific cues. For instance, oscillation in the brain and liver circadian clocks are entrained by light and feeding schedules, respectively.

5.2.4. *How is the SCN clock connected to tissue and cellular functions?*

As mentioned earlier, cells of most peripheral tissues harbor cell-autonomous circadian oscillators, which are synchronized to environmental cues through rhythmic signals from the SCN. Given that most cells lack photosensitivity, this implies that specialized circuits have evolved to communicate circadian information. The question is: how does SCN convey signals to peripheral clocks? It transmits the signals to peripheral oscillators through (1) both sympathetic and parasympathetic innervations, with a dominant sympathetic tone, (2) hormonal signals (such as glucocorticoids, insulin, IGF1), (3) temperature, and (4) local signals.

The major environmental cues driving the activation of peripheral oscillators are feeding–fasting cycles, oxygen levels, the variations of hormones and metabolites in response to feeding, and changes in body temperature. These external cues synchronize the circadian clock of an organism to its environment (Figure 5.16).

Mammals experience circadian rhythms in core body temperature with a fluctuation of 1–4°C, regulated by the SCN. This brings up the following question: does body temperature entrain peripheral oscillators? Culture of Rat-1 cell line can be entrained to temperature fluctuations, and they displayed circadian rhythms of gene expression. Similar observations have been reported in peripheral tissues *in vivo*.

Unlike peripheral oscillators, SCN rhythms *in vivo* are unaltered by environmental temperature changes, indicating that SCN is resistant to temperature cycles in the circadian range. This phenomenon is not a cell-autonomous property of SCN neurons. However, it should be mentioned that another study did not reproduce these observations. Further work is required to clarify this issue.

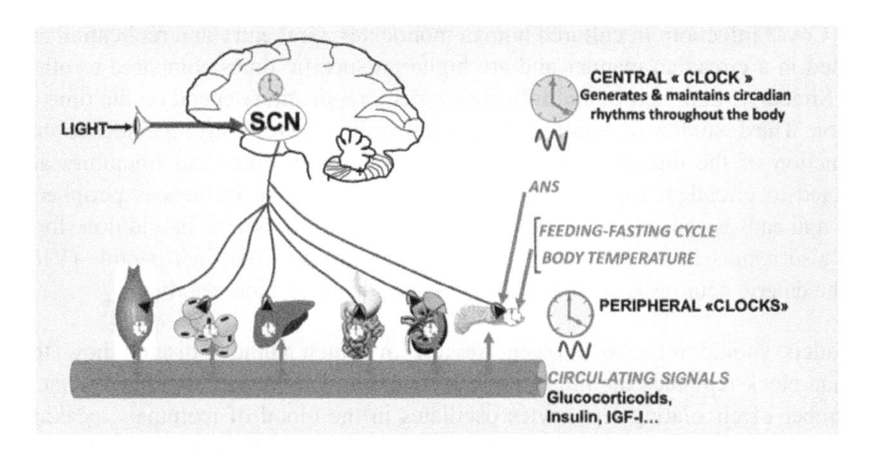

Figure 5.16. *Pathways of peripheral clock entrainment. The central clock (SCN) relays temporal information to peripheral clocks through the autonomic nervous system (ANS), body temperature, circulating signals (hormones) and feeding*

5.2.4.1. *Circadian clocks and the immune system*

A growing number of research studies provide evidence that both innate and adaptive immune systems are highly circadian regulated and exhibit 24-hour rhythmicity. We now know that immune cell migration and the function of T-cells and macrophages are regulated by the circadian clock. It should be emphasized that most circadian studies in the immune system have been focused on the role of BMAL1, because it is the only component of the clock machinery whose absence abrogates rhythmicity.

Some key circadian immune concepts should be mentioned because they underscore the role of circadian rhythms in (1) cytokine secretion from stimulated macrophages, (2) the number of leukocytes in blood, peaking during the behavioral rest phase, (3) the mobilization of leukocytes from the bone marrow into blood, and their homing to tissues, (4) the interactions of the host and the microbiota to maintain metabolic homeostasis of the host, and (5) metabolite production by microbiota to regulate the host rhythm.

Three studies dealing with the circadian rhythm and its relation to the immune system will be briefly discussed. First, the SCN through circadian signals synchronizes the rhythmic synthesis and secretion of catecholamines from the adrenergic branch of the sympathetic nervous system, which then affect the peripheral control of immune cell trafficking. Second, studies illustrating the immune responses to pathogens underscore the circadian rhythms that play a crucial role in disease progression, and it is exemplified in the context of the

SARS-Cov-2 infection. In cultured human monocytes, viral entry and replication are regulated in a circadian manner and are higher at specific times compared to other times. Similarly, both herpes and influenza virus's replication depend on the time of infection. Third, studies illustrating the gut immunity. There is strong evidence that the function of the digestive system, gut microbiota and intestinal immunity are connected to circadian rhythms. The timing of food intake influences peripheral clocks and causes changes in microbiota and nutrition absorption. In addition, food intake also results in the rhythmic secretion of vasoactive intestinal peptide (VIP) from the enteric neurons, which in turn influences the microbiome activity.

Readers should refer to a recent review in which authors discuss how the circadian clock regulates the migration and function of immune cells. For instance, the number of circulating leukocytes oscillates in the blood of mammals, peaking during the diurnal phase in humans. The circadian clocks also modulate leukocyte migration across the body, channeling the number of leukocytes at specific sites throughout the day (Wang et al. 2022).

5.2.4.2. *Circadian control of metabolism*

Metabolism involves a network of biochemical reactions that aim to regulate the whole-body energy homeostasis. It consists of anabolic and catabolic reactions which store and generate energy, respectively. Given that the circadian clock is present in almost every cell, it was hypothesized that it controls, in a cyclic manner, cellular metabolism in peripheral tissues as well as circulating blood metabolite levels. The molecular link between circadian clocks and metabolism has been elucidated when it was reported that mice carrying a mutation in the *Clock* gene lead to a disruption of glucose and lipid homeostasis and obesity. Altered feeding rhythms in these Clock mutant mice exhibit a marked attenuation of the diurnal feeding rhythm. In recent years, the role of peripheral clocks in the regulation of glucose homeostasis in the liver, skeletal muscle, adipose tissue, gut and pancreas has been extensively investigated. Emphasis was put on glucose because all tissues rely on glucose metabolism to meet their energy demand.

In the pancreas, the circadian clocks in β-cells and α-cells coordinate diurnal blood levels of insulin and glucagon, respectively. These two hormones play a central role in the regulation of blood glucose level. Furthermore, the sensitivity of the β-cell to glucose oscillates throughout the day.

In skeletal muscle and fat, insulin-stimulated glucose uptake is imposed by their respective circadian oscillators. The latter also regulate diurnal expression of glucose transporters (GLUTs). Given that muscle plays a primordial role in the regulation of the whole-body metabolism in response to feeding–fasting and physical activity, investigators have explored their influence on circadian biology. Exercise is likely the most robust environmental cue and it interacts with the circadian rhythm and the core clock of the skeletal muscle.

The liver is an organ with a broad repertoire of physiological functions, and in the fasting state glucagon increases the blood glucose level. It signals this biological response via its diurnal expressed receptor which stimulates glycogenolysis and gluconeogenesis pathways. The synthesis of glucose via gluconeogenesis occurs when the glucose availability is low, and it is stimulated by glucocorticoids. This process is regulated by CRY proteins through repression of the glucocorticoid receptor. Members of the nuclear receptor family exhibit circadian rhythms of transcription, within the liver and other tissues that are metabolically active. These receptors are regulators of clock function, thus providing a mechanism by which signals of metabolic status can modulate rhythmicity. For instance, glucocorticoid receptors stimulate the transcription of Per and other clock genes. Glucose metabolism is also regulated by REV-ERBA α. The liver has many metabolic functions, including carbohydrates, lipids, amino acids and hormones, and the central regulatory nodes of the metabolic pathways are under circadian control (Reinke and Asher 2016).

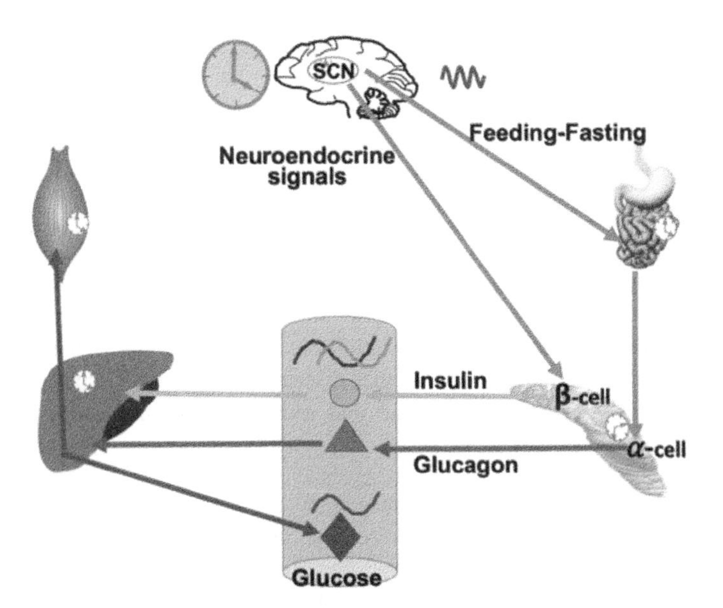

Figure 5.17. *Circadian control of glucose metabolism. The suprachiasmatic nucleus (SCN) coordinates the peripheral clocks of the β-cells and α-cells of the pancreas through neuroendocrine signals. It coordinates the diurnal blood levels of insulin and glucagon*

The nutrient flux in the bloodstream varies considerably depending on the time of day, which is explained by the fact that food is consumed during the activity phase and energy stores are mobilized to buffer the decline in nutrient levels. Hence, glucose, amino acids and lipids oscillate in the blood. As mentioned earlier, all tissues rely on glucose metabolism to meet their energy demand; it therefore requires a tight control of its circulating levels through a circadian regulation. The SCN is the principal driver of circadian blood glucose fluctuations by synchronizing food intake to the active phases, but it also requires co-operation between clocks in various organs to maintain glucose homeostasis. The question is: do these concepts apply to other blood metabolites?

There are two fuel-sensing molecules, AMPK and SIRT1, which emerge as modulators of the circadian clock function in mammals. Thus, SIRT1, which displays a histone deacetylase (HDAC) activity, has been reported to counteract the histone acetyltransferase (HAT) activity of CLOCK. With respect to AMPK, it directly regulates the central clock mechanism by phosphorylating and disrupting a major clock component CRY1, which results in circadian and gluconeogenic gene reprogramming (Lamia et al. 2009). Additionally, they regulate each other, since AMPK enhances the SIRT1 activity by increasing cellular NAD^+ levels and SIRT1 may cause AMPK phosphorylation. It strongly suggests that they are at the crossroads of nutritional status and circadian regulation (Bass and Takahashi 2010). Metabolites and energy sensors can directly regulate circadian clock function (Sato et al. 2019).

Perturbations of the clock and metabolic disease

In human subjects who experience disturbances in daily life, such as rotating shift work, working at night, or a sedentary life, this results in uncoupling of the clock from natural day–night cycles. It triggers an elevated blood glucose and lipid levels as well increased body mass. In addition, they are at a higher risk of developing type 2 diabetes and cardiovascular diseases. Long-term disruption of the clock can impact the gastrointestinal tract and may lead to colorectal cancer or metabolic syndrome.

Hence, our environment can influence our circadian biology, leading to misalignment (i.e. difference between outside time and circadian time, for example, during shift work) of our circadian rhythms. For a comprehensive recent overview of the relationship between circadian biology and metabolism, the reader is referred to Dollet et al. (2021).

The influence of feeding–fasting and physical activity on circadian biology has been explored, and exercise is likely the most robust environmental cue. The latter interacts with circadian rhythm and the core clock of the skeletal muscle.

Exercise improves metabolic health; investigators focused on its interaction with the molecular clock in skeletal muscle to elucidate how the clock communicates with cellular physiology. Animal models have been created, but unfortunately, they do not fully recapitulate human physiology (Gabriel and Zierath 2019).

In the fasting state, in the liver, glucagon increases blood glucose levels. It signals this biological response via its diurnal expressed receptor which stimulates glycogenolysis and gluconeogenesis pathways.

The cellular redox state, which reflects the metabolic status of the cell, can impact the circadian system. Changes in cell metabolism also impact the cellular redox state, leading to fluctuations of intracellular NAD^+ levels, which reflects that the metabolic status of the cell can modulate the rhythmicity. NAD^+ levels fluctuate substantially in a circadian manner, linking the peripheral clock to the transcriptional regulation of metabolism by epigenetic mechanisms involving SIRT1 (histone deacetylase).

5.2.4.3. *Clocks and the endocrine system*

Among hormones that fluctuate throughout the day are glucocorticoids, growth hormone, IGF1, insulin, glucagon, oxyntomodulin, prolactin, thyroid hormones and sex hormones. These hormones potentially feedback phase information to the circadian system (Figure 5.15).

With respect to the control of glucocorticoid (GC) secretion described in section 4.3.1, it was mentioned that the hypothalamic–pituitary–adrenal (HPA) axis stimulates their release from the adrenal cortex. Their secretion is synchronized to daily solar cycles in a classical circadian pattern. It is maximally secreted just prior to awakening and then gradually decreases to low levels during sleep.

Recent studies indicate that effector neurons from the SCN project to the PVN of the hypothalamus. Thus, through the PVN, the SCN would add another level of regulation of pituitary and peripheral hormone releases. Regarding the HPA axis, it has been established that the SCN and the ANS (autonomic nervous system) influence its activity and therefore coordinate GC release. In addition, the SCN plays a protective role against the overexposure of cells to high blood GC concentrations in the morning. It synchronizes elevated GC levels with low GC sensitivity of the target cells. However, this mechanism is overruled by stress-induced GC secretion or prolonged stress response in the evening hours when the GC sensitivity is high. In humans, it leads to the development of chronic stress-related hypercortisolism. See Kazakou et al. (2022).

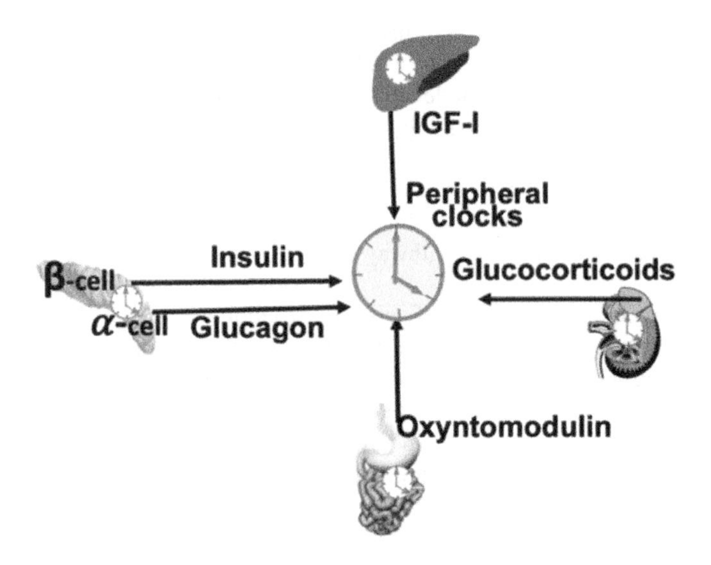

Figure 5.18. *The different hormone secretions listed above potentially feedback phase information to the circadian system*

To elicit a cellular response, GCs bind to their nuclear receptors. The latter exhibit circadian rhythms of transcription, within the liver and other tissues metabolically active. These receptors are regulators of clock function, thus providing a mechanism by which signals of metabolic status can modulate rhythmicity. For instance, glucocorticoid receptors stimulate the transcription of Per and other clock genes (Figure 5.15).

5.2.4.4. *Circadian clocks and reproduction*

Successful reproduction relies on hormonal and physiological processes. For instance, specific temporal organization of hormone release is required for ovulation, estrus and fertilization. These critical processes are coordinated by the circadian clock, and any disturbances in this coordination may result in poor or even an absence of fertility.

Neuronal SCN output has been extensively investigated in rodents. Effector neurons from the SCN project to gonadotropic-releasing hormone (GnRH) neurons, which constitute the final output that regulates the precise timing of the pre-ovulatory LH surge and ovulation. To reinforce the importance of the role of the molecular clock in regulating GnRH secretion, mice with clock-deficient genotypes (Clock mutant) display severe changes in estrogen and progesterone levels.

Similarly, Per2 mutant mice exhibit an irregular estrus cycle. These findings highlight the importance of the clock system in reproduction.

5.2.4.5. *Circadian rhythms and microbiota*

A complex community of microbes known as the intestinal microbiota is present in the distal gut of mammals. In the human, dietary glycans, such as plant cell-wall polymers, represent the major food source for the microbiota. The latter exhibits circadian rhythms which modulate many important aspects of host physiology, namely, the host's metabolism. The microbial circadian dynamic is tightly associated with nutrient availability. In the intestine, dietary glycans are transformed by Firmicutes which subsequently decline in abundance, allowing Bacteroidetes to expand bacteria in response to host glycans present in the mucus of the gut.

Gut bacteria break down digestion-resistant carbohydrates to produce short-chain fatty acid (SCFA) metabolites such as acetate, propionate and butyrate. These metabolites display diurnal rhythmicity, and their levels also impose the diurnal inhibitory effect of SCFAs on intestinal ghrelin release. These effects are lost in Bmal1-/- mice but can be restored by night-time restricted feeding. These findings point out that feeding time restores the rhythmicity independently of Bmal1. In addition, restricted feeding at night-time enhances the expression of colonic clock genes (Clock), which likely contributes to the restoration of rhythmicity (see Segers and Depoortere (2021)).

When treating mice with acetate or butyrate, it causes a significant change in the expression of the clock genes. Hepatic *Bmal1* expression was suppressed and *Per2* increased in the dark phase for high-fat diet in mice. It was not observed when the gut was devoid of microbiota, demonstrating that it was microbiota-mediated. In humans and mice, diet can alter gut microbiota, promoting obesity. In mice, a high-fat diet leads to a blunted microbial diurnal rhythmicity, which can be restored by a time-restricted feeding of this diet.

As mentioned earlier, there are many lifestyle factors in modern society that may contribute to the current metabolic problems such as diabetes and obesity. Disruptions in eating patterns, diet and sleep influence both the circadian timing system and the gut microbiome. It has been suggested that physiological adaptations to circadian disruption should be associated with gut microbial adaptation to time-of-day patterns to prevent metabolic consequences to the host in terms of the efficiency of energy utilization and extraction. The microbiota serves as a mediator for the interplay between the host's diet and circadian rhythmicity.

It should be kept in mind that most of these results were obtained in mouse models, and little is known about the microbial influence on circadian rhythms in humans.

Circadian disruption

For biological clocks to be effective, they must accurately keep time and adjust to environmental signs. It therefore requires SCN control of peripheral oscillators, and a loss of the SCN results in peripheral desynchronization of circadian clocks (Yoo et al. 2004). However, a genetic disruption of a circadian clock mechanism specifically in the hepatocyte of a mouse results in arrhythmicity of most hepatic transcripts. Notably, a subset of transcripts among these continues to cycle, despite the absence of a functional liver clock. This indicates that rhythmic gene expression can be driven by both local intracellular clocks and extracellular systemic cues (Kornmann et al. 2007).

It has been suggested that in some pathologies, the rhythm of the clock machinery would be disrupted, leading to impairment of the circadian control of some functions. Given the interplay between circadian clocks in cells of all tissues and organs, we can expect that a defect can impact the whole-body homeostasis.

One of the best examples of circadian disruption has been observed in human subjects who experience disturbances in daily life, such as rotating shift work or working at night. It results in an uncoupling of the clock from natural day–night cycles, leading to a misalignment, as mentioned earlier. It triggers an elevated blood glucose and lipid levels as well as increased body mass. In addition, they are at a higher risk of developing type 2 diabetes and cardiovascular diseases. Interestingly, when obesity and type 2 diabetes develop, rhythms of the molecular clocks in pancreatic islets are disrupted, which concomitantly dysregulates insulin and glucagon secretion. Furthermore, long-term disruption of the clock can impact the gastrointestinal tract and may lead to colorectal cancer or metabolic syndrome. Hence, the pathology of some metabolic diseases might be explained in great part by the perturbation of our circadian rhythms. For a comprehensive recent overview of the relationship between circadian biology and metabolism, the reader is referred to Dollet et al. (2021).

Recent studies have demonstrated that exercise is likely the most robust environmental cue interacting with circadian rhythm and the core clock of the skeletal muscle. It would therefore have a potential therapeutic role that makes it possible to readjust circadian rhythms (see the review by Gabriel and Zierath (2019).

5.2.5. *Avian circadian clock*

The role of the circadian clock in annual cycles of birds has been known for some time (Rowan 1926). It is more complex than that of mammals, involving the pineal gland, the retina and the SCN, which are the pacemakers that regulate peripheral tissues. A model of interactions between these pacemakers has been proposed and suggests that oscillators in the pineal gland and SCN stabilize and amplify each other through the emission of signal(s) by both components that is/are sensed by peripheral tissues (Gwinner and Brandstatter 2001). Though SCN is the primordial pacemaker, the mammalian pineal gland also participates to a lesser extent in circadian organization. Peripheral rhythms have not been explored in birds as extensively as in mammals, but most tissues express oscillator genes.

Circadian rhythms are also regulated by a set of genes, collectively called "clock genes", which are similar to those described in living organisms from *Drosophila* to humans. Their products dynamically interact with the rhythmic patterns of transcription, translation feedback loop, mentioned earlier.

In addition to photoreceptors shared by all vertebrate classes, non-mammalian vertebrates express photopigments critical for the entrainment of both circadian and circannual cycles. In birds, they are mainly located in the pineal gland. In the 1930s, in an elegant study carried out on ducks, it was shown that eyes are not necessary for the seasonal control of reproduction since after enucleation, to make them blind, they continue to exhibit reproductive responses to changing photoperiod (Benoit and Assenmacher 1954). Subsequent research has identified four photoreceptive structures, including the pineal gland. When the gland is surgically removed, birds become arrhythmic, demonstrating that the pineal gland is necessary for self-sustained circadian rhythmicity. In fact, the pineal gland is part of the central core of the avian circadian system, which consists of the pineal gland, the SCN and the retina. They express robust molecular clockworks, but each of them is not capable of self-sustained circadian rhythms. Exposure to photoperiods of longer than 11.5 h/day results in the rapid stimulation of the hypothalamo–pituitary–gonadal axis, causing development and growth of testes and ovarian follicles.

5.3. Conclusion

Over the last two decades, a plethora of studies have focused on circadian clocks, which set the timing for many circadian rhythms, regulating a wide spectrum of physiological and behavioral processes. Although in all living organisms, circadian clocks are composed of multiple oscillators, which are coordinated in different ways, these discoveries uncovered the complexity in the circadian system. It raised several questions regarding the synchronization and coherence of rhythms at the cellular level as well as the architecture of circadian clocks at the systemic levels. For instance, how to understand the coupling between circadian and metabolic systems? In mammals, the master pacemaker of the circadian clock had been assigned to the SCN, but that view has been challenged after the description of extra-SCN brain clocks and oscillators. To further understand the complex connections and interactions between the mammalian circadian clock and other biological systems will require the development of mathematical models of the circadian clock. Then, modelers and biologists will clear up the complexity of clock component interactions.

Concluding Remarks

The aim of this book was to provide an overview of the main concepts that have been established over the last century, in the three major disciplines of biology: endocrinology, neurology and immunology. The development of new technologies has allowed researchers to address both new and unsolved issues, which then led to spectacular advances in these three fields. New mechanisms were discovered and gave rise to a wide diversity of concepts. For space reasons, some of them are not reported. However, they are briefly mentioned and discussed in this section, especially concepts which are emerging, unexpected or evolving.

Studies carried out on endocannabinoids and bile acids (BAs) led to new and unexpected concepts. The concept of the endocannabinoid (EC) system was outlined in 1995 by DiMarzo and Fontana. Since then, a plethora of studies have shown that the concept is continuously evolving, which is linked to the discovery of additional cellular and subcellular localizations of CB1 receptors. These studies indicate the presence of ubiquitous EC signaling, which broadens its role in the regulation of key metabolic, hormonal and neuronal processes. Recent discoveries have expanded this system – known as the endocannabinoidome – that includes several mediators that are biochemically related to the ECs. Additional concepts have emerged in the EC signaling, and they should be taken into consideration. They are extensively reviewed in Busquets-Garcia et al. (2022).

Bile has been studied for centuries for its therapeutic properties, and it dates back to the Egyptian period (in the Ebers Papyrus, 1,550 BCE). But the molecular aspects by which the bile acids (BAs), components of the bile, impact health started to be investigated over the last two centuries. The development of new technologies has revealed that BAs can spill over into the systemic circulation and signal energy availability to almost all organs through their receptors. Therefore, BAs function as hormones and emerged as postprandial messengers that fine-tune whole-body metabolism. Furthermore, BAs counteract different processes contributing to the onset of obesity, which has led to the concept that they could be used as biomarkers for disease progression.

Over the past decade, historical enteroendocrine cell (EEC) concepts have changed. The EECs of the gut represent the largest endocrine network, and we consider that the EEC biology evolved from a unihormonal cell type to a complex plurihormonal dynamic cell. Furthermore, there is evidence that several hormones do not exert their actions through a single receptor as classically described, but as a prohormone or multiple truncated forms of peptides. The cleavage gave rise to the concept of enzymatic "inactivation" of ECC hormones.

The history of cell niche concept is steadily evolving in a variety of disciplines, including hematopoiesis oncology and stem cell biology. The biochemical characteristics and regulation of stem cell niche are under intense investigation. The term "niche" is used to refer to a conceptual analysis of a specialized microenvironment where stem cells reside prior to self-renewal or differentiation. When the first cancer stem cells (CSCs) were described in 1994 by Lapidot et al., these cells were also characterized by their ability to self-renew, differentiate and initiate tumors (Lapidot et al. 1994). Given the similarities between normal stem cells and CSCs, the niche concept also applied. Prior to this observation, investigators were interested in the interactions between cancer cells and their microenvironment. The first experiments documenting these processes were carried out by Virchow in 1863. Then, in 1889, S. Paget, who was studying organ preferences of breast cancer metastasis, highlighted the relationships between the tumor and its microenvironment. In the 1980s, it was reported that the tumor microenvironment (TME) exerts its influence on the process of metastatic colonization; it gave rise to the concept of TME, which refers to the role of the microenvironment in the process. In this context, it is worth noting that the ability of CSCs to remain quiescent should be linked to the concept of "angiogenic dormancy". This occurs when tumors cannot expand due to decreased vascularization. Consequently, its mass is stable over time, as a fraction of dying cells equals the dividing ones. Taken together, these concepts add new layers of complexity in our understanding of tumor cell growth and in the metastatic process.

Changing concepts are exemplified by hematopoietic lineage commitment. There are three models of stem cell lineage commitment. The classic hematopoietic model suggests lineage commitment associated with branching transitions. The second model suggests that hematopoietic stem cell (HSC) differentiation occurs via defined progenitor populations. The third one suggests that the predominant fate of HSCs in native hematopoiesis is megakaryopoiesis and there may be multiple lineage trajectories that generate equivalent mature hematopoietic cell types (Loughran et al. 2020).

Regarding the regulation of adipose thermogenesis, it was thought that in mammals, UCP1 was the sole regulator of the process. But the concept has been dismissed after genetic ablation of UCP1. It is now acknowledged that alternative pathways are activated to enhance heat production in adipose tissue (Auger and Kajimura 2022).

Over the past several years, studies have focused on two aspects of the expansion of fat mass in obese patients, occurring either by hypertrophic expansion of adipocytes or by the formation of new adipocytes (adipogenesis). In the latter process, evidence indicated the appearance of newly formed "beige" adipocytes within white adipose tissue (WAT). This therefore raises the question of where these cells come from. Two concepts were proposed: 1) beige cells come from a pool of precursors and 2) white adipocytes can convert into beige cells (transdifferentiation) (Herz and Kiefer 2019). There is evidence from lineage tracing studies for both theories. WAT browning represents an intriguing concept in humans, and it strongly suggests the induction of thermogenic adipocyte (beige) in fat depots. This transformation of WAT into thermogenic fat depot raises questions about the potential therapeutic implications.

Finally, the central dogma of biology held that genetic information flowed from DNA to RNA to protein. We now know that transcription also generates non-protein-coding RNA species which dominate the genomic output of the higher organisms. Furthermore, reverse transcriptase can synthesize DNA based on an RNA template, highlighting the possibility of a reverse flow of genetic information.

In conclusion, this book provides an overview of major discoveries which laid out a great diversity of concepts that have emerged over the last century. It should be emphasized that many fields of research on biology have come from a long way since they started, and remarkable advances in technology have enabled researchers to progress in areas that were an unexpected field of knowledge half a century ago. In the future, it is difficult to predict what direction should be taken by scientists. We all know that in biology there are still many species of molecules and chemicals in the cell which remain mostly hidden and they likely have important effects in biological processes. Currently, our tools cannot physically visualize these "dark matter" species, and we therefore cannot address the question of how many of them are in the cell and their respective roles (Ross 2016). Obviously, researchers should be receptive to new ideas, theories and concepts that can bring novel opportunities in all fields of research. The challenge is often to figure out what would be the most fruitful topic to investigate.

References

Abram, C.L. and Lowell, C.A. (2009). The ins and outs of leukocyte integrin signaling. *Annu Rev Immunol*, 27, 339–362.

Adams, F. (1891). *The Genuine Works of Hippocrates*, Volume 1. W. Wood & Co, New York.

Adeva-Andany, M.M., Gonzalez-Lucan, M., Donapetry-Garcia, C., Fernandez-Fernandez, C., Ameneiros-Rodriguez, E. (2016). Glycogen metabolism in humans. *BBA Clin*, 5, 85–100.

Afzelius, B. (1970). Brown adipose tissue: Its gross anatomy, histology and cytology. *Brown Adipose Tissue*, 1–32.

Agarwal, P., Morriseau, T.S., Kereliuk, S.M., Doucette, C.A., Wicklow, B.A., Dolinsky, V.W. (2018). Maternal obesity, diabetes during pregnancy and epigenetic mechanisms that influence the developmental origins of cardiometabolic disease in the offspring. *Crit Rev Clin Lab Sci*, 55, 71–101.

Agnati, L.F., Zoli, M., Strömberg, I., Fuxe, K. (1995). Intercellular communication in the brain: Wiring versus volume transmission. *Neuroscience*, 69, 711–726.

Ahlquist, R.P. (1973). Adrenergic receptors: A personal and practical view. *Perspect Biol Med*, 17, 119–122.

Al Massadi, O., López, M., Tschöp, M., Diéguez, C., Nogueiras, R. (2017). Current understanding of the hypothalamic ghrelin pathways inducing appetite and adiposity. *Trends Neurosci*, 40, 167–180.

Albright, F. and Reifenstein Jr., E.C. (1948). *The Parathyroid Glands and Metabolic Bone Disease: Selected Studies*. Williams & Wilkins, Baltimore.

Albright, F., Butler, A.M., Bloomberg, E. (1937). Rickets resistant to vitamin D therapy. *Am J Diseases Children*, 54, 529–547.

Algarra, I., Cabrera, T., Garrido, F. (2000). The HLA crossroad in tumor immunology. *Hum Immunol*, 61, 65–73.

Allard, J.B. and Duan, C. (2018). IGF-binding proteins: Why do they exist and why are there so many? *Front Endocrinol (Lausanne)*, 9, 117.

Allfrey, V.G., Faulkner, R., Mirsky, A.E. (1964). Acetylation and methylation of histones and their possible role in the regulation of RNA synthesis. *Proc Natl Acad Sci USA*, 51, 786–794.

Allsopp, R.C., Vaziri, H., Patterson, C., Goldstein, S., Younglai, E.V., Futcher, A.B., Greider, C.W., Harley, C.B. (1992). Telomere length predicts replicative capacity of human fibroblasts. *Proc Natl Acad Sci USA*, 89, 10114–10118.

Allum, F. and Grundberg, E. (2020). Capturing functional epigenomes for insight into metabolic diseases. *Mol Metab*, 38, 100936.

Alonso, J.L. and Goldmann, W.H. (2016). Cellular mechanotransduction. *AIMS Biophysics*, 3, 50–62.

Altindis, E., Cai, W., Sakaguchi, M., Zhang, F., Guoxiao, W., Liu, F., De Meyts, P., Gelfanov, V., Pan, H., Dimarchi, R. et al. (2018). Viral insulin-like peptides activate human insulin and IGF-1 receptor signaling: A paradigm shift for host-microbe interactions. *Proc Natl Acad Sci USA*, 115, 2461–2466.

Altman, J. (1962). Are new neurons formed in the brains of adult mammals? *Science*, 135, 1127–1128.

Anand, B.K. and Brobeck, J.R. (1951). Localization of a "feeding center" in the hypothalamus of the rat. *Proc Soc Exp Biol Med*, 77, 323–325.

Anand, B., Chhina, G., Sharma, K., Dua, S., Singh, B. (1964). Activity of single neurons in the hypothalamic feeding centers: Effect of glucose. *Am J Physiol-Legacy Content*, 207, 1146–1154.

Anderson, E.H., Ruegsegger, M.A., Murugesan, G., Kottke-Marchant, K., Marchant, R.E. (2004). Extracellular matrix-like surfactant polymers containing arginine-glycine-aspartic acid (RGD) peptides. *Macromol Biosci*, 4, 766–775.

Andjelkovic, M., Alessi, D.R., Meier, R., Fernandez, A., Lamb, N.J., Frech, M., Cron, P., Cohen, P., Lucocq, J.M., Hemmings, B.A. (1997). Role of translocation in the activation and function of protein kinase B. *J Biol Chem*, 272, 31515–31524.

Andrew, A., Kramer, B., Rawdon, B.B. (1998). The origin of gut and pancreatic neuroendocrine (APUD) cells – The last word? *J Pathol*, 186, 117–8.

Andrewes, C. (1971). Francis Peyton Rous (1879–1070). *Biogr Mem Fellows Royal Soc*, 17, 643–662.

Anklesaria, P., Teixidó, J., Laiho, M., Pierce, J.H., Greenberger, J.S., Massagué, J. (1990). Cell-cell adhesion mediated by binding of membrane-anchored transforming growth factor alpha to epidermal growth factor receptors promotes cell proliferation. *Proc Natl Acad Sci USA*, 87, 3289–8293.

Arble, D.M., Copinschi, G., Vitaterna, M.H., Van Cauter, E., Turek, F.W. (2012). Circadian rhythms in neuroendocrine systems. In *Handbook of Neuroendocrinology*, Levine, G.F.W.P.E. (ed.). Academic Press, San Diego.

Argiles, J.M., Busquets, S., Felipe, A., Lopez-Soriano, F.J. (2005). Molecular mechanisms involved in muscle wasting in cancer and ageing: Cachexia versus sarcopenia. *Int J Biochem Cell Biol*, 37, 1084–1104.

Arkins, S., Johnson, R.W., Minshall, C., Dantzer, R., Kelley, K.W. (2010). Immunophysiology: The interaction of hormones, lymphohemopoietic cytokines, and the neuroimmune axis. *Compr Physiol*, 469–495.

Árnadóttir, J. and Chalfie, M. (2010). Eukaryotic mechanosensitive channels. *Annu Rev Biophys*, 39, 111–137.

Arnold, F. (1942). Pseudohypoparathyroidism – An example of “Seabright-Bantam” syndrome. *Endocrinology*, 3, 922–932.

Arnon, T.I., Markel, G., Mandelboim, O. (2006). Tumor and viral recognition by natural killer cells receptors. *Semin Cancer Biol*, 16, 348–358.

Atkinson, M.A. (1997). Molecular mimicry and the pathogenesis of insulin-dependent diabetes mellitus: Still just an attractive hypothesis. *Ann Med*, 29, 393–399.

Auger, C. and Kajimura, S. (2022). Adipose tissue remodeling in pathophysiology. *Annu Rev Pathol: Mech Disease*, 18.

Avery, O.T., Macleod, C.M., McCarty, M. (1944). Studies on the chemical nature of the substance inducing transformation of pneumococcal types: Induction of transformation by a desoxyribonucleic acid fraction isolated from pneumococcus type III. *J Exp Med*, 79, 137–158.

Avitsur, R., Kavelaars, A., Heijnen, C., Sheridan, J.F. (2005). Social stress and the regulation of tumor necrosis factor-alpha secretion. *Brain Behav Immun*, 19, 311–317.

Azuma, M. (2019). Co-signal molecules in T-cell activation. In *Co-signal Molecules in T Cell Activation: Immune Regulation in Health and Disease*, Azuma, M. and Yagita, H. (eds). Springer, Singapore.

Bacakova, L., Zarubova, J., Travnickova, M., Musilkova, J., Pajorova, J., Slepicka, P., Kasalkova, N.S., Svorcik, V., Kolska, Z., Motarjemi, H. et al. (2018). Stem cells: Their source, potency and use in regenerative therapies with focus on adipose-derived stem cells – A review. *Biotechnol Adv*, 36, 1111–1126.

Baccala, R., Gonzalez-Quintial, R., Lawson, B.R., Stern, M.E., Kono, D.H., Beutler, B., Theofilopoulos, A.N. (2009). Sensors of the innate immune system: Their mode of action. *Nat Rev Rheumatol*, 5, 448–456.

Baeuerle, P.A. and Baltimore, D. (1988). I kappa B: A specific inhibitor of the NF-kappa B transcription factor. *Science*, 242, 540–546.

Balsalobre, A., Damiola, F., Schibler, U. (1998). A serum shock induces circadian gene expression in mammalian tissue culture cells. *Cell*, 93, 929–937.

Banerjee, S., Sengupta, K., Dhar, K., Mehta, S., D'amore, P.A., Dhar, G., Banerjee, S.K. (2006). Breast cancer cells secreted platelet-derived growth factor-induced motility of vascular smooth muscle cells is mediated through neuropilin-1. *Mol Carcinog*, 45, 871–880.

Bankaitis, E.D., Ha, A., Kuo, C.J., Magness, S.T. (2018). Reserve stem cells in intestinal homeostasis and injury. *Gastroenterology*, 155, 1348–1361.

Banting, F.G., Campbell, W.R., Fletcher, A.A. (1923). Further clinical experience with insulin (pancreatic extracts) in the treatment of diabetes mellitus. *British Medical Journal*, 1, 8–12.

Baraille, F., Planchais, J., Dentin, R., Guilmeau, S., Postic, C. (2015). Integration of ChREBP-mediated glucose sensing into whole body metabolism. *Physiology (Bethesda)*, 30, 428–437.

Bard, J.B. (2008). Waddington's legacy to developmental and theoretical biology. *Biol Theory*, 3, 188–197.

Baribeau, D.A. and Anagnostou, E. (2015). Oxytocin and vasopressin: Linking pituitary neuropeptides and their receptors to social neurocircuits. *Front Neurosci*, 9, 335.

Barker, D.J., Osmond, C., Golding, J., Kuh, D., Wadsworth, M.E. (1989). Growth in utero, blood pressure in childhood and adult life, and mortality from cardiovascular disease. *Bmj*, 298, 564–567.

Barker, D.J., Gluckman, P.D., Godfrey, K.M., Harding, J.E., Owens, J.A., Robinson, J.S. (1993). Fetal nutrition and cardiovascular disease in adult life. *Lancet*, 341, 938–941.

Barker, N., Van Es, J.H., Kuipers, J., Kujala, P., Van Den Born, M., Cozijnsen, M., Haegebarth, A., Korving, J., Begthel, H., Peters, P.J. et al. (2007). Identification of stem cells in small intestine and colon by marker gene Lgr5. *Nature*, 449, 1003–1007.

Barlan, K. and Gelfand, V.I. (2017). Microtubule-based transport and the distribution, tethering, and organization of organelles. *Cold Spring Harb Perspect Biol*, 9, a025817.

Barlow, J.P. and Solomon, T.P. (2018). Do skeletal muscle-secreted factors influence the function of pancreatic beta-cells? *Am J Physiol Endocrinol Metab*, 314, E297–E307.

Barlow, D.P., Stöger, R., Herrmann, B.G., Saito, K., Schweifer, N. (1991). The mouse insulin-like growth factor type-2 receptor is imprinted and closely linked to the Tme locus. *Nature*, 349, 84–87.

Barr, M.L. and Bertram, E.G. (1949). A morphological distinction between neurones of the male and female, and the behaviour of the nucleolar satellite during accelerated nucleoprotein synthesis. *Nature*, 163, 676.

Bascos, N.A.D. and Landry, S.J. (2019). A history of molecular chaperone structures in the protein data bank. *Int J Mol Sci*, 20, 6195.

Bashyam, M.D., Animireddy, S., Bala, P., Naz, A., George, S.A. (2019). The Yin and Yang of cancer genes. *Gene*, 704, 121–133.

Bass, J. and Takahashi, J.S. (2010). Circadian integration of metabolism and energetics. *Science*, 330, 1349–1354.

Bassett, J.H. and Williams, G.R. (2008). Critical role of the hypothalamic-pituitary-thyroid axis in bone. *Bone*, 43, 418–426.

Bastepe, M. and Jüppner, H. (2003). Pseudohypoparathyroidism and mechanisms of resistance toward multiple hormones: Molecular evidence to clinical presentation. *The Journal of Clinical Endocrinology & Metabolism*, 88, 4055–4058.

Bates, S.E. (2020). Epigenetic therapies for cancer. *N Engl J Med*, 383, 650–663.

Bayliss, W.M. and Starling, E.H. (1902). The mechanism of pancreatic secretion. *J Physiol*, 28, 325–353.

Baylor, D.A., Lamb, T.D., Yau, K.W. (1979). Responses of retinal rods to single photons. *J Physiol*, 288, 613–634.

Beadle, G.W. and Tatum, E.L. (1941). Genetic control of biochemical reactions in neurospora. *Proc Natl Acad Sci USA*, 27, 499–506.

Begemann, K., Neumann, A.M., Oster, H. (2020). Regulation and function of extra-SCN circadian oscillators in the brain. *Acta Physiol (Oxf)*, 229, e13446.

Bennett, S. (1996). A brief history of automatic control. *IEEE Control Systems Magazine*, 16, 17–25.

Benninger, R.K., Zhang, M., Head, W.S., Satin, L.S., Piston, D.W. (2008). Gap junction coupling and calcium waves in the pancreatic islet. *Biophysical Journal*, 95, 5048–5061.

Benoit, J. and Assenmacher, I. (1954). Comparative sensitivity of superficial and deep receptors in photosexual reflex in duck. *Comptes rendus hebdomadaires des seances de l'Academie des sciences*, 239, 105–107.

Berg, G., Rybakova, D., Fischer, D., Cernava, T., Verges, M.C., Charles, T., Chen, X., Cocolin, L., Eversole, K., Corral, G.H. et al. (2020). Microbiome definition re-visited: Old concepts and new challenges. *Microbiome*, 8, 103.

Bergman, R.N. and Bucolo, R.J. (1974). Interaction of insulin and glucose in the control of hepatic glucose balance. *Am J Physiol*, 227, 1314–1322.

Bernard, C. (1849). Chiens rendus diabetiques. *C R Soc Bio*, 1, 60.

Bernard, C. (1879). *Leçons sur les phénomènes de la vie commune aux animaux et aux végétaux*. Baillière, Paris.

Berwick, D.C., Hers, I., Heesom, K.J., Moule, S.K., Tavareá, J.M. (2002). The identification of ATP-citrate lyase as a protein kinase B (Akt) substrate in primary adipocytes. *J Biol Chem*, 277, 33895–33900.

Besedovsky, H.O. and Del Rey, A. (1996). Immune-neuro-endocrine interactions: Facts and hypotheses. *Endocr Rev*, 17, 64–102.

Besedovsky, H.O., Del Rey, A., Sorkin, E., Dinarello, C. (1986). Immunoregulatory feedback between interleukin-1 and glucocorticoid hormones. *Science*, 233, 652–654.

Betancourt, A.M., Eltoum, I.A., Desmond, R.A., Russo, J., Lamartiniere, C.A. (2010). In utero exposure to bisphenol A shifts the window of susceptibility for mammary carcinogenesis in the rat. *Environ Health Perspect*, 118, 1614–1619.

Bhutani, N., Brady, J.J., Damian, M., Sacco, A., Corbel, S.Y., Blau, H.M. (2010). Reprogramming towards pluripotency requires AID-dependent DNA demethylation. *Nature*, 463, 1042–1047.

Bianconi, E., Piovesan, A., Facchin, F., Beraudi, A., Casadei, R., Frabetti, F., Vitale, L., Pelleri, M.C., Tassani, S., Piva, F. et al. (2013). An estimation of the number of cells in the human body. *Ann Hum Biol*, 40, 463–471.

Bikle, D.D. (2021). The free hormone hypothesis: When, why, and how to measure the free hormone levels to assess vitamin D, thyroid, sex hormone, and cortisol status. *JBMR Plus*, 5, e10418.

Blackburn, E.H. (2001). Switching and signaling at the telomere. *Cell*, 106, 661–673.

Blackburn, E.H., Greider, C.W., Szostak, J.W. (2006). Telomeres and telomerase: The path from maize, Tetrahymena and yeast to human cancer and aging. *Nat Med*, 12, 1133–1138.

Blalock, J.E. and Smith, E.M. (1980). Human leukocyte interferon: Structural and biological relatedness to adrenocorticotropic hormone and endorphins. *Proc Natl Acad Sci*, 77, 5972–5974.

Blanco, A.M., Calo, J., Soengas, J.L. (2021). The gut-brain axis in vertebrates: Implications for food intake regulation. *J Exp Biol*, 224, jeb231571.

Blau, J.E. and Collins, M.T. (2015). The PTH-vitamin D-FGF23 axis. *Rev Endocr Metab Disord*, 16, 165–174.

Bluestone, J.A., Herold, K., Eisenbarth, G. (2010). Genetics, pathogenesis and clinical interventions in type 1 diabetes. *Nature*, 464, 1293–1300.

Blunt, J.W., Deluca, H.F., Schnoes, H.K. (1968). 25–hydroxycholecalciferol. A biologically active metabolite of vitamin D3. *Biochemistry*, 7, 3317–3322.

Bodnar, A.G., Ouellette, M., Frolkis, M., Holt, S.E., Chiu, C.P., Morin, G.B., Harley, C.B., Shay, J.W., Lichtsteiner, S., Wright, W.E. (1998). Extension of life-span by introduction of telomerase into normal human cells. *Science*, 279, 349–352.

Böhm, S.K., Grady, E.F., Bunnett, N.W. (1997). Regulatory mechanisms that modulate signalling by G-protein-coupled receptors. *Biochem J*, 322, 1–18.

Bohr, C., Hasselbalch, K., Krogh, A. (1904). Ueber einen in biologischer Beziehung wichtigen Einfluss, den die Kohlensäurespannung des Blutes auf dessen Sauerstoffbindung übt1. *Skandinavisches Archiv Für Physiologie*, 16, 402–412.

Bonilla, F.A. and Oettgen, H.C. (2010). Adaptive immunity. *Journal of Allergy and Clinical Immunology*, 125, S33–S40.

Borg, W.P., Sherwin, R.S., During, M.J., Borg, M.A., Shulman, G.I. (1995). Local ventromedial hypothalamus glucopenia triggers counterregulatory hormone release. *Diabetes*, 44, 180–184.

Börjesson, S.I. and Elinder, F. (2008). Structure, function, and modification of the voltage sensor in voltage-gated ion channels. *Cell Biochem Biophys*, 52, 149–174.

Boucher, J., Kleinridders, A., Kahn, C.R. (2014). Insulin receptor signaling in normal and insulin-resistant states. *Cold Spring Harb Perspect Biol*, 6(1), a009191.

Boulais, P.E. and Frenette, P.S. (2015). Making sense of hematopoietic stem cell niches. *Blood*, 125, 2621–2629.

Boulanger, M.J., Chow, D.-C., Brevnova, E.E., Garcia, K.C. (2003). Hexameric structure and assembly of the interleukin-6/IL-6 α-receptor/gp130 complex. *Science*, 300, 2101–2104.

Boulton, T.G., Nye, S.H., Robbins, D.J., Ip, N.Y., Radziejewska, E., Morgenbesser, S.D., Depinho, R.A., Panayotatos, N., Cobb, M.H., Yancopoulos, G.D. (1991). ERKs: A family of protein-serine/threonine kinases that are activated and tyrosine phosphorylated in response to insulin and NGF. *Cell*, 65, 663–675.

Bove-Fenderson, E. and Mannstadt, M. (2018). Hypocalcemic disorders. *Best Pract Res Clin Endocrinol Metab*, 32, 639–656.

Bowman, W.C. (2006). Neuromuscular block. *Br J Pharmacol*, 147(Suppl 1), S277–S286.

Bowman, W.C. (2013). *Pharmacology of Neuromuscular Function*. Butterworth-Heinemann, Boston.

Breasted, J.H. (1905). *Egypt through the Stereoscope: A Journey through the Land of the Pharaohs*. Underwood & Underwood, New York.

Breer, H., Eberle, J., Frick, C., Haid, D., Widmayer, P. (2012). Gastrointestinal chemosensation: Chemosensory cells in the alimentary tract. *Histochem Cell Biol*, 138, 13–24.

Brink, R.A. (1956). A genetic change associated with the R locus in maize which is directed and potentially reversible. *Genetics*, 41, 872.

Brook, C.G., Lloyd, J.K., Wolf, O.H. (1972). Relation between age of onset of obesity and size and number of adipose cells. *Br Med J*, 2, 25–27.

Brooks, G.A. (1998). Mammalian fuel utilization during sustained exercise. *Comp Biochem Physiol B Biochem Mol Biol*, 120, 89–107.

Brooks, G.A. and Mercier, J. (1994). Balance of carbohydrate and lipid utilization during exercise: The "crossover" concept. *J Appl Physiol (1985)*, 76, 2253–2261.

Brotto, M. and Bonewald, L. (2015). Bone and muscle: Interactions beyond mechanical. *Bone*, 80, 109–114.

Broussard, S.R., McCusker, R.H., Novakofski, J.E., Strle, K., Shen, W.H., Johnson, R.W., Dantzer, R., Kelley, K.W. (2004). IL-1beta impairs insulin-like growth factor i-induced differentiation and downstream activation signals of the insulin-like growth factor i receptor in myoblasts. *J Immunol*, 172, 7713–7720.

Brown, E.M. and Macleod, R.J. (2001). Extracellular calcium sensing and extracellular calcium signaling. *Physiol Rev*, 81, 239–297.

Brown, E.M., Gamba, G., Riccardi, D., Lombardi, M., Butters, R., Kifor, O., Sun, A., Hediger, M.A., Lytton, J., Hebert, S.C. (1993). Cloning and characterization of an extracellular Ca(2+)-sensing receptor from bovine parathyroid. *Nature*, 366, 575–580.

Brownell, J.E., Zhou, J., Ranalli, T., Kobayashi, R., Edmondson, D.G., Roth, S.Y., Allis, C.D. (1996). Tetrahymena histone acetyltransferase A: A homolog to yeast Gcn5p linking histone acetylation to gene activation. *Cell*, 84, 843–851.

Bucay, N., Sarosi, I., Dunstan, C.R., Morony, S., Tarpley, J., Capparelli, C., Scully, S., Tan, H.L., Xu, W., Lacey, D.L. et al. (1998). Osteoprotegerin-deficient mice develop early onset osteoporosis and arterial calcification. *Genes Dev*, 12, 1260–1268.

Buchman, T.G. (2002). The community of the self. *Nature*, 420, 246–251.

Bulynko, Y.A. and O'Malley, B.W. (2011). Nuclear receptor coactivators: Structural and functional biochemistry. *Biochemistry*, 50, 313–328.

Burcelin, R., Crivelli, V., Perrin, C., Da Costa, A., Mu, J., Kahn, B.B., Birnbaum, M.J., Kahn, C.R., Vollenweider, P., Thorens, B. (2003). GLUT4, AMP kinase, but not the insulin receptor, are required for hepatoportal glucose sensor–stimulated muscle glucose utilization. *The Journal of Clinical Investigation*, 111, 1555–1562.

Burgdorf, S. and Kurts, C. (2008). Endocytosis mechanisms and the cell biology of antigen presentation. *Curr Opin Immunol*, 20, 89–95.

Burgoyne, R.D. (2007). Neuronal calcium sensor proteins: Generating diversity in neuronal Ca2+ signalling. *Nat Rev Neurosci*, 8, 182–193.

Burgoyne, R.D., Helassa, N., McCue, H.V., Haynes, L.P. (2019). Calcium sensors in neuronal function and dysfunction. *Cold Spring Harb Perspect Biol*, 11(5), a035154.

Burnet, M. (1957). Cancer: A biological approach. III. Viruses associated with neoplastic conditions. IV. Practical applications. *Br Med J*, 1, 841–847.

Busquets-Garcia, A., Bolanos, J.P., Marsicano, G. (2022). Metabolic messengers: Endocannabinoids. *Nat Metab*, 4, 848–855.

Butt, R.L. and Volkoff, H. (2019). Gut microbiota and energy homeostasis in fish. *Front Endocrinol*, 10, 9.

Buttgereit, F. and Brand, M.D. (1995). A hierarchy of ATP-consuming processes in mammalian cells. *Biochem J*, 312(Pt 1), 163–167.

Buzzai, M., Bauer, D.E., Jones, R.G., Deberardinis, R.J., Hatzivassiliou, G., Elstrom, R.L., Thompson, C.B. (2005). The glucose dependence of Akt-transformed cells can be reversed by pharmacologic activation of fatty acid beta-oxidation. *Oncogene*, 24, 4165–4173.

Cahill Jr., G.F. (2006). Fuel metabolism in starvation. *Annu Rev Nutr*, 26, 1–22.

Cannon, W.B. (1929). Organization for physiological homeostasis. *Physiological Reviews*, 9, 399–431.

Cao, Y., Yao, Z., Sarkar, D., Lawrence, M., Sanchez, G.J., Parker, M.H., Macquarrie, K.L., Davison, J., Morgan, M.T., Ruzzo, W.L. et al. (2010). Genome-wide MyoD binding in skeletal muscle cells: A potential for broad cellular reprogramming. *Dev Cell*, 18, 662–674.

Care, A.D., Sherwood, L.M., Potts Jr., J.T., Aurbach, G.D. (1966). Perfusion of the isolated parathyroid gland of the goat and sheep. *Nature*, 209, 55–57.

Carling, D., Zammit, V.A., Hardie, D.G. (1987). A common bicyclic protein kinase cascade inactivates the regulatory enzymes of fatty acid and cholesterol biosynthesis. *FEBS Lett*, 223, 217–222.

Carlsson, A., Lindqvist, M., Magnusson, T., Waldeck, B. (1958). On the presence of 3-hydroxytyramine in brain. *Science*, 127, 471.

Carpenter, K.J. and Sutherland, B. (1995). Eijkman's contribution to the discovery of vitamins. *J Nutr*, 125, 155–163.

Carpenter, G., Lembach, K.J., Morrison, M.M., Cohen, S. (1975). Characterization of the binding of 125-I-labeled epidermal growth factor to human fibroblasts. *J Biol Chem*, 250, 4297–4304.

Carpenter, G., King Jr., L., Cohen, S. (1978). Epidermal growth factor stimulates phosphorylation in membrane preparations in vitro. *Nature*, 276, 409–410.

Carroll, B. and Dunlop, E.A. (2017). The lysosome: A crucial hub for AMPK and mTORC1 signalling. *Biochem J*, 474, 1453–1466.

Cartaud, J., Benedetti, E.L., Cohen, J.B., Meunier, J.-C., Changeux, J.-P. (1973). Presence of a lattice structure in membrane fragments rich in nicotinic receptor protein from the electric organ of Torpedo marmorata. *FEBS Lett*, 33, 109–113.

Castillo-Armengol, J., Fajas, L., Lopez-Mejia, I.C. (2019). Inter-organ communication: A gatekeeper for metabolic health. *EMBO Rep*, 20, e47903.

de Castro, F. (1929). Über die Struktur und Innervation des Glomus caroticum beim Menschen und bei den Säugetieren. *Zeitschrift für Anatomie und Entwicklungsgeschichte*, 89, 250–265.

de Castro, F. (2009). Towards the sensory nature of the carotid body: Hering, de Castro and Heymans. *Front Neuroanat*, 3, 1–11.

Cattaneo, F., Guerra, G., Parisi, M., De Marinis, M., Tafuri, D., Cinelli, M., Ammendola, R. (2014). Cell-surface receptors transactivation mediated by g protein-coupled receptors. *Int J Mol Sci*, 15, 19700–19728.

Cavaillès, V., Dauvois, S., Danielian, P.S., Parker, M.G. (1994). Interaction of proteins with transcriptionally active estrogen receptors. *Proc Natl Acad Sci USA*, 91, 10009–10013.

Chalfie, M. (2009). Neurosensory mechanotransduction. *Nat Rev Mol Cell Biol*, 10, 44–52.

Chamouni, A., Schreiweis, C., Oury, F. (2015). Bone, brain & beyond. *Rev Endocr Metab Disord*, 16, 99–113.

Chan, S.J. and Steiner, D.F. (1977). Preproinsulin, a new precursor in insulin biosynthesis. *Trends Biochem Sci*, 2, 254–256.

Chang, J.R., Ghafouri, M., Mukerjee, R., Bagashev, A., Chabrashvili, T., Sawaya, B.E. (2012). Role of p53 in neurodegenerative diseases. *Neurodegener Dis*, 9, 68–80.

Changeux, J.P. (1961). The feedback control mechanisms of biosynthetic L-threonine deaminase by L-isoleucine. *Cold Spring Harb Symp Quant Biol*, 26, 313–318.

Changeux, J.P. (2012). The nicotinic acetylcholine receptor: The founding father of the pentameric ligand-gated ion channel superfamily. *J Biol Chem*, 287, 40207–40215.

Changeux, J.P. (2020). Discovery of the first neurotransmitter receptor: The acetylcholine nicotinic receptor. *Biomolecules*, 10(4).

Chavez-Abiega, S., Mos, I., Centeno, P.P., Elajnaf, T., Schlattl, W., Ward, D.T., Goedhart, J., Kallay, E. (2020). Sensing extracellular calcium – An insight into the structure and function of the calcium-sensing receptor (CaSR). *Adv Exp Med Biol*, 1131, 1031–1063.

Chen, L. and Flies, D.B. (2013). Molecular mechanisms of T cell co-stimulation and co-inhibition. *Nat Rev Immunol*, 13, 227–242.

Cherrington, A.D. (1999). Banting Lecture 1997. Control of glucose uptake and release by the liver in vivo. *Diabetes*, 48, 1198–1214.

Cheung, H.H., Lee, T.L., Rennert, O.M., Chan, W.Y. (2009). DNA methylation of cancer genome. *Birth Defects Res C Embryo Today*, 87, 335–350.

Chin, D. and Means, A.R. (2000). Calmodulin: A prototypical calcium sensor. *Trends Cell Biol*, 10, 322–328.

Chock, P.B., Rhee, S.G., Stadtman, E.R. (1980). Interconvertible enzyme cascades in cellular regulation. *Annu Rev Biochem*, 49, 813–841.

Choudhuri, S. (2011). From Waddington's epigenetic landscape to small noncoding RNA: Some important milestones in the history of epigenetics research. *Toxicol Mech Methods*, 21, 252–274.

Chovatiya, R. and Medzhitov, R. (2014). Stress, inflammation, and defense of homeostasis. *Mol Cell*, 54, 281–288.

Chow, L.T., Gelinas, R.E., Broker, T.R., Roberts, R.J. (1977). An amazing sequence arrangement at the 5' ends of adenovirus 2 messenger RNA. *Cell*, 12, 1–8.

Chu, D.M., Antony, K.M., Ma, J., Prince, A.L., Showalter, L., Moller, M., Aagaard, K.M. (2016). The early infant gut microbiome varies in association with a maternal high-fat diet. *Genome Med*, 8, 77.

Chuaire, L. (2006). Telomere and Telomerase: Brief review of a history initiated by Hermann Müller and Barbara McClintock. *Colombia Médica*, 37, 332–335.

Clarke, I.J. (2015). Hypothalamus as an endocrine organ. *Compr Physiol*, 5, 217–253.

Clemmons, D.R. (2016). Role of IGF binding proteins in regulating metabolism. *Trends Endocrinol Metab*, 27, 375–391.

Clevers, H. (2011). The cancer stem cell: Premises, promises and challenges. *Nat Med*, 17, 313–319.

Cline, G.W., Magnusson, I., Rothman, D.L., Petersen, K.F., Laurent, D., Shulman, G.I. (1997). Mechanism of impaired insulin-stimulated muscle glucose metabolism in subjects with insulin-dependent diabetes mellitus. *J Clin Invest*, 99, 2219–2224.

Clos, J., Crépel, F., Legrand, C., Legrand, J., Rabie, A., Vigouroux, E. (1974). Thyroid physiology during the postnatal period in the rat: A study of the development of thyroid function and of the morphogenetic effects of thyroxine with special reference to cerebellar maturation. *Gen Comp Endocrinol*, 23, 178–192.

Coffman, R.L. (2006). Origins of the T(H)1–T(H)2 model: A personal perspective. *Nat Immunol*, 7, 539–541.

Cohen, S. (1962). Isolation of a mouse submaxillary gland protein accelerating incisor eruption and eyelid opening in the new-born animal. *J Biol Chem*, 237, 1555–1562.

Cohen, S. and Levi-Montalcini, R. (1957). Purification and properties of a nerve growth-promoting factor isolated from mouse sarcoma 180. *Cancer Res*, 17, 15–20.

Cohen, S., Bigazzi, P.E., Yoshida, T. (1974). Similarities of T cell function in cell-mediated immunity and antibody production. *Cell Immunol*, 12, 150–159.

Coleman, D.L. and Hummel, K.P. (1969). Effects of parabiosis of normal with genetically diabetic mice. *Am J Physiol*, 217, 1298–1304.

Colombo, C.V., Gnugnoli, M., Gobbini, E., Longhese, M.P. (2020). How do cells sense DNA lesions? *Biochem Soc Trans*, 48, 677–691.

Comninos, A.N., Jayasena, C.N., Dhillo, W.S. (2014). The relationship between gut and adipose hormones, and reproduction. *Hum Reprod Update*, 20, 153–174.

Connolly, K., Cho, Y.H., Duan, R., Fikes, J., Gregorio, T., Lafleur, D.W., Okoye, Z., Salcedo, T.W., Santiago, G., Ullrich, S. et al. (2001). In vivo inhibition of Fas ligand-mediated killing by TR6, a Fas ligand decoy receptor. *J Pharmacol Exp Ther*, 298, 25–33.

Conti, M., Richter, W., Mehats, C., Livera, G., Park, J.Y., Jin, C. (2003). Cyclic AMP-specific PDE4 phosphodiesterases as critical components of cyclic AMP signaling. *J Biol Chem*, 278, 5493–5496.

Cook, K.S., Min, H.Y., Johnson, D., Chaplinsky, R.J., Flier, J.S., Hunt, C.R., Spiegelman, B.M. (1987). Adipsin: A circulating serine protease homolog secreted by adipose tissue and sciatic nerve. *Science*, 237, 402–405.

Cooper, S.J. (2008). From Claude Bernard to Walter Cannon. Emergence of the concept of homeostasis. *Appetite*, 51, 419–427.

Copp, D.H. (1963). Calcitonin – A new hormone from the parathyroid which lowers blood calcium. *Oral Surg Oral Med Oral Pathol*, 16, 872–877.

Copp, D.H. and Cheney, B. (1962). Calcitonin – A hormone from the parathyroid which lowers the calcium-level of the blood. *Nature*, 193, 381–382.

Cori, C.F. and Cori, G.T. (1946). Carbohydrate metabolism. *Annual Review of Biochemistry*, 15, 193–218.

Cornish, J., Callon, K.E., Bava, U., Kamona, S.A., Cooper, G.J., Reid, I.R. (2001). Effects of calcitonin, amylin, and calcitonin gene-related peptide on osteoclast development. *Bone*, 29, 162–168.

Cortese, R., Lu, L., Yu, Y., Ruden, D., Claud, E.C. (2016). Epigenome-microbiome crosstalk: A potential new paradigm influencing neonatal susceptibility to disease. *Epigenetics*, 11, 205–215.

Cowley, M.A., Smart, J.L., Rubinstein, M., Cerdán, M.G., Diano, S., Horvath, T.L., Cone, R.D., Low, M.J. (2001). Leptin activates anorexigenic POMC neurons through a neural network in the arcuate nucleus. *Nature*, 411, 480–484.

Coyle, E.F., Jeukendrup, A.E., Wagenmakers, A.J., Saris, W.H. (1997). Fatty acid oxidation is directly regulated by carbohydrate metabolism during exercise. *Am J Physiol*, 273, E268–E275.

Crick, F. (1974). The double helix: A personal view. *Nature*, 248, 766–769.

Cropley, J.E., Suter, C.M., Beckman, K.B., Martin, D.I. (2006). Germ-line epigenetic modification of the murine A vy allele by nutritional supplementation. *Proc Natl Acad Sci USA*, 103, 17308–17312.

Cross, D.A., Alessi, D.R., Cohen, P., Andjelkovich, M., Hemmings, B.A. (1995). Inhibition of glycogen synthase kinase-3 by insulin mediated by protein kinase B. *Nature*, 378, 785–789.

Cui, N., Hu, M., Khalil, R.A. (2017). Biochemical and biological attributes of matrix metalloproteinases. *Prog Mol Biol Transl Sci*, 147, 1–73.

Dai, Z., Ramesh, V., Locasale, J.W. (2020). The evolving metabolic landscape of chromatin biology and epigenetics. *Nat Rev Genet*, 21, 737–753.

Dale, H.H. (1935). The Harveian oration on some epochs in medical research. *British Medical Journal*, 2, 771.

Dale, H.H., Feldberg, W., Vogt, M. (1936). Release of acetylcholine at voluntary motor nerve endings. *The Journal of Physiology*, 86, 353–380.

Damian, R.T. (1964). Molecular mimicry: Antigen sharing by parasite and host and its consequences. *Am Nat*, 98, 129–149.

Danilova, N., Sakamoto, K.M., Lin, S. (2008). p53 family in development. *Mech Dev*, 125, 919–931.

Daub, H., Weiss, F.U., Wallasch, C., Ullrich, A. (1996). Role of transactivation of the EGF receptor in signalling by G-protein-coupled receptors. *Nature*, 379, 557–560.

Davie, J.R. (2003). Inhibition of histone deacetylase activity by butyrate. *J Nutr*, 133, 2485s–2493s.

Davis, R.L., Weintraub, H., Lassar, A.B. (1987). Expression of a single transfected cDNA converts fibroblasts to myoblasts. *Cell*, 51, 987–1000.

Dawson, A. (1979). Oxidation of cytosolic NADH formed during aerobic metabolism in mammalian cells. *Trends Biochem Sci*, 4, 171–176.

Day, F.L., Jorissen, R.N., Lipton, L., Mouradov, D., Sakthianandeswaren, A., Christie, M., Li, S., Tsui, C., Tie, J., Desai, J. et al. (2013). PIK3CA and PTEN gene and exon mutation-specific clinicopathologic and molecular associations in colorectal cancer. *Clin Cancer Res*, 19, 3285–3296.

De, M. (1929). Observation botanique. *Histoire de L'Académie Royale des Sciences Paris*, 35–36.

De Larco, J.E. and Todaro, G.J. (1978). Growth factors from murine sarcoma virus-transformed cells. *Proc Natl Acad Sci USA*, 75, 4001–4005.

De Lorenzo, A., Romano, L., Di Renzo, L., Di Lorenzo, N., Cenname, G., Gualtieri, P. (2020). Obesity: A preventable, treatable, but relapsing disease. *Nutrition*, 71, 110615.

De Mendoza, D. and Pilon, M. (2019). Control of membrane lipid homeostasis by lipid-bilayer associated sensors: A mechanism conserved from bacteria to humans. *Progr Lipid Res*, 76, 100996.

De Meyts, P., Roth, J., Neville Jr., D.M., Gavin 3rd, J.R., Lesniak, M.A. (1973). Insulin interactions with its receptors: Experimental evidence for negative cooperativity. *Biochem Biophys Res Commun*, 55, 154–161.

De Wulf, H. and Hers, H.G. (1967). The stimulation of glycogen synthesis and of glycogen synthetase in the liver by the administration of glucose. *Eur J Biochem*, 2, 50–56.

Dechiara, T.M., Robertson, E.J., Efstratiadis, A. (1991). Parental imprinting of the mouse insulin-like growth factor II gene. *Cell*, 64, 849–859.

Defilippi, P., Rosso, A., Dentelli, P., Calvi, C., Garbarino, G., Tarone, G., Pegoraro, L., Brizzi, M.F. (2005). β1 integrin and IL-3R coordinately regulate STAT5 activation and anchorage-dependent proliferation. *J Cell Biol*, 168, 1099–1108.

Degenhardt, K., Mathew, R., Beaudoin, B., Bray, K., Anderson, D., Chen, G., Mukherjee, C., Shi, Y., Gélinas, C., Fan, Y. et al. (2006). Autophagy promotes tumor cell survival and restricts necrosis, inflammation, and tumorigenesis. *Cancer Cell*, 10, 51–64.

Deichmann, U. (2016). Epigenetics: The origins and evolution of a fashionable topic. *Dev Biol*, 416, 249–254.

Delcourt, N., Bockaert, J., Marin, P. (2007). GPCR-jacking: From a new route in RTK signalling to a new concept in GPCR activation. *Trends Pharmacol Sci*, 28, 602–607.

Deleo, A.B., Jay, G., Appella, E., Dubois, G.C., Law, L.W., Old, L.J. (1979). Detection of a transformation-related antigen in chemically induced sarcomas and other transformed cells of the mouse. *Proc Natl Acad Sci USA*, 76, 2420–2424.

Deluca, H.F. (1975). The kidney as an endocrine organ involved in the function of vitamin D. *Am J Med*, 58, 39–47.

Deluca, H.F. (2014). History of the discovery of vitamin D and its active metabolites. *Bonekey Rep*, 3, 479.

Demaria, J.E., Nagy, G.M., Lerant, A.A., Fekete, M.I., Levenson, C.W., Freeman, M.E. (2000). Dopamine transporters participate in the physiological regulation of prolactin. *Endocrinology*, 141, 366–374.

Deng, G. and Sui, G. (2013). Noncoding RNA in oncogenesis: A new era of identifying key players. *Int J Mol Sci*, 14, 18319–18349.

Deng, J., Shimamura, T., Perera, S., Carlson, N.E., Cai, D., Shapiro, G.I., Wong, K.K., Letai, A. (2007). Proapoptotic BH3-only BCL-2 family protein BIM connects death signaling from epidermal growth factor receptor inhibition to the mitochondrion. *Cancer Res*, 67, 11867–11875.

Díaz, M., Garde, E., Lopez-Bermejo, A., de Zegher, F., Ibañez, L. (2013). Differential DNA methylation profile in infants born small-for-gestational-age: Association with markers of adiposity and insulin resistance from birth to age 24 months. *BMJ Open Diabetes Res Care*, 8, e001402.

Dibble, C.C. and Manning, B.D. (2013). Signal integration by mTORC1 coordinates nutrient input with biosynthetic output. *Nat Cell Biol*, 15, 555–564.

Dick, F.A., Goodrich, D.W., Sage, J., Dyson, N.J. (2018). Non-canonical functions of the RB protein in cancer. *Nat Rev Cancer*, 18, 442–451.

Di Liberto, V., Borroto-Escuela, D.O., Frinchi, M., Verdi, V., Fuxe, K., Belluardo, N., Mudò, G. (2017). Existence of muscarinic acetylcholine receptor (mAChR) and fibroblast growth factor receptor (FGFR) heteroreceptor complexes and their enhancement of neurite outgrowth in neural hippocampal cultures. *Biochim Biophys Acta Gen Subj*, 1861, 235–245.

Di Liberto, V., Mudo, G., Belluardo, N. (2019). Crosstalk between receptor tyrosine kinases (RTKs) and G protein-coupled receptors (GPCR) in the brain: Focus on heteroreceptor complexes and related functional neurotrophic effects. *Neuropharmacology*, 152, 67–77.

Dimova, D.K. and Dyson, N.J. (2005). The E2F transcriptional network: Old acquaintances with new faces. *Oncogene*, 24, 2810–2826.

Dinarello, C.A. (2007). Historical insights into cytokines. *Eur J Immunol*, 37(Suppl 1), S34–S45.

Dockray, G.J. (2014). Gastrointestinal hormones and the dialogue between gut and brain. *J Physiol*, 592, 2927–2941.

Dodd, G.T. and Tiganis, T. (2017). Insulin action in the brain: Roles in energy and glucose homeostasis. *J Neuroendocrinol*. doi: 10.1111/jne.12513.

Dohlman, H.G. (2002). Diminishing returns. *Nature*, 418, 591.

Dolinoy, D.C. (2008). The agouti mouse model: An epigenetic biosensor for nutritional and environmental alterations on the fetal epigenome. *Nutr Rev*, 66(Suppl 1), S7–S11.

Dolinoy, D.C., Weidman, J.R., Waterland, R.A., Jirtle, R.L. (2006). Maternal genistein alters coat color and protects Avy mouse offspring from obesity by modifying the fetal epigenome. *Environ Health Perspect*, 114, 567–572.

Dollet, L., Pendergrast, L.A., Zierath, J.R. (2021). The role of the molecular circadian clock in human energy homeostasis. *Curr Opin Lipidol*, 32, 16–23.

Domene, H.M. and Fierro-Carrion, G. (2018). Genetic disorders of GH action pathway. *Growth Horm IGF Res*, 38, 19–23.

Dominguez-Villar, M. and Hafler, D.A. (2018). Regulatory T cells in autoimmune disease. *Nat Immunol*, 19, 665–673.

Donato Jr., J., Cravo, R.M., Frazão, R., Gautron, L., Scott, M.M., Lachey, J., Castro, I.A., Margatho, L.O., Lee, S., Lee, C. et al. (2011). Leptin's effect on puberty in mice is relayed by the ventral premammillary nucleus and does not require signaling in Kiss1 neurons. *J Clin Invest*, 121, 355–368.

Drucker, D.J. (2016). Evolving concepts and translational relevance of enteroendocrine cell biology. *J Clin Endocrinol Metab*, 101, 778–786.

Duca, F.A. and Yue, J.T.Y. (2014). Fatty acid sensing in the gut and the hypothalamus: In vivo and in vitro perspectives. *Mol Cell Endocrinol*, 397, 23–33.

Ducy, P., Amling, M., Takeda, S., Priemel, M., Schilling, A.F., Beil, F.T., Shen, J., Vinson, C., Rueger, J.M., Karsenty, G. (2000). Leptin inhibits bone formation through a hypothalamic relay: A central control of bone mass. *Cell*, 100, 197–207.

Duez, H. and Staels, B. (2008). The nuclear receptors Rev-erbs and RORs integrate circadian rhythms and metabolism. *Diab Vasc Dis Res*, 5, 82–88.

Dunn, G.P., Bruce, A.T., Ikeda, H., Old, L.J., Schreiber, R.D. (2002). Cancer immunoediting: From immunosurveillance to tumor escape. *Nat Immunol*, 3, 991–998.

Eagle, H. (1955). Nutrition needs of mammalian cells in tissue culture. *Science*, 122, 501–514.

Eaves, C.J. (2015). Hematopoietic stem cells: Concepts, definitions, and the new reality. *Blood*, 125, 2605–2613.

Eckardt, K., Görgens, S.W., Raschke, S., Eckel, J. (2014). Myokines in insulin resistance and type 2 diabetes. *Diabetologia*, 57, 1087–1099.

Edgerton, D.S., Kraft, G., Smith, M.S., Moore, L.M., Farmer, B., Scott, M., Moore, M.C., Nauck, M.A., Cherrington, A.D. (2019). Effect of portal glucose sensing on incretin hormone secretion in a canine model. *Am J Physiol Endocrinol Metab*, 317, E244–e249.

Edwards, G.M., Wilford, F.H., Liu, X., Hennighausen, L., Djiane, J., Streuli, C.H. (1998). Regulation of mammary differentiation by extracellular matrix involves protein-tyrosine phosphatases. *J Biol Chem*, 273, 9495–9500.

Ehrlich, P. (1900). Croonian lecture. On immunity with special reference to cell life. *Proc Royal Soc London*, 66, 424–448.

Ehrlich, P. (1909). On the current state of cancer research. *Ned Tijdschr Geneeskd*, 5, 273–290.

Eichel, K. and Von Zastrow, M. (2018). Subcellular organization of GPCR signaling. *Trends Pharmacol Sci*, 39, 200–208.

Eimar, H., Tamimi, I., Murshed, M., Tamimi, F. (2013). Cholinergic regulation of bone. *J Musculoskelet Neuronal Interact*, 13, 124–132.

Eknoyan, G. (2004). Emergence of the concept of endocrine function and endocrinology. *Adv Chronic Kidney Dis*, 11, 371–376.

Elchebly, M., Payette, P., Michaliszyn, E., Cromlish, W., Collins, S., Loy, A.L., Normandin, D., Cheng, A., Himms-Hagen, J., Chan, C.C. et al. (1999). Increased insulin sensitivity and obesity resistance in mice lacking the protein tyrosine phosphatase-1B gene. *Science*, 283, 1544–1548.

Elenkov, I.J. (2004). Glucocorticoids and the Th1/Th2 balance. *Ann NY Acad Sci*, 1024, 138–146.

Elenkov, I.J., Wilder, R.L., Chrousos, G.P., Vizi, E.S. (2000). The sympathetic nerve – An integrative interface between two supersystems: The brain and the immune system. *Pharmacol Rev*, 52, 595–638.

Elliott, T.R. (1905). The action of adrenalin. *J Physiol*, 32, 401–467.

Elstrom, R.L., Bauer, D.E., Buzzai, M., Karnauskas, R., Harris, M.H., Plas, D.R., Zhuang, H., Cinalli, R.M., Alavi, A., Rudin, C.M. et al. (2004). Akt stimulates aerobic glycolysis in cancer cells. *Cancer Res*, 64, 3892–3899.

Endo, T.A., Masuhara, M., Yokouchi, M., Suzuki, R., Sakamoto, H., Mitsui, K., Matsumoto, A., Tanimura, S., Ohtsubo, M., Misawa, H. et al. (1997). A new protein containing an SH2 domain that inhibits JAK kinases. *Nature*, 387, 921–924.

Eriksson, J.G., Lindi, V., Uusitupa, M., Forsén, T.J., Laakso, M., Osmond, C., Barker, D.J. (2002). The effects of the Pro12Ala polymorphism of the peroxisome proliferator-activated receptor-gamma2 gene on insulin sensitivity and insulin metabolism interact with size at birth. *Diabetes*, 51, 2321–2324.

Esteller, M. (2008). Epigenetics in cancer. *N Engl J Med*, 358, 1148–1159.

Evans, H.M. and Long, J. (1921). The effect of the anterior lobe administered intraperitoneally upon growth maturation, and oestrus cycles of the rat. *Anat Rec*, 21, 62–63.

Evans, W.H. and Martin, P.E. (2002). Gap junctions: Structure and function. *Mol Membr Biol*, 19, 121–136.

Fan, X., Kraynak, J., Knisely, J.P.S., Formenti, S.C., Shen, W.H. (2020). PTEN as a guardian of the genome: Pathways and targets. *Cold Spring Harb Perspect Med*, 10, a036194.

Fang, J.Y., Cheng, Z.H., Chen, Y.X., Lu, R., Yang, L., Zhu, H.Y., Lu, L.G. (2004). Expression of Dnmt1, demethylase, MeCP2 and methylation of tumor-related genes in human gastric cancer. *World J Gastroenterol*, 10, 3394–3398.

Farooqi, S. and O'Rahilly, S. (2006). Genetics of obesity in humans. *Endocr Rev*, 27, 710–718.

Farooqi, I.S., Jebb, S.A., Langmack, G., Lawrence, E., Cheetham, C.H., Prentice, A.M., Hughes, I.A., McCamish, M.A., O'Rahilly, S. (1999). Effects of recombinant leptin therapy in a child with congenital leptin deficiency. *N Engl J Med*, 341, 879–884.

Fasshauer, M. and Blüher, M. (2015). Adipokines in health and disease. *Trends Pharmacol Sci*, 36, 461–470.

Fearon, E.R. (2011). Molecular genetics of colorectal cancer. *Annu Rev Pathol*, 6, 479–507.

Feinberg, A.P. and Vogelstein, B. (1983). Hypomethylation distinguishes genes of some human cancers from their normal counterparts. *Nature*, 301, 89–92.

Feliciello, A., Gottesman, M.E., Avvedimento, E.V. (2001). The biological functions of A-kinase anchor proteins. *J Mol Biol*, 308, 99–114.

Ferland-McCollough, D., Fernandez-Twinn, D.S., Cannell, I.G., David, H., Warner, M., Vaag, A.A., Bork-Jensen, J., Brøns, C., Gant, T.W., Willis, A.E. et al. (2012). Programming of adipose tissue miR-483–3p and GDF-3 expression by maternal diet in type 2 diabetes. *Cell Death Differ*, 19, 1003–1012.

Filhoulaud, G., Guilmeau, S., Dentin, R., Girard, J., Postic, C. (2013). Novel insights into ChREBP regulation and function. *Trends Endocrinol Metab*, 24, 257–268.

Fink, G. (1976). The development of the releasing factor concept. *Clin Endocrinol*, 5, s245–s260.

Finlay, C.A., Hinds, P.W., Levine, A.J. (1989). The p53 proto-oncogene can act as a suppressor of transformation. *Cell*, 57, 1083–1093.

Firth, S.M. and Baxter, R.C. (2002). Cellular actions of the insulin-like growth factor binding proteins. *Endocr Rev*, 23, 824–854.

Fischer, E.H. and Krebs, E.G. (1955). Conversion of phosphorylase b to phosphorylase a in muscle extracts. *J Biol Chem*, 216, 121–132.

Fischer, E.H., Graves, D.J., Crittenden, E.R., Krebs, E.G. (1959). Structure of the site phosphorylated in the phosphorylase b to a reaction. *J Biol Chem*, 234, 1698–1704.

Fisher, J.W. (1983). Control of erythropoietin production. *Proc Soc Exp Biol Med*, 173, 289–305.

Flajolet, M., Wang, Z., Futter, M., Shen, W., Nuangchamnong, N., Bendor, J., Wallach, I., Nairn, A.C., Surmeier, D.J., Greengard, P. (2008). FGF acts as a co-transmitter through adenosine A(2A) receptor to regulate synaptic plasticity. *Nat Neurosci*, 11, 1402–1409.

Fletcher, W.M. (1907). Lactic acid in amphibian muscle. *J Physiol*, 35, 247–309.

Flier, J.S., Kahn, C.R., Jarrett, D.B., Roth, J. (1977). Autoantibodies to the insulin receptor. Effect on the insulin-receptor interaction in IM-9 lymphocytes. *J Clin Invest*, 60, 784–794.

Folkman, J. (2003). Fundamental concepts of the angiogenic process. *Curr Mol Med*, 3, 643–651.

Fong, A.P., Yao, Z., Zhong, J.W., Cao, Y., Ruzzo, W.L., Gentleman, R.C., Tapscott, S.J. (2012). Genetic and epigenetic determinants of neurogenesis and myogenesis. *Dev Cell*, 22, 721–735.

Fontaine, C., Dubois, G., Duguay, Y., Helledie, T., Vu-Dac, N., Gervois, P., Soncin, F., Mandrup, S., Fruchart, J.C., Fruchart-Najib, J. et al. (2003). The orphan nuclear receptor Rev-Erbalpha is a peroxisome proliferator-activated receptor (PPAR) gamma target gene and promotes PPARgamma-induced adipocyte differentiation. *J Biol Chem*, 278, 37672–37680.

Ford, C.E., Hamerton, J.L., Barnes, D.W., Loutit, J.F. (1956). Cytological identification of radiation-chimaeras. *Nature*, 177, 452–454.

Fraga, M.F., Ballestar, E., Villar-Garea, A., Boix-Chornet, M., Espada, J., Schotta, G., Bonaldi, T., Haydon, C., Ropero, S., Petrie, K. et al. (2005). Loss of acetylation at Lys16 and trimethylation at Lys20 of histone H4 is a common hallmark of human cancer. *Nat Genet*, 37, 391–400.

Frakes, A.E. and Dillin, A. (2017). The UPR(ER): Sensor and coordinator of organismal homeostasis. *Mol Cell*, 66, 761–771.

Franklin, T.B., Russig, H., Weiss, I.C., Gräff, J., Linder, N., Michalon, A., Vizi, S., Mansuy, I.M. (2010). Epigenetic transmission of the impact of early stress across generations. *Biol Psychiatry*, 68, 408–415.

Fraser, R. and Lin, C.J. (2016). Epigenetic reprogramming of the zygote in mice and men: On your marks, get set, go! *Reproduction*, 152, R211–R222.

Fredholm, B. and Rosell, S. (1968). Effects of adrenergic blocking agents on lipid mobilization from canine subcutaneous adipose tissue after sympathetic nerve stimulation. *J Pharmacol Exp Ther*, 159, 1–7.

Frost, H.M. (1990). Skeletal structural adaptations to mechanical usage (SATMU) 2: Redefining Wolff's law: The remodeling problem. *Anat Rec*, 226, 414–422.

Fuchs, T.L., Bell, S.E., Chou, A., Gill, A.J. (2019). Revisiting the significance of prominent C cells in the thyroid. *Endocr Pathol*, 30, 113–117.

Fujii, H. (2007). Mechanisms of signal transduction from receptors of type I and type II cytokines. *J Immunotoxicol*, 4, 69–76.

Fujisawa, T. and Filippakopoulos, P. (2017). Functions of bromodomain-containing proteins and their roles in homeostasis and cancer. *Nat Rev Mol Cell Biol*, 18, 246–262.

Fukumoto, S. and Shimizu, Y. (2011). Fibroblast growth factor 23 as a phosphotropic hormone and beyond. *J Bone Miner Metab*, 29, 507–514.

Furth, J., Kahn, M.C., Breedis, C. (1937). The transmission of leukemia of mice with a single cell. *Am J Cancer*, 31, 276–282.

Gaasbeek, A. and Meinders, A.E. (2005). Hypophosphatemia: An update on its etiology and treatment. *Am J Med*, 118, 1094–1101.

Gabriel, B.M. and Zierath, J.R. (2019). Circadian rhythms and exercise – Re-setting the clock in metabolic disease. *Nat Rev Endocrinol*, 15, 197–206.

Gal-Yam, E.N., Saito, Y., Egger, G., Jones, P.A. (2008). Cancer epigenetics: Modifications, screening, and therapy. *Annu Rev Med*, 59, 267–280.

Garaulet, M., Ordovás, J.M., Gómez-Abellán, P., Martínez, J.A., Madrid, J.A. (2011). An approximation to the temporal order in endogenous circadian rhythms of genes implicated in human adipose tissue metabolism. *J Cell Physiol*, 226, 2075–2080.

García, J.M., Silva, J., Peña, C., Garcia, V., Rodríguez, R., Cruz, M.A., Cantos, B., Provencio, M., España, P., Bonilla, F. (2004). Promoter methylation of the PTEN gene is a common molecular change in breast cancer. *Genes Chromosom Cancer*, 41, 117–124.

Gaston, S. and Menaker, M. (1968). Pineal function: The biological clock in the sparrow? *Science*, 160, 1125–1127.

Gendrel, A.V. and Heard, E. (2014). Noncoding RNAs and epigenetic mechanisms during X-chromosome inactivation. *Annu Rev Cell Dev Biol*, 30, 561–580.

Giancotti, F.G. and Ruoslahti, E. (1999). Integrin signaling. *Science*, 285, 1028–1033.

Giancotti, F.G. and Tarone, G. (2003). Positional control of cell fate through joint integrin/receptor protein kinase signaling. *Annu Rev Cell Dev Biol*, 19, 173–206.

Gibbs, J., Young, R.C., Smith, G.P. (1973). Cholecystokinin decreases food intake in rats. *J Comp Physiol Psychol*, 84, 488–495.

Ginsberg, M.H. (2014). Integrin activation. *BMB Rep*, 47, 655–659.

Giudice, J. and Taylor, J.M. (2017). Muscle as a paracrine and endocrine organ. *Curr Opin Pharmacol*, 34, 49–55.

Glass, C.K. and Rosenfeld, M.G. (2000). The coregulator exchange in transcriptional functions of nuclear receptors. *Genes Dev*, 14, 121–141.

Glinka, Y. and Prud'homme, G.J. (2008). Neuropilin-1 is a receptor for transforming growth factor beta-1, activates its latent form, and promotes regulatory T cell activity. *J Leukoc Biol*, 84, 302–310.

Gluckman, P.D., Hanson, M.A., Buklijas, T., Low, F.M., Beedle, A.S. (2009). Epigenetic mechanisms that underpin metabolic and cardiovascular diseases. *Nat Rev Endocrinol*, 5, 401–408.

Godfrey, K.M., Sheppard, A., Gluckman, P.D., Lillycrop, K.A., Burdge, G.C., McLean, C., Rodford, J., Slater-Jefferies, J.L., Garratt, E., Crozier, S.R. et al. (2011). Epigenetic gene promoter methylation at birth is associated with child's later adiposity. *Diabetes*, 60, 1528–1534.

Goh, L.K. and Sorkin, A. (2013). Endocytosis of receptor tyrosine kinases. *Cold Spring Harb Perspect Biol*, 5, a017459.

Goldstein, M.S. (1961). Humoral nature of the hypoglycemic factor of muscular work. *Diabetes*, 10, 232–234.

Goldstein, J.L. and Brown, M.S. (2009). The LDL receptor. *Arteriosclerosis Thrombosis Vasc Biol*, 29, 431–438.

Goldstein, D.S. and Kopin, I.J. (2007). Evolution of concepts of stress. *Stress*, 10, 109–120.

Gomarasca, M., Banfi, G., Lombardi, G. (2020). Myokines: The endocrine coupling of skeletal muscle and bone. *Adv Clin Chem*, 94, 155–218.

Gomes, A.S., Ramos, H., Soares, J., Saraiva, L. (2018). p53 and glucose metabolism: An orchestra to be directed in cancer therapy. *Pharmacol Res*, 131, 75–86.

Goodier, M.R., Jonjić, S., Riley, E.M., Juranić Lisnić, V. (2018). CMV and natural killer cells: Shaping the response to vaccination. *Eur J Immunol*, 48, 50–65.

Govindan, M.V., Devic, M., Green, S., Gronemeyer, H., Chambon, P. (1985). Cloning of the human glucocorticoid receptor cDNA. *Nucleic Acids Res*, 13, 8293–8304.

Grabmayr, H., Romanin, C., Fahrner, M. (2020). STIM proteins: An ever-expanding family. *Int J Mol Sci*, 22, 378.

Gregory, K.J., Nguyen, E.D., Reiff, S.D., Squire, E.F., Stauffer, S.R., Lindsley, C.W., Meiler, J., Conn, P.J. (2013). Probing the metabotropic glutamate receptor 5 ($mGlu_5$) positive allosteric modulator (PAM) binding pocket: Discovery of point mutations that engender a "molecular switch" in PAM pharmacology. *Mol Pharmacol*, 83, 991–1006.

Greider, C.W. and Blackburn, E.H. (1985). Identification of a specific telomere terminal transferase activity in Tetrahymena extracts. *Cell*, 43, 405–413.

Gronemeyer, H. and Moras, D. (1995). Nuclear receptors. How to finger DNA. *Nature*, 375, 190–191.

Gruson, D., Buglioni, A., Burnett Jr., J.C. (2014). PTH: Potential role in management of heart failure. *Clin Chim Acta*, 433, 290–296.

Guan, K.L. and Xiong, Y. (2011). Regulation of intermediary metabolism by protein acetylation. *Trends Biochem Sci*, 36, 108–116.

Gudlur, A., Zeraik, A.E., Hirve, N., Hogan, P.G. (2020). STIM calcium sensing and conformational change. *J Physiol*, 598, 1695–1705.

Guerrero-Bosagna, C., Settles, M., Lucker, B., Skinner, M.K. (2010). Epigenetic transgenerational actions of vinclozolin on promoter regions of the sperm epigenome. *PLoS ONE*, 5.

Guillemin, R. (1978). Control of adenohypophysial functions by peptides of the central nervous system. *Harvey Lect*, 71, 71–131.

Guillemin, R. (2005). Hypothalamic hormones aka hypothalamic releasing factors. *J Endocrinol*, 184, 11–28.

Guillemin, R., Sakiz, E., Ward, D.N. (1965). Further purification of TSH-releasing factor (TRF) from sheep hypothalamic tissues, with observations on the amino acid composition. *Proc Soc Exp Biol Med*, 118, 1132–1137.

Guillou, A., Romanò, N., Steyn, F., Abitbol, K., Le Tissier, P., Bonnefont, X., Chen, C., Mollard, P., Martin, A.O. (2015). Assessment of lactotroph axis functionality in mice: Longitudinal monitoring of PRL secretion by ultrasensitive-ELISA. *Endocrinology*, 156, 1924–1930.

Guo, H.F. and Vander Kooi, C.W. (2015). Neuropilin functions as an essential cell surface receptor. *J Biol Chem*, 290, 29120–29126.

Gupta, R.A., Shah, N., Wang, K.C., Kim, J., Horlings, H.M., Wong, D.J., Tsai, M.C., Hung, T., Argani, P., Rinn, J.L. et al. (2010). Long non-coding RNA HOTAIR reprograms chromatin state to promote cancer metastasis. *Nature*, 464, 1071–1076.

Gustafsson, J.A. (2016). Historical overview of nuclear receptors. *J Steroid Biochem Mol Biol*, 157, 3–6.

Guzman, F., Fazeli, Y., Khuu, M., Salcido, K., Singh, S., Benavente, C.A. (2020). Retinoblastoma tumor suppressor protein roles in epigenetic regulation. *Cancers (Basel)*, 12.

Gwinner, E. and Brandstatter, R. (2001). Complex bird clocks. *Philos Trans Royal Soc London Ser B Biol Sci*, 356, 1801–1810.

Haeckel, E. (1868). *Natürliche Schöpfungsgeschichte. Gemeinverständliche wissenschaftliche Vorträge über die Entwickelungslehre im Allgemeinen und diejenige von Darwin, Goethe und Lamarck im Besonderen*. Reimer, Berlin.

Haissaguerre, M., Saucisse, N., Cota, D. (2014). Influence of mTOR in energy and metabolic homeostasis. *Mol Cell Endocrinol*, 397, 67–77.

Haj, F.G., Verveer, P.J., Squire, A., Neel, B.G., Bastiaens, P.I. (2002). Imaging sites of receptor dephosphorylation by PTP1B on the surface of the endoplasmic reticulum. *Science*, 295, 1708–1711.

Halachmi, S., Marden, E., Martin, G., Mackay, H., Abbondanza, C., Brown, M. (1994). Estrogen receptor-associated proteins: Possible mediators of hormone-induced transcription. *Science*, 264, 1455–1458.

Hall, B., Limaye, A., Kulkarni, A.B. (2009). Overview: Generation of gene knockout mice. *Curr Protoc Cell Biol*, 19.12.1–19.12.17.

Hamer, H.M., Jonkers, D., Venema, K., Vanhoutvin, S., Troost, F.J., Brummer, R.J. (2008). Review article: The role of butyrate on colonic function. *Aliment Pharmacol Ther*, 27, 104–119.

Hamm, L.L., Nakhoul, N., Hering-Smith, K.S. (2015). Acid-base homeostasis. *Clin J Am Soc Nephrol*, 10, 2232–2242.

Hamrick, M.W. (2011). A role for myokines in muscle-bone interactions. *Exerc Sport Sci Rev*, 39, 43–47.

Han, H., Wolff, E.M., Liang, G. (2012). Epigenetic alterations in bladder cancer and their potential clinical implications. *Adv Urol*, (2012), 546917.

Han, Y., You, X., Xing, W., Zhang, Z., Zou, W. (2018). Paracrine and endocrine actions of bone – The functions of secretory proteins from osteoblasts, osteocytes, and osteoclasts. *Bone Res*, 6, 16.

Handa, R.J., Corbier, P., Shryne, J.E., Schoonmaker, J.N., Gorski, R.A. (1985). Differential effects of the perinatal steroid environment on three sexually dimorphic parameters of the rat brain. *Biol Reprod*, 32, 855–864.

Hannan, F.M., Kallay, E., Chang, W., Brandi, M.L., Thakker, R.V. (2018). The calcium-sensing receptor in physiology and in calcitropic and noncalcitropic diseases. *Nat Rev Endocrinol*, 15, 33–51.

Hanoune, J. and Defer, N. (2001). Regulation and role of adenylyl cyclase isoforms. *Annu Rev Pharmacol Toxicol*, 41, 145–174.

Hansen, J.S., Pedersen, B.K., Xu, G., Lehmann, R., Weigert, C., Plomgaard, P. (2016). Exercise-induced secretion of FGF21 and follistatin are blocked by pancreatic clamp and impaired in type 2 diabetes. *J Clin Endocrinol Metab*, 101, 2816–2825.

Hansford, S. and Huntsman, D.G. (2014). Boveri at 100: Theodor Boveri and genetic predisposition to cancer. *J Pathol*, 234, 142–145.

Hanson, M.A. and Gluckman, P.D. (2014). Early developmental conditioning of later health and disease: Physiology or pathophysiology? *Physiol Rev*, 94, 1027–1076.

Hardy, J.D. (1953). Control of heat loss and heat production in physiologic temperature regulation. *Harvey Lect*, 49, 242–270.

Harley, C.B., Futcher, A.B., Greider, C.W. (1990). Telomeres shorten during ageing of human fibroblasts. *Nature*, 345, 458–460.

Harno, E., Gali Ramamoorthy, T., Coll, A.P., White, A. (2018). POMC: The physiological power of hormone processing. *Physiol Rev*, 98, 2381–2430.

Harris, G.W. (1955). *Neural Control of the Pituitary Gland*. Edward Arnold, London.

Harris, H., Miller, O.J., Klein, G., Worst, P., Tachibana, T. (1969). Suppression of malignancy by cell fusion. *Nature*, 223, 363–368.

Hassler, M.R. and Egger, G. (2012). Epigenomics of cancer – Emerging new concepts. *Biochimie*, 94, 2219–2230.

Hatting, M., Tavares, C.D.J., Sharabi, K., Rines, A.K., Puigserver, P. (2018). Insulin regulation of gluconeogenesis. *Ann NY Acad Sci*, 1411, 21–35.

Haussler, M.R. and Norman, A.W. (1969). Chromosomal receptor for a vitamin D metabolite. *Proc Natl Acad Sci USA*, 62, 155–162.

Hawley, S.A., Ross, F.A., Gowans, G.J., Tibarewal, P., Leslie, N.R., Hardie, D.G. (2014). Phosphorylation by Akt within the ST loop of AMPK-α1 down-regulates its activation in tumour cells. *Biochem J*, 459, 275–287.

Hay, D.L. and Pioszak, A.A. (2016). Receptor activity-modifying proteins (RAMPs): New insights and roles. *Annu Rev Pharmacol Toxicol*, 56, 469–487.

Hayden, M.S. and Ghosh, S. (2004). Signaling to NF-kappaB. *Genes Dev*, 18, 2195–2224.

Hayflick, L. and Moorhead, P.S. (1961). The serial cultivation of human diploid cell strains. *Exp Cell Res*, 25, 585–621.

Hebbes, T.R., Thorne, A.W., Crane-Robinson, C. (1988). A direct link between core histone acetylation and transcriptionally active chromatin. *EMBO J*, 7, 1395–1402.

Heijmans, B.T., Tobi, E.W., Stein, A.D., Putter, H., Blauw, G.J., Susser, E.S., Slagboom, P.E., Lumey, L.H. (2008). Persistent epigenetic differences associated with prenatal exposure to famine in humans. *Proc Natl Acad Sci USA*, 105, 17046–17049.

Hemmings, B.A. and Restuccia, D.F. (2012). PI3K-PKB/Akt pathway. *Cold Spring Harb Perspect Biol*, 4, a011189.

Hench, P.S., Kendall, E.C., Slocumb, C.H., Polley, H.F. (1949). Adrenocortical hormone in arthritis: Preliminary report. *Ann Rheum Dis*, 8, 97.

Hensch, T.K. (2004). Critical period regulation. *Annu Rev Neurosci*, 27, 549–579.

Herbison, A.E. (2008). Estrogen positive feedback to gonadotropin-releasing hormone (GnRH) neurons in the rodent: The case for the rostral periventricular area of the third ventricle (RP3V). *Brain Res Rev*, 57, 277–287.

Herbison, A.E. (2020). A simple model of estrous cycle negative and positive feedback regulation of GnRH secretion. *Front Neuroendocrinol*, 57, 100837.

Herceg, Z. and Vaissière, T. (2011). Epigenetic mechanisms and cancer: An interface between the environment and the genome. *Epigenetics*, 6, 804–819.

Herman, J.G., Merlo, A., Mao, L., Lapidus, R.G., Issa, J.P., Davidson, N.E., Sidransky, D., Baylin, S.B. (1995). Inactivation of the CDKN2/p16/MTS1 gene is frequently associated with aberrant DNA methylation in all common human cancers. *Cancer Res*, 55, 4525–4530.

Herman, J.G., Umar, A., Polyak, K., Graff, J.R., Ahuja, N., Issa, J.P., Markowitz, S., Willson, J.K., Hamilton, S.R., Kinzler, K.W. et al. (1998). Incidence and functional consequences of hMLH1 promoter hypermethylation in colorectal carcinoma. *Proc Natl Acad Sci USA*, 95, 6870–6875.

Hers, H.G. (1976). The control of glycogen metabolism in the liver. *Annu Rev Biochem*, 45, 167–189.

Hers, H.G. and Hue, L. (1983). Gluconeogenesis and related aspects of glycolysis. *Annu Rev Biochem*, 52, 617–653.

Herz, C.T. and Kiefer, F.W. (2019). Adipose tissue browning in mice and humans. *J Endocrinol*, 241, R97–R109.

Hetherington, A. and Ranson, S. (1940). Hypothalamic lesions and adiposity in the rat. *Anat Rec*, 78, 149–172.

Hevesy, G.V. and Paneth, F. (1913). Die Löslichkeit des Bleisulfids und Bleichromats. *Zeitschrift für anorganische Chemie*, 82, 323–328.

Hinnebusch, A.G. (2005). Translational regulation of GCN4 and the general amino acid control of yeast. *Annu Rev Microbiol*, 59, 407–450.

Hirsch, J. and Batchelor, B. (1976). Adipose tissue cellularity in human obesity. *Clin Endocrinol Metab*, 5, 299–311.

Hirsch, P.F. and Munson, P.L. (1969). Thyrocalcitonin. *Physiol Rev*, 49, 548–622.

Hjort, L., Martino, D., Grunnet, L.G., Naeem, H., Maksimovic, J., Olsson, A.H., Zhang, C., Ling, C., Olsen, S.F., Saffery, R. et al. (2018). Gestational diabetes and maternal obesity are associated with epigenome-wide methylation changes in children. *JCI Insight*, 3, e122572.

Ho, B.B. and Bergwitz, C. (2021). FGF23 signalling and physiology. *J Mol Endocrinol*, 66, R23–R32.

Ho, S.M., Tang, W.Y., Belmonte De Frausto, J., Prins, G.S. (2006). Developmental exposure to estradiol and bisphenol A increases susceptibility to prostate carcinogenesis and epigenetically regulates phosphodiesterase type 4 variant 4. *Cancer Res*, 66, 5624–5632.

Ho, J., Cruise, E.S., Dowling, R.J.O., Stambolic, V. (2020). PTEN nuclear functions. *Cold Spring Harb Perspect Med*, 10(5), a036079.

Hoffmann, A., Natoli, G., Ghosh, G. (2006). Transcriptional regulation via the NF-kappaB signaling module. *Oncogene*, 25, 6706–6716.

Holliday, R. and Pugh, J.E. (1975). DNA modification mechanisms and gene activity during development. *Science*, 187, 226–232.

Holloszy, J.O., Kohrt, W.M., Hansen, P.A. (1998). The regulation of carbohydrate and fat metabolism during and after exercise. *Front Biosci*, 3, D1011–D1027.

Hotamisligil, G.S. (2006). Inflammation and metabolic disorders. *Nature*, 444, 860–867.

Hotamisligil, G.S., Shargill, N., Spiegelman, B. (1993). Adipose expression of tumor necrosis factor-alpha: Direct role in obesity-linked insulin resistance. *Science*, 259, 87–91.

Hotchkiss, R.D. (1948). The quantitative separation of purines, pyrimidines, and nucleosides by paper chromatography. *J Biol Chem*, 175, 315–332.

Houslay, M.D. (2010). Underpinning compartmentalised cAMP signalling through targeted cAMP breakdown. *Trends Biochem Sci*, 35, 91–100.

Houssay, B.A. (1936). The hypophysis and metabolism. *New Eng J Med*, 214, 961–985.

Houssay, B.A. (1943). History of hypophysial diabetes. In *Essays in Biology in Honor of Herbert M. Evans*. University of California Press, Berkeley.

Howard, E.J., Lam, T.K., Duca, F.A. (2022). The gut microbiome: Connecting diet, glucose homeostasis, and disease. *Annu Rev Med*, 73, 469–481.

Huang, C.H., Yu, X., Liao, W.B. (2018). The expensive-tissue hypothesis in vertebrates: Gut microbiota effect, a review. *Int J Mol Sci*, 19(6), 1792.

Huang, Q., Kahn, C.R., Altindis, E. (2019). Viral hormones: Expanding dimensions in endocrinology. *Endocrinology*, 160, 2165–2179.

Hue, L. and Taegtmeyer, H. (2009). The Randle cycle revisited: A new head for an old hat. *Am J Physiol Endocrinol Metab*, 297, E578–E591.

Huebner, R.J. and Todaro, G.J. (1969). Oncogenes of RNA tumor viruses as determinants of cancer. *Proc Natl Acad Sci*, 64, 1087–1094.

Humpton, T.J. and Vousden, K.H. (2016). Regulation of cellular metabolism and hypoxia by p53. *Cold Spring Harb Perspect Med*, 6(7), a026146.

Hunt, T. (2013). On the regulation of protein phosphatase 2A and its role in controlling entry into and exit from mitosis. *Adv Biol Regul*, 53, 173–178.

Hunter, T. and Sefton, B.M. (1980). Transforming gene product of Rous sarcoma virus phosphorylates tyrosine. *Proc Natl Acad Sci USA*, 77, 1311–1315.

Hurt, E.M., Thomas, S.B., Peng, B., Farrar, W.L. (2006). Reversal of p53 epigenetic silencing in multiple myeloma permits apoptosis by a p53 activator. *Cancer Biol Ther*, 5, 1154–1160.

Husted, A.S., Trauelsen, M., Rudenko, O., Hjorth, S.A., Schwartz, T.W. (2017). GPCR-mediated signaling of metabolites. *Cell Metab*, 25, 777–796.

Hynes, R.O. (1976). Cell surface proteins and malignant transformation. *Biochim Biophys Acta*, 458, 73–107.

Hynes, R.O. (1992). Integrins: Versatility, modulation, and signaling in cell adhesion. *Cell*, 69, 11–25.

Hynes, R.O. (2009). The extracellular matrix: Not just pretty fibrils. *Science*, 326, 1216–1219.

Ilonen, J., Lempainen, J., Veijola, R. (2019). The heterogeneous pathogenesis of type 1 diabetes mellitus. *Nat Rev Endocrinol*, 15, 635–650.

Imai, S.-I., Armstrong, C.M., Kaeberlein, M., Guarente, L. (2000). Transcriptional silencing and longevity protein Sir2 is an NAD-dependent histone deacetylase. *Nature*, 403, 795–800.

Inagami, T. (1998). A memorial to Robert Tiegerstedt. *Hypertension*, 32, 953–957.

Irannejad, R. and Von Zastrow, M. (2014). GPCR signaling along the endocytic pathway. *Curr Opin Cell Biology*, 27, 109–116.

Iroz, A., Couty, J.-P., Postic, C. (2015). Hepatokines: Unlocking the multi-organ network in metabolic diseases. *Diabetologia*, 58, 1699–1703.

Irvine, D.J., Purbhoo, M.A., Krogsgaard, M., Davis, M.M. (2002). Direct observation of ligand recognition by T cells. *Nature*, 419, 845–849.

Isaacs, A., Lindenmann, J., Andrewes, C.H. (1957). Virus interference I. The interferon. *Proc Royal Soc London Ser B Biol Sci*, 147, 258–267.

Ishida, Y., Kondo, T., Takayasu, T., Iwakura, Y., Mukaida, N. (2004). The essential involvement of cross-talk between IFN-γ and TGF-β in the skin wound-healing process. *J Immunol*, 172, 1848–1855.

Ito, Y., Koessler, T., Ibrahim, A.E., Rai, S., Vowler, S.L., Abu-Amero, S., Silva, A.L., Maia, A.T., Huddleston, J.E., Uribe-Lewis, S. et al. (2008). Somatically acquired hypomethylation of IGF2 in breast and colorectal cancer. *Hum Mol Genet*, 17, 2633–2643.

Ivaska, J., Bosca, L., Parker, P.J. (2003). PKCepsilon is a permissive link in integrin-dependent IFN-gamma signalling that facilitates JAK phosphorylation of STAT1. *Nat Cell Biol*, 5, 363–369.

Ivell, R., Schmale, H., Richter, D. (1983). Vasopressin and oxytocin precursors as model preprohormones. *Neuroendocrinology*, 37, 235–240.

Jacklet, J.W. and Geronimo, J. (1971). Circadian rhythm: Population of interacting neurons. *Science*, 174, 299–302.

Jacob, F. and Monod, J. (1961). Genetic regulatory mechanisms in the synthesis of proteins. *Journal of Molecular Biology*, 3, 318–356.

Jacobson, L.O., Simmons, E.L., Marks, E.K., Eldredge, J.H. (1951). Recovery from radiation injury. *Science*, 113, 510–511.

Jayasena, C.N., Nijher, G.M., Chaudhri, O.B., Murphy, K.G., Ranger, A., Lim, A., Patel, D., Mehta, A., Todd, C., Ramachandran, R. et al. (2009). Subcutaneous injection of kisspeptin-54 acutely stimulates gonadotropin secretion in women with hypothalamic amenorrhea, but chronic administration causes tachyphylaxis. *J Clin Endocrinol Metab*, 94, 4315–4323.

Jeltsch, A. and Jurkowska, R.Z. (2014). New concepts in DNA methylation. *Trends Biochem Sci*, 39, 310–318.

Jenner, E. (1801). On the origin of the vaccine inoculation. *Medical Phys J*, 5, 505.

Jensen, E.V. (1966). Mechanism of estrogen action in relation to carcinogenesis. *Proceedings. Canadian Cancer Conference*, 143–165.

Jensen-Cody, S.O. and Potthoff, M.J. (2020). Hepatokines and metabolism: Deciphering communication from the liver. *Mol Metab*, 44, 101138.

Jin, P., Jan, L.Y., Jan, Y.-N. (2020). Mechanosensitive ion channels: Structural features relevant to mechanotransduction mechanisms. *Annu Rev Neurosci*, 43, 207–229.

Johannsen, W. (1909). *Elemente der exakten Erblichkeitslehre*. Fischer, Jena.

Jonasch, E., Futreal, P.A., Davis, I.J., Bailey, S.T., Kim, W.Y., Brugarolas, J., Giaccia, A.J., Kurban, G., Pause, A., Frydman, J. et al. (2012). State of the science: An update on renal cell carcinoma. *Mol Cancer Res*, 10, 859–880.

Jones, P.A. and Baylin, S.B. (2007). The epigenomics of cancer. *Cell*, 128, 683–692.

Jordt, S.-E. and Julius, D. (2002). Molecular basis for species-specific sensitivity to "hot" chili peppers. *Cell*, 108, 421–430.

Joulia-Ekaza, D. and Cabello, G. (2006). Myostatin regulation of muscle development: Molecular basis, natural mutations, physiopathological aspects. *Exp Cell Res*, 312, 2401–2414.

Juric, D., Castel, P., Griffith, M., Griffith, O.L., Won, H.H., Ellis, H., Ebbesen, S.H., Ainscough, B.J., Ramu, A., Iyer, G. et al. (2015). Convergent loss of PTEN leads to clinical resistance to a PI(3)Kα inhibitor. *Nature*, 518, 240–244.

Kaidanovich-Beilin, O. and Woodgett, J.R. (2011). GSK-3: Functional insights from cell biology and animal models. *Front Mol Neurosci*, 4, 40.

Kampourakis, K. and Stern, F. (2018). Reconsidering the meaning of concepts in biology: Why distinctions are so important. *Bioessays*, 40, e1800148.

Kandel, E.S. and Hay, N. (1999). The regulation and activities of the multifunctional serine/threonine kinase Akt/PKB. *Exp Cell Res*, 253, 210–229.

Kanwal, R. and Gupta, S. (2012). Epigenetic modifications in cancer. *Clin Genet*, 81, 303–311.

Kapcala, L.P., Chautard, T., Eskay, R.L. (1995). The protective role of the hypothalamic-pituitary-adrenal axis against lethality produced by immune, infectious, and inflammatory stress. *Ann NY Acad Sci*, 771, 419–437.

Kapiloff, M.S., Schillace, R.V., Westphal, A.M., Scott, J.D. (1999). mAKAP: An A-kinase anchoring protein targeted to the nuclear membrane of differentiated myocytes. *J Cell Sci*, 112(Pt 16), 2725–2736.

Kaplan, D.H., Shankaran, V., Dighe, A.S., Stockert, E., Aguet, M., Old, L.J., Schreiber, R.D. (1998). Demonstration of an interferon gamma-dependent tumor surveillance system in immunocompetent mice. *Proc Natl Acad Sci USA*, 95, 7556–7561.

Karin, M. and Ben-Neriah, Y. (2000). Phosphorylation meets ubiquitination: The control of NF-[kappa]B activity. *Annu Rev Immunol*, 18, 621–663.

Karin, M., Yamamoto, Y., Wang, Q.M. (2004). The IKK NF-kappa B system: A treasure trove for drug development. *Nat Rev Drug Discov*, 3, 17–26.

Karlsson, F.H., Tremaroli, V., Nookaew, I., Bergström, G., Behre, C.J., Fagerberg, B., Nielsen, J., Bäckhed, F. (2013). Gut metagenome in European women with normal, impaired and diabetic glucose control. *Nature*, 498, 99–103.

Kasai, C., Sugimoto, K., Moritani, I., Tanaka, J., Oya, Y., Inoue, H., Tameda, M., Shiraki, K., Ito, M., Takei, Y. et al. (2015). Comparison of the gut microbiota composition between obese and non-obese individuals in a Japanese population, as analyzed by terminal restriction fragment length polymorphism and next-generation sequencing. *BMC Gastroenterol*, 15, 100.

Kasakura, S. and Lowenstein, L. (1965). A factor stimulating DNA synthesis derived from the medium of leukocyte cultures. *Nature*, 208, 794–795.

Katakami, H., Arimura, A., Frohman, L.A. (1985). Involvement of hypothalamic somatostatin in the suppression of growth hormone secretion by central corticotropin-releasing factor in conscious male rats. *Neuroendocrinology*, 41, 390–393.

Katz, E. and Streuli, C.H. (2007). The extracellular matrix as an adhesion checkpoint for mammary epithelial function. *Int J Biochem Cell Biol*, 39, 715–726.

Kaufmann, S.H.E. and Schaible, U.E. (2005). 100th anniversary of Robert Koch's Nobel Prize for the discovery of the tubercle bacillus. *Trends in Microbiology*, 13, 469–475.

Kaufmann, T., Schlipf, S., Sanz, J., Neubert, K., Stein, R., Borner, C. (2003). Characterization of the signal that directs Bcl-xL, but not Bcl-2, to the mitochondrial outer membrane. *J Cell Biol*, 160, 53–64.

Kawabata, H. (2019). Transferrin and transferrin receptors update. *Free Radic Biol Med*, 133, 46–54.

Kawai, K. and Takahashi, M. (2020). Intracellular RET signaling pathways activated by GDNF. *Cell Tissue Res*, 382, 113–123.

Kazakou, P., Nicolaides, N.C., Chrousos, G.P. (2022). Basic concepts and hormonal regulators of the stress system. *Horm Res Paediatr*, 96(1), 8–16.

Kazerounian, S. and Lawler, J. (2018). Integration of pro- and anti-angiogenic signals by endothelial cells. *J Cell Commun Signal*, 12, 171–179.

Kechagia, J.Z., Ivaska, J., Roca-Cusachs, P. (2019). Integrins as biomechanical sensors of the microenvironment. *Nat Rev Mol Cell Biol*, 20, 457–473.

Kellaway, P. (1945). The Nobel Prize in physiology and medicine 1944. *Can Med Assoc J*, 53, 57.

Kelley, D.E. (2005). Skeletal muscle fat oxidation: Timing and flexibility are everything. *J Clin Invest*, 115, 1699–1702.

Kemp, R.G. and Foe, L.G. (1983). Allosteric regulatory properties of muscle phosphofructokinase. *Mol Cell Biochem*, 57, 147–154.

Kendall, E.C. (1953). Hormones of the adrenal cortex. *Bull NY Acad Med*, 29, 91.

Kennedy, G.C. (1953). The role of depot fat in the hypothalamic control of food intake in the rat. *Proc R Soc Lond B Biol Sci*, 140, 578–596.

Kennett, J.E. and McKee, D.T. (2012). Oxytocin: An emerging regulator of prolactin secretion in the female rat. *J Neuroendocrinol*, 24, 403–412.

Kim, J.-W. and Dang, C.V. (2005). Multifaceted roles of glycolytic enzymes. *Trends Biochem Sci*, 30, 142–150.

Kim, M.P. and Lozano, G. (2018). Mutant p53 partners in crime. *Cell Death Differ*, 25, 161–168.

Kim, N.W., Piatyszek, M.A., Prowse, K.R., Harley, C.B., West, M.D., Ho, P.L., Coviello, G.M., Wright, W.E., Weinrich, S.L., Shay, J.W. (1994). Specific association of human telomerase activity with immortal cells and cancer. *Science*, 266, 2011–2015.

Kim, M.-S., Pak, Y.K., Jang, P.-G., Namkoong, C., Choi, Y.-S., Won, J.-C., Kim, K.-S., Kim, S.-W., Kim, H.-S., Park, J.-Y. (2006). Role of hypothalamic Foxo1 in the regulation of food intake and energy homeostasis. *Nat Neurosci*, 9, 901–906.

Kim, J.M., Rasmussen, J.P., Rudensky, A.Y. (2007). Regulatory T cells prevent catastrophic autoimmunity throughout the lifespan of mice. *Nat Immunol*, 8, 191–197.

Kim, E., Goraksha-Hicks, P., Li, L., Neufeld, T.P., Guan, K.L. (2008). Regulation of TORC1 by Rag GTPases in nutrient response. *Nat Cell Biol*, 10, 935–945.

Kirby, T.J. and Lammerding, J. (2018). Emerging views of the nucleus as a cellular mechanosensor. *Nat Cell Biol*, 20, 373–381.

Klein Herenbrink, C., Sykes, D.A., Donthamsetti, P., Canals, M., Coudrat, T., Shonberg, J., Scammells, P.J., Capuano, B., Sexton, P.M., Charlton, S.J. et al. (2016). The role of kinetic context in apparent biased agonism at GPCRs. *Nat Commun*, 7, 10842.

Klionsky, D.J. (2008). Autophagy revisited: A conversation with Christian de Duve. *Autophagy*, 4, 740–743.

Klok, M.D., Jakobsdottir, S., Drent, M.L. (2007). The role of leptin and ghrelin in the regulation of food intake and body weight in humans: A review. *Obes Rev*, 8, 21–34.

Knochelmann, H.M., Dwyer, C.J., Bailey, S.R., Amaya, S.M., Elston, D.M., Mazza-McCrann, J.M., Paulos, C.M. (2018). When worlds collide: Th17 and Treg cells in cancer and autoimmunity. *Cell Mol Immunol*, 15, 458–469.

Knudson Jr., A.G. (1971). Mutation and cancer: Statistical study of retinoblastoma. *Proc Natl Acad Sci USA*, 68, 820–823.

Kong, Y.Y., Yoshida, H., Sarosi, I., Tan, H.L., Timms, E., Capparelli, C., Morony, S., Oliveira-Dos-Santos, A.J., Van, G., Itie, A. et al. (1999). OPGL is a key regulator of osteoclastogenesis, lymphocyte development and lymph-node organogenesis. *Nature*, 397, 315–323.

Koopman, F.A., Van Maanen, M.A., Vervoordeldonk, M.J., Tak, P.P. (2017). Balancing the autonomic nervous system to reduce inflammation in rheumatoid arthritis. *J Int Med*, 282, 64–75.

Koren, O., Goodrich, J.K., Cullender, T.C., Spor, A., Laitinen, K., Bäckhed, H.K., Gonzalez, A., Werner, J.J., Angenent, L.T., Knight, R. et al. (2012). Host remodeling of the gut microbiome and metabolic changes during pregnancy. *Cell*, 150, 470–480.

Kornberg, H. (1965). Anaplerotic sequences in microbial metabolism. *Angew Chem Int Ed Engl*, 4, 558–565.

Kornmann, B., Schaad, O., Bujard, H., Takahashi, J.S., Schibler, U. (2007). System-driven and oscillator-dependent circadian transcription in mice with a conditionally active liver clock. *PLoS Biol*, 5, e34.

Köse, M. (2017). GPCRs and EGFR – Cross-talk of membrane receptors in cancer. *Bioorg Med Chem Lett*, 27, 3611–3620.

Koshiishi, N., Chong, J.M., Fukasawa, T., Ikeno, R., Hayashi, Y., Funata, N., Nagai, H., Miyaki, M., Matsumoto, Y., Fukayama, M. (2004). p300 gene alterations in intestinal and diffuse types of gastric carcinoma. *Gastric Cancer*, 7, 85–90.

Kotas, M.E. and Medzhitov, R. (2015). Homeostasis, inflammation, and disease susceptibility. *Cell*, 160, 816–827.

Krebs, E.G., and Beavo, J.A. (1979). Phosphorylation-dephosphorylation of enzymes. *Annual Rev Biochem*, 48, 923–959.

Ktistakis, N.T. (2017). In praise of M. Anselmier who first used the term "autophagie" in 1859. *Autophagy*, 13, 2015–2017.

Kubota, H., Noguchi, R., Toyoshima, Y., Ozaki, Y., Uda, S., Watanabe, K., Ogawa, W., Kuroda, S. (2012). Temporal coding of insulin action through multiplexing of the AKT pathway. *Mol Cell*, 46, 820–832.

Kucharski, R., Maleszka, J., Foret, S., Maleszka, R. (2008). Nutritional control of reproductive status in honeybees via DNA methylation. *Science*, 319, 1827–1830.

Kuhne, W. (1864). *Untersuchungen uber das Protoplasma und die Contractilitat*. W. Engelmann, Leipzig.

Kunitz, M. and Northrop, J.H. (1935). Crystalline chymo-trypsin and chymo-trypsinogen I: Isolation, crystallization, and general properties of a new proteolytic enzyme and its precursor. *J Gen Physiol*, 18, 433–458.

Kurt, B. and Kurtz, A. (2015). Plasticity of renal endocrine function. *Am J Physiol Regul Integr Comp Physiol*, 308, R455–R466.

Kwabi-Addo, B., Wang, S., Chung, W., Jelinek, J., Patierno, S.R., Wang, B.D., Andrawis, R., Lee, N.H., Apprey, V., Issa, J.P. et al. (2010). Identification of differentially methylated genes in normal prostate tissues from African American and Caucasian men. *Clin Cancer Res*, 16, 3539–3547.

Kwon, M.J., Jang, B., Yi, J.Y., Han, I.O., Oh, E.S. (2012). Syndecans play dual roles as cell adhesion receptors and docking receptors. *FEBS Lett*, 586, 2207–2211.

Labrie, F. (1991). Intracrinology. *Mol Cell Endocrinol*, 78, C113–C118.

Labrie, F. (2010). DHEA, important source of sex steroids in men and even more in women. *Prog Brain Res*, 182, 97–148.

Lacy, P. (2015). Editorial: Secretion of cytokines and chemokines by innate immune cells. *Front Immunol*, 6.

Ladurner, A.G. (2006). Rheostat control of gene expression by metabolites. *Mol Cell*, 24, 1–11.

Lafoya, B., Munroe, J.A., Miyamoto, A., Detweiler, M.A., Crow, J.J., Gazdik, T., Albig, A.R. (2018). Beyond the matrix: The many non-ECM ligands for integrins. *Int J Mol Sci*, 19(2), 449.

Lainez, N.M. and Coss, D. (2019). Leukemia inhibitory factor represses GnRH gene expression via cFOS during inflammation in male mice. *Neuroendocrinology*, 108, 291–307.

Lamarck, J. (1963). *Zoological Philosophy: An Exposition with Regard to the Natural History of Animals*. Harner Publishing C, New York.

Lamia, K.A., Sachdeva, U.M., Ditacchio, L., Williams, E.C., Alvarez, J.G., Egan, D.F., Vasquez, D.S., Juguilon, H., Panda, S., Shaw, R.J. (2009). AMPK regulates the circadian clock by cryptochrome phosphorylation and degradation. *Science*, 326, 437–440.

Lamy, C.M., Sanno, H., Labouèbe, G., Picard, A., Magnan, C., Chatton, J.-Y., Thorens, B. (2014). Hypoglycemia-activated GLUT2 neurons of the nucleus tractus solitarius stimulate vagal activity and glucagon secretion. *Cell Metab*, 19, 527–538.

Landreth, K.S. (2002). Critical windows in development of the rodent immune system. *Hum Exp Toxicol*, 21, 493–498.

Lane, D.P. and Crawford, L.V. (1979). T antigen is bound to a host protein in SV40-transformed cells. *Nature*, 278, 261–263.

Lang, C.H., Hong-Brown, L., Frost, R.A. (2005). Cytokine inhibition of JAK-STAT signaling: A new mechanism of growth hormone resistance. *Pediatr Nephrol*, 20, 306–312.

Langer, S. and Lehmann, J. (1988). Presynaptic receptors on catecholamine neurones. In *Catecholamines I*, Trendelenburg, U. and Weiner, N. (eds). Springer, Berlin.

Langley, J.N. (1905). On the reaction of cells and of nerve-endings to certain poisons, chiefly as regards the reaction of striated muscle to nicotine and to curari. *J Physiol*, 33, 374–413.

Langley, J.N. (1906). Croonian Lecture, 1906: On nerve endings and on special excitable substances in cells. *Proc Royal Soc London Ser B Containing Papers of a Biological Character*, 78, 170–194.

Lapidot, T., Sirard, C., Vormoor, J., Murdoch, B., Hoang, T., Caceres-Cortes, J., Minden, M., Paterson, B., Caligiuri, M.A., Dick, J.E. (1994). A cell initiating human acute myeloid leukaemia after transplantation into SCID mice. *Nature*, 367, 645–648.

Laplante, M. and Sabatini, D.M. (2012). mTOR signaling in growth control and disease. *Cell*, 149, 274–293.

Laron, Z., Pertzelan, A., Mannheimer, S. (1966). Genetic pituitary dwarfism with high serum concentation of growth hormone – A new inborn error of metabolism? *Isr J Med Sci*, 2, 152–155.

Larue, C.G. and Le Magnen, J. (1972). The olfactory control of meal pattern in rats. *Physiol Behav*, 9, 817–821.

Latko, M., Czyrek, A., Porebska, N., Kucinska, M., Otlewski, J., Zakrzewska, M., Opalinski, L. (2019). Cross-talk between fibroblast growth factor receptors and other cell surface proteins. *Cells*, 8(5), 455.

Latorre, R., Sternini, C., De Giorgio, R., Greenwood-Van Meerveld, B. (2016). Enteroendocrine cells: A review of their role in brain-gut communication. *Neurogastroenterol Motil*, 28, 620–630.

Latva-Rasku, A., Honka, M.-J., Stančáková, A., Koistinen, H.A., Kuusisto, J., Guan, L., Manning, A.K., Stringham, H., Gloyn, A.L., Lindgren, C.M. (2018). A partial loss-of-function variant in AKT2 is associated with reduced insulin-mediated glucose uptake in multiple insulin-sensitive tissues: A genotype-based callback positron emission tomography study. *Diabetes*, 67, 334–342.

Lavoie, H., Gagnon, J., Therrien, M. (2020). ERK signalling: A master regulator of cell behaviour, life and fate. *Nat Rev Mol Cell Biol*, 21, 607–632.

Lee, S.-K. and Lorenzo, J.A. (1999). Parathyroid hormone stimulates TRANCE and inhibits osteoprotegerin messenger ribonucleic acid expression in murine bone marrow cultures: Correlation with osteoclast-like cell formation. *Endocrinology*, 140, 3552–3561.

Legler, J. (2010). Epigenetics: An emerging field in environmental toxicology. *Integr Environ Assess Manag*, 6, 314–315.

Leibbrandt, A. and Penninger, J.M. (2009). RANKL/RANK as key factors for osteoclast development and bone loss in arthropathies. In *Molecular Mechanisms of Spondyloarthropathies*, López-Larrea, C. and Díaz-Peña, R. (eds). Springer, New York.

Lemmon, M.A. and Schlessinger, J. (2010). Cell signaling by receptor tyrosine kinases. *Cell*, 141, 1117–1134.

Lemmon, M.A., Freed, D.M., Schlessinger, J., Kiyatkin, A. (2016). The dark side of cell signaling: Positive roles for negative regulators. *Cell*, 164, 1172–1184.

Lenstra, T.L., Rodriguez, J., Chen, H., Larson, D.R. (2016). Transcription dynamics in living cells. *Annu Rev Biophys*, 45, 25–47.

Lepage, R. and Albert, C. (2006). Fifty years of development in the endocrinology laboratory. *Clin Biochem*, 39, 542–557.

Levin, B.E. (2006). Metabolic sensing neurons and the control of energy homeostasis. *Physiol Behav*, 89, 486–489.

Levine, A.J. (1993). The tumor suppressor genes. *Annu Rev Biochem*, 62, 623–651.

Levine, A.J. and Puzio-Kuter, A.M. (2010). The control of the metabolic switch in cancers by oncogenes and tumor suppressor genes. *Science*, 330, 1340–1344.

Levine, A.J., Tomasini, R., McKeon, F.D., Mak, T.W., Melino, G. (2011). The p53 family: Guardians of maternal reproduction. *Nat Rev Mol Cell Biol*, 12, 259–265.

Levis, R.W. (1989). Viable deletions of a telomere from a Drosophila chromosome. *Cell*, 58, 791–801.

Li, X., Liu, H., Qin, L., Tamasi, J., Bergenstock, M., Shapses, S., Feyen, J.H., Notterman, D.A., Partridge, N.C. (2007). Determination of dual effects of parathyroid hormone on skeletal gene expression in vivo by microarray and network analysis. *J Biol Chem*, 282, 33086–33097.

Li, Y., Nishihara, E., Kakudo, K. (2011). Hashimoto's thyroiditis: Old concepts and new insights. *Curr Opin Rheumatol*, 23, 102–107.

Lillycrop, K.A., Phillips, E.S., Jackson, A.A., Hanson, M.A., Burdge, G.C. (2005). Dietary protein restriction of pregnant rats induces and folic acid supplementation prevents epigenetic modification of hepatic gene expression in the offspring. *J Nutr*, 135, 1382–1386.

Lin, W.W. and Hsieh, S.L. (2011). Decoy receptor 3: A pleiotropic immunomodulator and biomarker for inflammatory diseases, autoimmune diseases and cancer. *Biochem Pharmacol*, 81, 838–847.

Lindahl, T. and Barnes, D.E. (2000). Repair of endogenous DNA damage. *Cold Spring Harb Symp Quant Biol*, 65, 127–133.

Lipmann, F. (1969). Einar Lundsgaard. *Science*, 164, 246–247.

Littman, D.R. and Rudensky, A.Y. (2010). Th17 and regulatory T cells in mediating and restraining inflammation. *Cell*, 140, 845–858.

Liu, E. and Weissman, B. (1992). Oncogenes and tumor suppressor genes. *Cancer Treat Res*, 63, 1–13.

Liu, S., Tang, W., Zhou, J., Stubbs, J.R., Luo, Q., Pi, M., Quarles, L.D. (2006). Fibroblast growth factor 23 is a counter-regulatory phosphaturic hormone for vitamin D. *J Am Soc Nephrol*, 17, 1305–1315.

Liu, L., Xu, Z., Zhong, L., Wang, H., Jiang, S., Long, Q., Xu, J., Guo, J. (2013). Prognostic value of EZH2 expression and activity in renal cell carcinoma: A prospective study. *PLoS One*, 8, e81484.

Lodish, M.B., Trivellin, G., Stratakis, C.A. (2016). Pituitary gigantism: Update on molecular biology and management. *Curr Opin Endocrinol Diab Obes*, 23, 72.

Loewi, O. (1921). Über humorale übertragbarkeit der Herznervenwirkung. *Pflüger's Archiv für die gesamte Physiologie des Menschen und der Tiere*, 189, 239–242.

Loewi, O. and Navratil, E. (1926). Über humorale Übertragbarkeit der Herznervenwirkung. *Pflüger's Archiv für die gesamte Physiologie des Menschen und der Tiere*, 214, 678–688.

Lombardi, G. (2019). Exercise-dependent modulation of bone metabolism and bone endocrine function: New findings and therapeutic perspectives. *Journal of Science in Sport and Exercise*, 1, 20–28.

López-Cotarelo, P., Escribano-Díaz, C., González-Bethencourt, I.L., Gómez-Moreira, C., Deguiz, M.L., Torres-Bacete, J., Gómez-Cabañas, L., Fernández-Barrera, J., Delgado-Martín, C., Mellado, M. et al. (2015). A novel MEK-ERK-AMPK signaling axis controls chemokine receptor CCR7–dependent survival in human mature dendritic cells. *J Biol Chem*, 290, 827–840.

López-Muñoz, F. and Alamo, C. (2009). Historical evolution of the neurotransmission concept. *J Neural Trans*, 116, 515–533.

Lorenz, K.Z. (1958). The evolution of behavior. *Sci Am*, 199, 67–74 passim.

Lotka, A.J. (1910). Contribution to the theory of periodic reactions. *J Phys Chem*, 14, 271–274.

Loughran, S.J., Haas, S., Wilkinson, A.C., Klein, A.M., Brand, M. (2020). Lineage commitment of hematopoietic stem cells and progenitors: Insights from recent single cell and lineage tracing technologies. *Exp Hematol*, 88, 1–6.

Lowrey, P.L. and Takahashi, J.S. (2011). Genetics of circadian rhythms in Mammalian model organisms. *Adv Genet*, 74, 175–230.

Lussana, F., Painter, R.C., Ocke, M.C., Buller, H.R., Bossuyt, P.M., Roseboom, T.J. (2008). Prenatal exposure to the Dutch famine is associated with a preference for fatty foods and a more atherogenic lipid profile. *Am J Clin Nutr*, 88, 1648–1652.

Luteijn, M.J. and Ketting, R.F. (2013). PIWI-interacting RNAs: From generation to transgenerational epigenetics. *Nat Rev Genet*, 14, 523–534.

Luttrell, L.M. (2006). Transmembrane signaling by G protein-coupled receptors. In *Transmembrane Signaling Protocols*, Ali, H. and Haribabu, B. (eds). Humana Press, Totowa.

Lwoff, A. (1962). *L'ordre biologique*. FeniXX, London.

Macallan, D.C., Fullerton, C.A., Neese, R.A., Haddock, K., Park, S.S., Hellerstein, M.K. (1998). Measurement of cell proliferation by labeling of DNA with stable isotope-labeled glucose: Studies in vitro, in animals, and in humans. *Proc Natl Acad Sci USA*, 95, 708–713.

Magnen, J. (2012). *Neurobiology of Feeding and Nutrition*. Academic Press, San Diego.

Mahmoud, S., Gharagozloo, M., Simard, C., Gris, D. (2019). Astrocytes maintain glutamate homeostasis in the CNS by controlling the balance between glutamate uptake and release. *Cells*, 8(2), 184.

de Mairan (1729). Observation botanique. *Histoire de L'Académie Royale des Sciences Paris*, 35–36.

Majno, G. (1991). *The Healing Hand: Man and Wound in the Ancient World.* Harvard University Press, Cambridge.

Maman, S. and Witz, I.P. (2018). A history of exploring cancer in context. *Nature Reviews Cancer*, 18, 359–376.

Mangelsdorf, D.J., Thummel, C., Beato, M., Herrlich, P., Schütz, G., Umesono, K., Blumberg, B., Kastner, P., Mark, M., Chambon, P. (1995). The nuclear receptor superfamily: The second decade. *Cell*, 83, 835.

Mangnall, D., Bird, N.C., Majeed, A.W. (2003). The molecular physiology of liver regeneration following partial hepatectomy. *Liver Int*, 23, 124–138.

Manning, B.D. and Cantley, L.C. (2007). AKT/PKB signaling: Navigating downstream. *Cell*, 129, 1261–1274.

Manning, B.D. and Toker, A. (2017). AKT/PKB signaling: Navigating the network. *Cell*, 169, 381–405.

Marcelo, K.L., Means, A.R., York, B. (2016). The Ca(2+)/calmodulin/CaMKK2 axis: Nature's metabolic CaMshaft. *Trends Endocrinol Metab*, 27, 706–718.

Margolis, B., Bellot, F., Honegger, A.M., Ullrich, A., Schlessinger, J., Zilberstein, A. (1990). Tyrosine kinase activity is essential for the association of phospholipase C-gamma with the epidermal growth factor receptor. *Mol Cell Biol*, 10, 435–441.

Marshall, C.J. (1995). Specificity of receptor tyrosine kinase signaling: Transient versus sustained extracellular signal-regulated kinase activation. *Cell*, 80, 179–185.

Martelli, A.M., Tabellini, G., Bressanin, D., Ognibene, A., Goto, K., Cocco, L., Evangelisti, C. (2012). The emerging multiple roles of nuclear Akt. *Biochim Biophys Acta*, 1823, 2168–2178.

Martin, F.L. (2013). Epigenetic influences in the aetiology of cancers arising from breast and prostate: A hypothesised transgenerational evolution in chromatin accessibility. *ISRN Oncol*, (2013), 624794.

Martin, T.J. and Sims, N.A. (2015). RANKL/OPG; critical role in bone physiology. *Rev Endocr Metab Disord*, 16, 131–139.

Martinac, B. (2014). The ion channels to cytoskeleton connection as potential mechanism of mechanosensitivity. *Biochim Biophys Acta*, 1838, 682–691.

Martino, F., Perestrelo, A.R., Vinarský, V., Pagliari, S., Forte, G. (2018). Cellular mechanotransduction: From tension to function. *Front Physiol*, 9, 824.

Masuyama, H. and Hiramatsu, Y. (2012). Effects of a high-fat diet exposure in utero on the metabolic syndrome-like phenomenon in mouse offspring through epigenetic changes in adipocytokine gene expression. *Endocrinology*, 153, 2823–2830.

Masuyama, H., Mitsui, T., Nobumoto, E., Hiramatsu, Y. (2015). The effects of high-fat diet exposure in utero on the obesogenic and diabetogenic traits through epigenetic changes in adiponectin and leptin gene expression for multiple generations in female mice. *Endocrinology*, 156, 2482–2491.

Matera, L. (1996). Endocrine, paracrine and autocrine actions of prolactin on immune cells. *Life Sci*, 59, 599–614.

Mathis, D. and Benoist, C. (2004). Back to central tolerance. *Immunity*, 20, 509–516.

Matsumoto, M., Pocai, A., Rossetti, L., Depinho, R.A., Accili, D. (2007). Impaired regulation of hepatic glucose production in mice lacking the forkhead transcription factor Foxo1 in liver. *Cell Metab*, 6, 208–216.

Mayer, J. (1953). Glucostatic mechanism of regulation of food intake. *New Eng J Med*, 249, 13–16.

Mayer, E.A. (2011). Gut feelings: The emerging biology of gut-brain communication. *Nat Rev Neurosci*, 12, 453–466.

McCarthy, M.M. and Arnold, A.P. (2011). Reframing sexual differentiation of the brain. *Nat Neurosci*, 14, 677–683.

McCollum, E.V., Pitz, W., Simmonds, N., Becker, J.E., Shipley, P.G., Bunting, R.W. (2002). The effect of additions of fluorine to the diet of the rat on the quality of the teeth. 1925. Studies on experimental rickets. XXI. An experimental demonstration of the existence of a vitamin which promotes calcium deposition. 1922. The effect of additions of fluorine to the diet of the rat on the quality of the teeth. 1925. *J Biol Chem*, 277, E8.

McCrimmon, R. (2009). Glucose sensing during hypoglycemia: Lessons from the lab. *Diab Care*, 32, 1357–1363.

McGarry, J.D. and Foster, D.W. (1977). Hormonal control of ketogenesis: Biochemical considerations. *Arch Intern Med*, 137, 495–501.

McGrath, J. and Solter, D. (1984). Completion of mouse embryogenesis requires both the maternal and paternal genomes. *Cell*, 37, 179–183.

McIlwraith, E.K. and Belsham, D.D. (2020). Hypothalamic reproductive neurons communicate through signal transduction to control reproduction. *Mol Cell Endocrinol*, 518, 110971.

Mcintyre, N., Holdsworth, C., Turner, D.F. (1964). New interpretation of oral glucose tolerance. *Lancet*, 284, 20–21.

McKay, L.I. and Cidlowski, J.A. (1999). Molecular control of immune/inflammatory responses: Interactions between nuclear factor-kappa B and steroid receptor-signaling pathways. *Endocr Rev*, 20, 435–459.

McLatchie, L.M., Fraser, N.J., Main, M.J., Wise, A., Brown, J., Thompson, N., Solari, R., Lee, M.G., Foord, S.M. (1998). RAMPs regulate the transport and ligand specificity of the calcitonin-receptor-like receptor. *Nature*, 393, 333–339.

McLean, F.C. (1957). The parathyroid hormone and bone. *Clin Orthop*, 9, 46–60.

Meex, R.C.R. and Watt, M.J. (2017). Hepatokines: Linking nonalcoholic fatty liver disease and insulin resistance. *Nat Rev Endocrinol*, 13, 509–520.

Menendez, D., Shatz, M., Resnick, M.A. (2013). Interactions between the tumor suppressor p53 and immune responses. *Curr Opin Oncol*, 25, 85–92.

Menkin, V. (1944). Chemical basis of fever. *Science*, 100, 337–338.

Mihaylova, M.M. and Shaw, R.J. (2011). The AMPK signalling pathway coordinates cell growth, autophagy and metabolism. *Nat Cell Biol*, 13, 1016–1023.

Mithieux, G. (2014). Metabolic effects of portal vein sensing. *Diabetes, Obesity and Metabolism*, 16, 56–60.

Mitrophanov, A.Y. and Groisman, E.A. (2008). Positive feedback in cellular control systems. *Bioessays*, 30, 542–555.

Miyamoto, S., Teramoto, H., Gutkind, J.S., Yamada, K.M. (1996). Integrins can collaborate with growth factors for phosphorylation of receptor tyrosine kinases and MAP kinase activation: Roles of integrin aggregation and occupancy of receptors. *J Cell Biol*, 135, 1633–1642.

Miyara, M., Gorochov, G., Ehrenstein, M., Musset, L., Sakaguchi, S., Amoura, Z. (2011). Human FoxP3+ regulatory T cells in systemic autoimmune diseases. *Autoimmun Rev*, 10, 744–755.

Mizokami, A., Kawakubo-Yasukochi, T., Hirata, M. (2017). Osteocalcin and its endocrine functions. *Biochem Pharmacol*, 132, 1–8.

Mizushima, N. (2018). A brief history of autophagy from cell biology to physiology and disease. *Nat Cell Biol*, 20, 521–527.

Mocellin, S., Panelli, M.C., Wang, E., Nagorsen, D., Marincola, F.M. (2003). The dual role of IL-10. *Trends Immunol*, 24, 36–43.

Moelling, K., Schad, K., Bosse, M., Zimmermann, S., Schweneker, M. (2002). Regulation of Raf-Akt cross-talk. *J Biol Chem*, 277, 31099–31106.

Moerman, E.J., Teng, K., Lipschitz, D.A., Lecka-Czernik, B. (2004). Aging activates adipogenic and suppresses osteogenic programs in mesenchymal marrow stroma/stem cells: The role of PPAR-γ2 transcription factor and TGF-β/BMP signaling pathways. *Aging Cell*, 3, 379–389.

Mohan, S., Richman, C., Guo, R., Amaar, Y., Donahue, L.R., Wergedal, J., Baylink, D.J. (2003). Insulin-like growth factor regulates peak bone mineral density in mice by both growth hormone-dependent and -independent mechanisms. *Endocrinology*, 144, 929–936.

Mohawk, J.A., Green, C.B., Takahashi, J.S. (2012). Central and peripheral circadian clocks in mammals. *Annu Rev Neurosci*, 35, 445–462.

Møller, N. and Jørgensen, J.O. (2009). Effects of growth hormone on glucose, lipid, and protein metabolism in human subjects. *Endocr Rev*, 30, 152–177.

Moncaut, N., Rigby, P.W., Carvajal, J.J. (2013). Dial M(RF) for myogenesis. *FEBS J*, 280, 3980–3990.

Monod, J. and Jacob, F. (1961). Teleonomic mechanisms in cellular metabolism, growth, and differentiation. *Cold Spring Harb Symp Quant Biol*, 26, 389–401.

Monod, J., Changeux, J.-P., Jacob, F. (1963). Allosteric proteins and cellular control systems. *J Mol Biol*, 6, 306–329.

Moodie, S.A., Willumsen, B.M., Weber, M.J., Wolfman, A. (1993). Complexes of Ras.GTP with Raf-1 and mitogen-activated protein kinase kinase. *Science*, 260, 1658–1661.

Moore, B. (1906). On the treatment of diabetus mellitus by acid extract of duodenal mucous membrane. *Biochem J*, 1, 28–38.

Moore, C.S. and Crocker, S.J. (2012). An alternate perspective on the roles of TIMPs and MMPs in pathology. *Am J Pathol*, 180, 12–16.

Moore, R.Y. and Eichler, V.B. (1972). Loss of a circadian adrenal corticosterone rhythm following suprachiasmatic lesions in the rat. *Brain Res*, 42(1), 201–206.

Moore, M.C., Coate, K.C., Winnick, J.J., An, Z., Cherrington, A.D. (2012). Regulation of hepatic glucose uptake and storage in vivo. *Adv Nutr*, 3, 286–294.

Morgan, H.D., Sutherland, H.G., Martin, D.I., Whitelaw, E. (1999). Epigenetic inheritance at the agouti locus in the mouse. *Nat Genet*, 23, 314–318.

Morishima, A., Grumbach, M.M., Simpson, E.R., Fisher, C., Qin, K. (1995). Aromatase deficiency in male and female siblings caused by a novel mutation and the physiological role of estrogens. *J Clin Endocrinol Metab*, 80, 3689–3698.

Moro, L., Venturino, M., Bozzo, C., Silengo, L., Altruda, F., Beguinot, L., Tarone, G., Defilippi, P. (1998). Integrins induce activation of EGF receptor: Role in MAP kinase induction and adhesion-dependent cell survival. *EMBO J*, 17, 6622–6632.

Morrison, D.K. (2012). MAP kinase pathways. *Cold Spring Harb Perspect Biol*, 4(11), a011254.

Moser, M. and Leo, O. (2010). Key concepts in immunology. *Vaccine*, 28(Suppl 3), C2–C13.

Mosmann, T.R. and Sad, S. (1996). The expanding universe of T-cell subsets: Th1, Th2 and more. *Immunol Today*, 17, 138–146.

Mousseau, T.A. and Fox, C.W. (1998). *Maternal Effects as Adaptations*. Oxford University Press, New York.

Murray, P., Higham, C., Clayton, P. (2015). 60 years of neuroendocrinology: The hypothalamo-GH axis: The past 60 years. *J Endocrinol*, 226, T123–T140.

Naftolin, F., Garcia-Segura, L.M., Horvath, T.L., Zsarnovszky, A., Demir, N., Fadiel, A., Leranth, C., Vondracek-Klepper, S., Lewis, C., Chang, A. et al. (2007). Estrogen-induced hypothalamic synaptic plasticity and pituitary sensitization in the control of the estrogen-induced gonadotrophin surge. *Reprod Sci*, 14, 101–116.

Nagata, S., Taira, H., Hall, A., Johnsrud, L., Streuli, M., Ecsödi, J., Boll, W., Cantell, K., Weissmann, C. (1980). Synthesis in *E. coli* of a polypeptide with human leukocyte interferon activity. *Nature*, 284, 316–320.

Nanney, D.L. (1958). Epigenetic control systems. *Proc Natl Acad Sci USA*, 44, 712.

Napetschnig, J. and Wu, H. (2013). Molecular basis of NF-κB signaling. *Annu Rev Biophys*, 42, 443–468.

Narasimha, A.M., Kaulich, M., Shapiro, G.S., Choi, Y.J., Sicinski, P., Dowdy, S.F. (2014). Cyclin D activates the Rb tumor suppressor by mono-phosphorylation. *Elife*, 3, e02872.

Nash, W.T., Teoh, J., Wei, H., Gamache, A., Brown, M.G. (2014). Know thyself: NK-cell inhibitory receptors prompt self-tolerance, education, and viral control. *Front Immunol*, 5, 175.

Newgard, C.B., Moore, S.V., Foster, D.W., McGarry, J.D. (1984). Efficient hepatic glycogen synthesis in refeeding rats requires continued carbon flow through the gluconeogenic pathway. *J Biol Chem*, 259, 6958–6963.

Newman, J.C. and Verdin, E. (2014). β–hydroxybutyrate: Much more than a metabolite. *Diab Res Clin Pract*, 106, 173–181.

Ni, Q., Mehta, S., Zhang, J. (2018). Live-cell imaging of cell signaling using genetically encoded fluorescent reporters. *FEBS J*, 285, 203–219.

Nicholls, D.G. and Locke, R.M. (1984). Thermogenic mechanisms in brown fat. *Physiol Rev*, 64, 1–64.

Nicolaides, N.C., Charmandari, E., Chrousos, G.P., Kino, T. (2014). Circadian endocrine rhythms: The hypothalamic–pituitary–adrenal axis and its actions. *Ann NY Acad Sci*, 1318, 71–80.

Nishizuka, Y. and Sakakura, T. (1969). Thymus and reproduction: Sex-linked dysgenesia of the gonad after neonatal thymectomy in mice. *Science*, 166, 753–755.

Noack, M. and Miossec, P. (2014). Th17 and regulatory T cell balance in autoimmune and inflammatory diseases. *Autoimmun Rev*, 13, 668–677.

Norcross, M.A. (1984). A synaptic basis for T-lymphocyte activation. *Ann Immunol (Paris)*, 135d, 113–134.

O'Connor, J.C., McCusker, R.H., Strle, K., Johnson, R.W., Dantzer, R., Kelley, K.W. (2008). Regulation of IGF-I function by proinflammatory cytokines: At the interface of immunology and endocrinology. *Cell Immunol*, 252, 91–110.

Oh, K.J., Han, H.S., Kim, M.J., Koo, S.H. (2013). CREB and FoxO1: Two transcription factors for the regulation of hepatic gluconeogenesis. *BMB Rep*, 46, 567–574.

Ohno, S., Kaplan, W.D., Kinosita, R. (1959). Formation of the sex chromatin by a single X-chromosome in liver cells of Rattus norvegicus. *Exp Cell Res*, 18, 415–418.

Old, L.J. and Boyse, E.A. (1964). Immunology of experimental tumors. *Annu Rev Med*, 15, 167–186.

Oldstone, M.B. (1987). Molecular mimicry and autoimmune disease. *Cell*, 50, 819–820.

O'Neill, L.A., Golenbock, D., Bowie, A.G. (2013). The history of Toll-like receptors – Redefining innate immunity. *Nat Rev Immunol*, 13, 453–460.

Onishi, M., Nosaka, T., Kitamura, T. (1998). Cytokine receptors: Structures and signal transduction. *Int Rev Immunol*, 16, 617–634.

Oomura, Y., Kimura, K., Ooyama, H., Maeno, T., Iki, M., Kuniyoshi, M. (1964). Reciprocal activities of the ventromedial and lateral hypothalamic areas of cats. *Science*, 143, 484–485.

Oppenheimer, J.H. (1968). Role of plasma proteins in the binding, distribution and metabolism of the thyroid hormones. *N Eng J Med*, 278, 1153–1162.

Ortega-Prieto, P. and Postic, C. (2019). Carbohydrate sensing through the transcription factor ChREBP. *Front Genet*, 10, 472.

Osbak, K.K., Colclough, K., Saint-Martin, C., Beer, N.L., Bellanné-Chantelot, C., Ellard, S., Gloyn, A.L. (2009). Update on mutations in glucokinase (GCK), which cause maturity-onset diabetes of the young, permanent neonatal diabetes, and hyperinsulinemic hypoglycemia. *Hum Mutat*, 30, 1512–1526.

O'Shea, J.J., Gadina, M., Siegel, R.M. (2019). Cytokines and cytokine receptors. In *Clinical Immunology Principles and Practice*, 5th edition, Rich, R.R. (ed.). Elsevier, Amsterdam.

Ouchi, N., Parker, J.L., Lugus, J.J., Walsh, K. (2011). Adipokines in inflammation and metabolic disease. *Nat Rev Immunol*, 11, 85–97.

Owen, O.E. (2005). Ketone bodies as a fuel for the brain during starvation. *Biochem Mol Biol Edu*, 33, 246–251.

Owen, O.E., Morgan, A.P., Kemp, H.G., Sullivan, J.M., Herrera, M.G., Cahill Jr., G.F. (1967). Brain metabolism during fasting. *J Clin Invest*, 46, 1589–1595.

Oyola, M.G. and Handa, R.J. (2017). Hypothalamic-pituitary-adrenal and hypothalamic–pituitary–gonadal axes: Sex differences in regulation of stress responsivity. *Stress (Amsterdam, Netherlands)*, 20, 476–494.

Paget, S. (1889). The distribution of secondary growths in cancer of the breast. *Lancet*, 133, 571–573.

Painter, R.C., De Rooij, S.R., Bossuyt, P.M., Phillips, D.I., Osmond, C., Barker, D.J., Bleker, O.P., Roseboom, T.J. (2006). Blood pressure response to psychological stressors in adults after prenatal exposure to the Dutch famine. *J Hypertens*, 24, 1771–1778.

Palmer, B.F. and Clegg, D.J. (2014). Oxygen sensing and metabolic homeostasis. *Mol Cell Endocrinol*, 397, 51–58.

Palmer, D.C. and Restifo, N.P. (2009). Suppressors of cytokine signaling (SOCS) in T cell differentiation, maturation, and function. *Trends Immunol*, 30, 592–602.

Papadimitriou, A. and Priftis, K.N. (2009). Regulation of the hypothalamic–pituitary–adrenal axis. *Neuroimmunomodulation*, 16, 265–271.

Parfitt, A.M. (2002). Misconceptions (2): Turnover is always higher in cancellous than in cortical bone. *Bone*, 30, 807–809.

Pariante, C.M. and Miller, A.H. (2001). Glucocorticoid receptors in major depression: Relevance to pathophysiology and treatment. *Biol Psychiatry*, 49, 391–404.

Parry, R.V., Chemnitz, J.M., Frauwirth, K.A., Lanfranco, A.R., Braunstein, I., Kobayashi, S.V., Linsley, P.S., Thompson, C.B., Riley, J.L. (2005). CTLA-4 and PD-1 receptors inhibit T-cell activation by distinct mechanisms. *Mol Cell Biol*, 25, 9543–9553.

Patel, P. and Woodgett, J.R. (2017). Glycogen synthase kinase 3: A kinase for all pathways? *Curr Top Dev Biol*, 123, 277–302.

Patrick, J. and Lindstrom, J. (1973). Autoimmune response to acetylcholine receptor. *Science*, 180, 871–872.

Paul, H.A., Bomhof, M.R., Vogel, H.J., Reimer, R.A. (2016). Diet-induced changes in maternal gut microbiota and metabolomic profiles influence programming of offspring obesity risk in rats. *Sci Rep*, 6, 20683.

Pavlova, N.N. and Thompson, C.B. (2016). The emerging hallmarks of cancer metabolism. *Cell Metab*, 23, 27–47.

Pawson, T. (1995). Protein modules and signalling networks. *Nature*, 373, 573–580.

Pawson, T. (2004). Specificity in signal transduction: From phosphotyrosine-SH2 domain interactions to complex cellular systems. *Cell*, 116, 191–203.

Pearse, A.G. (1969). The calcitonin secreting C cells and their relationship to the APUD cell series. *J Endocrinol*, 45(Suppl), 13–14.

Pedersen, B.K. and Febbraio, M.A. (2012). Muscles, exercise and obesity: Skeletal muscle as a secretory organ. *Nat Rev Endocrinol*, 8, 457–465.

Pedersen, B.K. and Hoffman-Goetz, L. (2000). Exercise and the immune system: Regulation, integration, and adaptation. *Physiol Rev*, 80, 1055–1081.

Pelletier, R.M., Layeghkhavidaki, H., Vitale, M.L. (2020). Glucose, insulin, insulin receptor subunits α and β in normal and spontaneously diabetic and obese ob/ob and db/db infertile mouse testis and hypophysis. *Reprod Biol Endocrinol*, 18, 25.

Penny, G.D., Kay, G.F., Sheardown, S.A., Rastan, S., Brockdorff, N. (1996). Requirement for Xist in X chromosome inactivation. *Nature*, 379, 131–137.

Petersen, K.F., Laurent, D., Rothman, D.L., Cline, G.W., Shulman, G.I. (1998). Mechanism by which glucose and insulin inhibit net hepatic glycogenolysis in humans. *J Clin Invest*, 101, 1203–1209.

Pfeifer, G.P., Steigerwald, S.D., Hansen, R.S., Gartler, S.M., Riggs, A.D. (1990). Polymerase chain reaction-aided genomic sequencing of an X chromosome-linked CpG island: Methylation patterns suggest clonal inheritance, CpG site autonomy, and an explanation of activity state stability. *Proc Natl Acad Sci USA*, 87, 8252–8256.

Pflüger, E. (1868). Ueber die Ursache der Athembewegungen, sowie der Dyspnoë und Apnoë. *Archiv für die gesamte Physiologie des Menschen und der Tiere*, 1, 61–106.

Phillipps, H.R., Yip, S.H., Grattan, D.R. (2020). Patterns of prolactin secretion. *Mol Cell Endocrinol*, 502, 110679.

Phillips, D.M. (1963). The presence of acetyl groups of histones. *Biochem J*, 87, 258–263.

Pilkis, S.J., El-Maghrabi, M.R., Claus, T.H. (1988). Hormonal regulation of hepatic gluconeogenesis and glycolysis. *Annu Rev Biochem*, 57, 755–783.

Pioszak, A.A. and Hay, D.L. (2020). RAMPs as allosteric modulators of the calcitonin and calcitonin-like class B G protein-coupled receptors. *Adv Pharmacol*, 88, 115–141.

Piro, A., Tagarelli, G., Lagonia, P., Tagarelli, A., Quattrone, A. (2010). Casimir Funk: His discovery of the vitamins and their deficiency disorders. *Ann Nutr Metab*, 57, 85–88.

Plass, C. and Soloway, P.D. (2002). DNA methylation, imprinting and cancer. *Eur J Hum Genet*, 10, 6–16.

Plass, C., Pfister, S.M., Lindroth, A.M., Bogatyrova, O., Claus, R., Lichter, P. (2013). Mutations in regulators of the epigenome and their connections to global chromatin patterns in cancer. *Nat Rev Genet*, 14, 765–780.

Pollack, Y., Stein, R., Razin, A., Cedar, H. (1980). Methylation of foreign DNA sequences in eukaryotic cells. *Proc Natl Acad Sci USA*, 77, 6463–6467.

Pompei, A., Cordisco, L., Amaretti, A., Zanoni, S., Matteuzzi, D., Rossi, M. (2007). Folate production by bifidobacteria as a potential probiotic property. *Appl Environ Microbiol*, 73, 179–185.

Popper, K. (2014). *Conjectures and Refutations: The Growth of Scientific Knowledge*. Routledge, London.

Poreba, E., Broniarczyk, J.K., Gozdzicka-Jozefiak, A. (2011). Epigenetic mechanisms in virus-induced tumorigenesis. *Clin Epigenet*, 2, 233–247.

Potten, C.S. and Loeffler, M. (1990). Stem cells: Attributes, cycles, spirals, pitfalls and uncertainties. Lessons for and from the crypt. *Development*, 110, 1001–1020.

Power, C. and Rasko, J.E. (2008). Whither prometheus' liver? Greek myth and the science of regeneration. *Ann Intern Med*, 149, 421–426.

Poyurovsky, M.V. and Prives, C. (2010). P53 and aging: A fresh look at an old paradigm. *Aging (Albany NY)*, 2, 380–382.

Pozo, K. and Goda, Y. (2010). Unraveling mechanisms of homeostatic synaptic plasticity. *Neuron*, 66, 337–351.

Preidis, G.A., Kim, K.H., Moore, D.D. (2017). Nutrient-sensing nuclear receptors PPARα and FXR control liver energy balance. *J Clin Invest*, 127, 1193–1201.

Presman, D.M., Ball, D.A., Paakinaho, V., Grimm, J.B., Lavis, L.D., Karpova, T.S., Hager, G.L. (2017). Quantifying transcription factor binding dynamics at the single-molecule level in live cells. *Methods*, 123, 76–88.

Priestley, J. (1772). XIX. Observations on different kinds of air. *Philos Trans Royal Soc London*, 62, 147–264.

Quiñones, M., Al-Massadi, O., Fernø, J., Nogueiras, R. (2014). Cross-talk between SIRT1 and endocrine factors: Effects on energy homeostasis. *Mol Cell Endocrinol*, 397, 42–50.

Rabinowitz, D., Klassen, G., Zierler, K. (1965). Effect of human growth hormone on muscle and adipose tissue metabolism in the forearm of man. *Journal of Clinical Investigation*, 44, 51–61.

Raff, M.C. (1996). Size control: The regulation of cell numbers in animal development. *Cell*, 86, 173–175.

Rakyan, V. and Whitelaw, E. (2003). Transgenerational epigenetic inheritance. *Curr Biol*, 13, R6.

Rakyan, V.K., Chong, S., Champ, M.E., Cuthbert, P.C., Morgan, H.D., Luu, K.V., Whitelaw, E. (2003). Transgenerational inheritance of epigenetic states at the murine Axin(Fu) allele occurs after maternal and paternal transmission. *Proc Natl Acad Sci USA*, 100, 2538–2543.

Ramón Y. and Cajal, S. (1894). The Croonian lecture – La fine structure des centres nerveux. *Proc Royal Soc London*, 55, 444–468.

Ranade, S.S., Syeda, R., Patapoutian, A. (2015). Mechanically activated ion channels. *Neuron*, 87, 1162–1179.

Randle, P.J., Garland, P.B., Hales, C.N., Newsholme, E.A. (1963). The glucose fatty-acid cycle. Its role in insulin sensitivity and the metabolic disturbances of diabetes mellitus. *Lancet*, 1, 785–789.

Randle, P.J., Priestman, D.A., Mistry, S.C., Halsall, A. (1994). Glucose fatty acid interactions and the regulation of glucose disposal. *J Cell Biochem*, 55(Suppl), 1–11.

Ranganath, L.R. (2008). The entero-insular axis: Implications for human metabolism. *Clin Chem Lab Med (CCLM)*, 46, 43–56.

Rankin, L.C. and Artis, D. (2018). Beyond host defense: Emerging functions of the immune system in regulating complex tissue physiology. *Cell*, 173, 554–567.

Ransohoff, R.M. and Brown, M.A. (2012). Innate immunity in the central nervous system. *J Clin Invest*, 122, 1164–1171.

Rao, T.C., Ma, V.P., Blanchard, A., Urner, T.M., Grandhi, S., Salaita, K., Mattheyses, A.L. (2020). EGFR activation attenuates the mechanical threshold for integrin tension and focal adhesion formation. *J Cell Sci*, 133(13), jcs238840.

Raschke, S., Eckardt, K., Bjørklund Holven, K., Jensen, J., Eckel, J. (2013). Identification and validation of novel contraction-regulated myokines released from primary human skeletal muscle cells. *PLoS One*, 8, e62008.

Rassoulzadegan, M., Grandjean, V., Gounon, P., Vincent, S., Gillot, I., Cuzin, F. (2006). RNA-mediated non-mendelian inheritance of an epigenetic change in the mouse. *Nature*, 441, 469–474.

Ravelli, G.P., Stein, Z.A., Susser, M.W. (1976). Obesity in young men after famine exposure in utero and early infancy. *N Engl J Med*, 295, 349–353.

Rehfeld, J.F. (1998). Processing of precursors of gastroenteropancreatic hormones: Diagnostic significance. *J Mol Med (Berl)*, 76, 338–345.

Rehfeld, J.F. (2017). Cholecystokinin – From local gut hormone to ubiquitous messenger. *Front Endocrinol*, 8(47).

Rehfeld, J.F. and Bundgaard, J.R. (2010). Cell-specific precursor processing. *Cell Pept Horm Synthesis Secretory Pathways*, 185–205.

Rehfeld, J.F., Bardram, L., Cantor, P., Cerman, J., Hilsted, L., Johnsen, A.H., Mogensen, N., Osdum, L. (1989). Peptide hormone expression and precursor processing. *Acta Oncol*, 28, 315–318.

Reichel, R.R. and Jacob, S.T. (1993). Control of gene expression by lipophilic hormones. *FASEB J*, 7, 427–436.

Reinke, H. and Asher, G. (2016). Circadian clock control of liver metabolic functions. *Gastroenterology*, 150, 574–580.

Ribeiro, F.F. and Xapelli, S. (2021). An overview of adult neurogenesis. *Adv Exp Med Biol*, 1331, 77–94.

Ricquier, D. and Kader, J.C. (1976). Mitochondrial protein alteration in active brown fat: A soidum dodecyl sulfate-polyacrylamide gel electrophoretic study. *Biochem Biophys Res Commun*, 73, 577–583.

Ritossa, F. (1962). A new puffing pattern induced by temperature shock and DNP in Drosophila. *Experientia*, 18, 571–573.

Robichaux III, W.G. and Cheng, X. (2018). Intracellular cAMP sensor EPAC: Physiology, pathophysiology, and therapeutics development. *Physiol Rev*, 98, 919–1053.

Robison, G.A., Butcher, R.W., Sutherland, E.W. (1967). Adenyl cyclase as an adrenergic receptor. *Ann NY Acad Sci*, 139, 703–723.

Robling, A.G. and Bonewald, L.F. (2020). The osteocyte: New insights. *Annu Rev Physiol*, 82, 485–506.

Robling, A.G., Castillo, A.B., Turner, C.H. (2006). Biomechanical and molecular regulation of bone remodeling. *Annu Rev Biomed Eng*, 8, 455–498.

Rochira, V., Zirilli, L., Maffei, L., Premrou, V., Aranda, C., Baldi, M., Ghigo, E., Aimaretti, G., Carani, C., Lanfranco, F. (2010). Tall stature without growth hormone: Four male patients with aromatase deficiency. *J Clin Endocrinol Metab*, 95, 1626–1633.

Roden, M., Price, T.B., Perseghin, G., Petersen, K.F., Rothman, D.L., Cline, G.W., Shulman, G.I. (1996). Mechanism of free fatty acid-induced insulin resistance in humans. *J Clin Invest*, 97, 2859–2865.

Rodgers, J.T., Lerin, C., Haas, W., Gygi, S.P., Spiegelman, B.M., Puigserver, P. (2005). Nutrient control of glucose homeostasis through a complex of PGC-1alpha and SIRT1. *Nature*, 434, 113–118.

Rodrigues, M., Kosaric, N., Bonham, C.A., Gurtner, G.C. (2019). Wound healing: A cellular perspective. *Physiol Rev*, 99, 665–706.

Roelfsema, F., Yang, R.J., Takahashi, P.Y., Erickson, D., Bowers, C.Y., Veldhuis, J.D. (2018). Aromatized estrogens amplify nocturnal growth hormone secretion in testosterone-replaced older hypogonadal men. *J Clin Endocrinol Metab*, 103, 4419–4427.

Rohde, K., Keller, M., La Cour Poulsen, L., Blüher, M., Kovacs, P., Böttcher, Y. (2019). Genetics and epigenetics in obesity. *Metabolism*, 92, 37–50.

Ronveaux, C.C., Tomé, D., Raybould, H.E. (2015). Glucagon-like peptide 1 interacts with ghrelin and leptin to regulate glucose metabolism and food intake through vagal afferent neuron signaling. *J Nutr*, 145, 672–680.

Ross, J.L. (2016). The dark matter of biology. *Biophysical Journal*, 111, 909–916.

Rous, P. (1911). A sarcoma of the fowl transmissible by an agent separable from the tumor cells. *J Exp Med*, 13, 397.

Rowan, W. (1926). *On Photoperiodism, Reproductive Periodicity, and the Annual Migrations of Birds and Certain Fishes*. Boston Society of Natural History, Boston.

Rowe, I., Treherne, J., Ashford, M. (1996). Activation by intracellular ATP of a potassium channel in neurones from rat basomedial hypothalamus. *J Physiol*, 490, 97–113.

Rucart, G. (1951). Experimental parathyroid hyperplasia from hypervitaminosis D2; physiological interpretation. *C R Seances Soc Biol Fil*, 145, 342–344.

Ruland, J. (2011). Return to homeostasis: Downregulation of NF-κB responses. *Nat Immunol*, 12, 709–714.

Ryu, H., Chung, M., Dobrzynski, M., Fey, D., Blum, Y., Lee, S.S., Peter, M., Kholodenko, B.N., Jeon, N.L., Pertz, O. (2015). Frequency modulation of ERK activation dynamics rewires cell fate. *Mol Syst Biol*, 11, 838.

Saito, A., Goldfine, I.D., Williams, J.A. (1981). Characterization of receptors for cholecystokinin and related peptides in mouse cerebral cortex. *J Neurochem*, 37, 483–490.

Salmon Jr., W.D. and Daughaday, W.H. (1957). A hormonally controlled serum factor which stimulates sulfate incorporation by cartilage in vitro. *J Lab Clin Med*, 49, 825–836.

Sánchez-Carbayo, M. (2012). Hypermethylation in bladder cancer: Biological pathways and translational applications. *Tumour Biol*, 33, 347–361.

Sanders, V.M., Baker, R.A., Ramer-Quinn, D.S., Kasprowicz, D.J., Fuchs, B.A., Street, N.E. (1997). Differential expression of the beta2–adrenergic receptor by Th1 and Th2 clones: Implications for cytokine production and B cell help. *J Immunol*, 158, 4200–4210.

Sanders, N.M., Dunn-Meynell, A.A., Levin, B.E. (2004). Third ventricular alloxan reversibly impairs glucose counterregulatory responses. *Diabetes*, 53, 1230–1236.

Santos, A.J.M., Lo, Y.H., Mah, A.T., Kuo, C.J. (2018). The intestinal stem cell niche: Homeostasis and adaptations. *Trends Cell Biol*, 28, 1062–1078.

Sarnat, H.B. (1983). The discovery, proof and reproof of neurosecretion (Speidel, 1917; Scharrer and Scharrer, 1934). *Can J Neurol Sci / J Can Sci Neurol*, 10, 208–212.

Sarnat, H.B. and Flores-Sarnat, L. (2021). Excitatory/inhibitory synaptic ratios in polymicrogyria and down syndrome help explain epileptogenesis in malformations. *Pediatr Neurol*, 116, 41–54.

Sassone-Corsi, P. (2012). Minireview: NAD+, a circadian metabolite with an epigenetic twist. *Endocrinology*, 153, 1–5.

Sato, S., Basse, A.L., Schonke, M., Chen, S., Samad, M., Altintas, A., Laker, R.C., Dalbram, E., Barres, R., Baldi, P. et al. (2019). Time of exercise specifies the impact on muscle metabolic pathways and systemic energy homeostasis. *Cell Metab*, 30, 92–110e4.

Saxton, R.A. and Sabatini, D.M. (2017). mTOR signaling in growth, metabolism, and disease. *Cell*, 169, 361–371.

Schenk, S., Hintermann, E., Bilban, M., Koshikawa, N., Hojilla, C., Khokha, R., Quaranta, V. (2003). Binding to EGF receptor of a laminin-5 EGF-like fragment liberated during MMP-dependent mammary gland involution. *J Cell Biol*, 161, 197–209.

Schiller, M., Ben-Shaanan, T.L., Rolls, A. (2020). Neuronal regulation of immunity: Why, how and where? *Nat Rev Immunol*, 21(1), 20–36.

Schonhoff, S.E., Giel-Moloney, M., Leiter, A.B. (2004). Minireview: Development and differentiation of gut endocrine cells. *Endocrinology*, 145, 2639–2644.

Schwabe, J.W. and Rhodes, D. (1991). Beyond zinc fingers: Steroid hormone receptors have a novel structural motif for DNA recognition. *Trends Biochem Sci*, 16, 291–296.

Sears, S.M. and Hewett, S.J. (2021). Influence of glutamate and GABA transport on brain excitatory/inhibitory balance. *Exp Biol Med (Maywood)*, 246, 1069–1083.

Segal, B.M., Dwyer, B.K., Shevach, E.M. (1998). An interleukin (IL)-10/IL-12 immunoregulatory circuit controls susceptibility to autoimmune disease. *J Exp Med*, 187, 537–546.

Segers, A. and Depoortere, I. (2021). Circadian clocks in the digestive system. *Nat Rev Gastroenterol Hepatol*, 18, 239–251.

Semenza, G.L. (2011). Regulation of metabolism by hypoxia-inducible factor 1. *Cold Spring Harb Symp Quant Biol*, 76, 347–353.

Semenza, G.L. (2013). HIF-1 mediates metabolic responses to intratumoral hypoxia and oncogenic mutations. *J Clin Invest*, 123, 3664–3671.

Sen, R. and Baltimore, D. (1986). Multiple nuclear factors interact with the immunoglobulin enhancer sequences. *Cell*, 46, 705–716.

Sender, R. and Milo, R. (2021). The distribution of cellular turnover in the human body. *Nat Med*, 27, 45–48.

Sender, R., Fuchs, S., Milo, R. (2016). Revised estimates for the number of human and bacteria cells in the body. *PLoS Biol*, 14, e1002533.

Seo, D.Y., Park, S.H., Marquez, J., Kwak, H.B., Kim, T.N., Bae, J.H., Koh, J.H., Han, J. (2021). Hepatokines as a molecular transducer of exercise. *J Clin Med*, 10(3), 385.

Seoane, J. and Gomis, R.R. (2017). TGF-beta family signaling in tumor suppression and cancer progression. *Cold Spring Harb Perspect Biol*, 9(12), a022277.

Sethi, S., Kong, D., Land, S., Dyson, G., Sakr, W.A., Sarkar, F.H. (2013). Comprehensive molecular oncogenomic profiling and miRNA analysis of prostate cancer. *Am J Transl Res*, 5, 200–211.

Severinsen, M.C.K. and Pedersen, B.K. (2020). Muscle-organ crosstalk: The emerging roles of myokines. *Endocr Rev*, 41(4), 594–609.

Shankaran, V., Ikeda, H., Bruce, A.T., White, J.M., Swanson, P.E., Old, L.J., Schreiber, R.D. (2001). IFNgamma and lymphocytes prevent primary tumour development and shape tumour immunogenicity. *Nature*, 410, 1107–1111.

Shen, W.H., Zhou, J.H., Broussard, S.R., Freund, G.G., Dantzer, R., Kelley, K.W. (2002). Proinflammatory cytokines block growth of breast cancer cells by impairing signals from a growth factor receptor. *Cancer Res*, 62, 4746–4756.

Sherr, C.J. and Roberts, J.M. (1999). CDK inhibitors: Positive and negative regulators of G1–phase progression. *Genes Dev*, 13, 1501–1512.

Shim, H., Dolde, C., Lewis, B.C., Wu, C.S., Dang, G., Jungmann, R.A., Dalla-Favera, R., Dang, C.V. (1997). c-Myc transactivation of LDH-A: Implications for tumor metabolism and growth. *Proc Natl Acad Sci USA*, 94, 6658–6663.

Shinki, T., Shimada, H., Wakino, S., Anazawa, H., Hayashi, M., Saruta, T., Deluca, H.F., Suda, T. (1997). Cloning and expression of rat 25–hydroxyvitamin D3–1alpha-hydroxylase cDNA. *Proc Natl Acad Sci USA*, 94, 12920–12925.

Sidossis, L.S., Stuart, C.A., Shulman, G.I., Lopaschuk, G.D., Wolfe, R.R. (1996). Glucose plus insulin regulate fat oxidation by controlling the rate of fatty acid entry into the mitochondria. *J Clin Invest*, 98, 2244–2250.

Siegfried, Z., Eden, S., Mendelsohn, M., Feng, X., Tsuberi, B.Z., Cedar, H. (1999). DNA methylation represses transcription in vivo. *Nat Genet*, 22, 203–206.

Silver, J., Russell, J., Sherwood, L.M. (1985). Regulation by vitamin D metabolites of messenger ribonucleic acid for preproparathyroid hormone in isolated bovine parathyroid cells. *Proc Natl Acad Sci USA*, 82, 4270–4273.

Silverman, J.A. (1985). Richard Morton, 1637–1698, limner of anorexia nervosa: His life and times. A tercentenary essay. *J Psychiatr Res*, 19, 83–88.

Silverman, M.N. and Sternberg, E.M. (2012). Glucocorticoid regulation of inflammation and its functional correlates: From HPA axis to glucocorticoid receptor dysfunction. *Ann NY Acad Sci*, 1261, 55–63.

Simonet, W.S., Lacey, D.L., Dunstan, C.R., Kelley, M., Chang, M.S., Lüthy, R., Nguyen, H.Q., Wooden, S., Bennett, L., Boone, T. et al. (1997). Osteoprotegerin: A novel secreted protein involved in the regulation of bone density. *Cell*, 89, 309–319.

Sims, N.A. and Martin, T.J. (2014). Coupling the activities of bone formation and resorption: A multitude of signals within the basic multicellular unit. *Bonekey Rep*, 3, 481.

Sinding, C. (1989). The history of resistant rickets: A model for understanding the growth of biomedical knowledge. *J Hist Biol*, 22, 461–495.

Singer, S.J. and Nicolson, G.L. (1972). The fluid mosaic model of the structure of cell membranes. *Science*, 175, 720–731.

Singh, P., Carraher, C., Schwarzbauer, J.E. (2010). Assembly of fibronectin extracellular matrix. *Annu Rev Cell Dev Biol*, 26, 397–419.

Singh, R., Letai, A., Sarosiek, K. (2019). Regulation of apoptosis in health and disease: The balancing act of BCL-2 family proteins. *Nat Rev Mol Cell Biol*, 20, 175–193.

Skinner, M.K. (2008). What is an epigenetic transgenerational phenotype? F3 or F2. *Reprod Toxicol*, 25, 2–6.

Smith, W. (1867). *Dictionary of Greek and Roman Biography and Mythology*. John Murray, London.

Smith, P.E. (1932). The secretory capacity of the anterior hypophysis as evidenced by the effect of partial hypophysectomies in rats. *Anat Rec*, 52, 191–207.

Smith, E.M. and Blalock, J.E. (1981). Human lymphocyte production of corticotropin and endorphin-like substances: Association with leukocyte interferon. *Proc Natl Acad Sci*, 78, 7530–7534.

Smith, G.P., Jerome, C., Cushin, B.J., Eterno, R., Simansky, K.J. (1981). Abdominal vagotomy blocks the satiety effect of cholecystokinin in the rat. *Science*, 213, 1036–1037.

Smith, E.P., Boyd, J., Frank, G.R., Takahashi, H., Cohen, R.M., Specker, B., Williams, T.C., Lubahn, D.B., Korach, K.S. (1994). Estrogen resistance caused by a mutation in the estrogen-receptor gene in a man. *N Engl J Med*, 331, 1056–1061.

Smyth, M.J., Dunn, G.P., Schreiber, R.D. (2006). Cancer immunosurveillance and immunoediting: The roles of immunity in suppressing tumor development and shaping tumor immunogenicity. *Adv Immunol*, 90, 1–50.

Sokoloff, L., Reivich, M., Kennedy, C., Des Rosiers, M.H., Patlak, C.S., Pettigrew, K.D., Sakurada, O., Shinohara, M. (1977). The [14C]deoxyglucose method for the measurement of local cerebral glucose utilization: Theory, procedure, and normal values in the conscious and anesthetized albino rat. *J Neurochem*, 28, 897–916.

Sonnenburg, J.L. and Bäckhed, F. (2016). Diet–microbiota interactions as moderators of human metabolism. *Nature*, 535, 56–64.

Sörensen, S.P.L. (1909). Uber die Messung und die Bedeutung der Wasserstoffionenkonzentration bei enzymatischen Prozessen. *Biochemische Zeitschrift*, 21, 131–200.

Soskin, S. (1940). The liver and carbohydrate metabolism. *Endocrinology*, 26, 297–308.

Spalding, K.L., Arner, E., Westermark, P.O., Bernard, S., Buchholz, B.A., Bergmann, O., Blomqvist, L., Hoffstedt, J., Naslund, E., Britton, T. et al. (2008). Dynamics of fat cell turnover in humans. *Nature*, 453, 783–787.

Spelsberg, T.C., Steggles, A.W., O'Malley, B.W. (1971). Progesterone-binding components of chick oviduct 3. Chromatin acceptor sites. *J Biol Chem*, 246, 4188–4197.

Spreckley, E. and Murphy, K.G. (2015). The L-cell in nutritional sensing and the regulation of appetite. *Front Nutr*, 2.

Starmer, J. and Magnuson, T. (2009). A new model for random X chromosome inactivation. *Development*, 136, 1–10.

Steegborn, C. (2014). Structure, mechanism, and regulation of soluble adenylyl cyclases – Similarities and differences to transmembrane adenylyl cyclases. *Biochim Biophys Acta*, 1842, 2535–2547.

Steele, R. (1959). Influences of glucose loading and of injected insulin on hepatic glucose output. *Ann NY Acad Sci*, 82, 420–430.

Stehelin, D., Varmus, H.E., Bishop, J.M., Vogt, P.K. (1976). DNA related to the transforming gene(s) of avian sarcoma viruses is present in normal avian DNA. *Nature*, 260, 170–173.

Steiner, D.F. (2011a). Adventures with insulin in the islets of Langerhans. *J Biol Chem*, 286, 17399–17421.

Steiner, D.F. (2011b). On the discovery of precursor processing. In *Proprotein Convertases*, Mbikay, M. and Seidah, N.G. (eds). Humana Press, Totowa.

Stern, J.H., Rutkowski, J.M., Scherer, P.E. (2016). Adiponectin, leptin, and fatty acids in the maintenance of metabolic homeostasis through adipose tissue crosstalk. *Cell Metab*, 23, 770–784.

Sternberg, E.M., Young 3rd, W.S., Bernardini, R., Calogero, A.E., Chrousos, G.P., Gold, P.W., Wilder, R.L. (1989). A central nervous system defect in biosynthesis of corticotropin-releasing hormone is associated with susceptibility to streptococcal cell wall-induced arthritis in Lewis rats. *Proc Natl Acad Sci USA*, 86, 4771–4775.

Stincic, T.L., Ronnekleiv, O.K., Kelly, M.J. (2018). Diverse actions of estradiol on anorexigenic and orexigenic hypothalamic arcuate neurons. *Horm Behav*, 104, 146–155.

Stine, Z.E., Walton, Z.E., Altman, B.J., Hsieh, A.L., Dang, C.V. (2015). MYC, metabolism, and cancer. *Cancer Discov*, 5, 1024–1039.

Stockard, C.R. (1921). Developmental rate and structural expression: An experimental study of twins, "double monsters" and single deformities, and the interaction among embryonic organs during their origin and development. *Am J Anat*, 28, 115–277.

Storlien, L., Oakes, N.D., Kelley, D.E. (2004). Metabolic flexibility. *Proc Nutr Soc*, 63, 363–368.

Streuli, C.H. and Akhtar, N. (2009). Signal co-operation between integrins and other receptor systems. *Biochem J*, 418, 491–506.

Surani, M.A., Barton, S.C., Norris, M.L. (1984). Development of reconstituted mouse eggs suggests imprinting of the genome during gametogenesis. *Nature*, 308, 548–550.

Surjit, M., Ganti, K.P., Mukherji, A., Ye, T., Hua, G., Metzger, D., Li, M., Chambon, P. (2011). Widespread negative response elements mediate direct repression by agonist-liganded glucocorticoid receptor. *Cell*, 145, 224–241.

Swann, J.B. and Smyth, M.J. (2007). Immune surveillance of tumors. *J Clin Invest*, 117, 1137–1146.

Sweeney, H.L. and Holzbaur, E.L.F. (2018). Motor proteins. *Cold Spring Harb Perspect Biol*, 10(5), a021931.

Syková, E. and Nicholson, C. (2008). Diffusion in brain extracellular space. *Physiol Rev*, 88, 1277–1340.

Takahashi, A. and Mizusawa, K. (2013). Posttranslational modifications of proopiomelanocortin in vertebrates and their biological significance. *Front Endocrinol (Lausanne)*, 4, 143.

Takahashi, K. and Yamanaka, S. (2006). Induction of pluripotent stem cells from mouse embryonic and adult fibroblast cultures by defined factors. *Cell*, 126, 663–676.

Takahashi, K., Okita, K., Nakagawa, M., Yamanaka, S. (2007). Induction of pluripotent stem cells from fibroblast cultures. *Nat Protoc*, 2, 3081–3089.

Takeda, S., Elefteriou, F., Levasseur, R., Liu, X., Zhao, L., Parker, K.L., Armstrong, D., Ducy, P., Karsenty, G. (2002). Leptin regulates bone formation via the sympathetic nervous system. *Cell*, 111, 305–317.

Takeshige, K., Baba, M., Tsuboi, S., Noda, T., Ohsumi, Y. (1992). Autophagy in yeast demonstrated with proteinase-deficient mutants and conditions for its induction. *J Cell Biol*, 119, 301–311.

Takeuchi, O. and Akira, S. (2010). Pattern recognition receptors and inflammation. *Cell*, 140, 805–820.

Tamkun, J.W., Desimone, D.W., Fonda, D., Patel, R.S., Buck, C., Horwitz, A.F., Hynes, R.O. (1986). Structure of integrin, a glycoprotein involved in the transmembrane linkage between fibronectin and actin. *Cell*, 46, 271–282.

Tammen, S.A., Friso, S., Choi, S.W. (2013). Epigenetics: The link between nature and nurture. *Mol Aspects Med*, 34, 753–764.

Taniguchi, T., Mantei, N., Schwarzstein, M., Nagata, S., Muramatsu, M., Weissmann, C. (1980). Human leukocyte and fibroblast interferons are structurally related. *Nature*, 285, 547–549.

Tannenbaum, G. and Epelbaum, J. (1999). Somatostatin. In *Handbook of Physiology (section 7); The Endocrine System – Volume V: Hormonal Control of Growth*, Kostyo, J.L. and Goodman, H.M. (eds). Oxford University Press, New York.

Tanti, J.F., Grémeaux, T., Van Obberghen, E., Le Marchand-Brustel, Y. (1994). Serine/threonine phosphorylation of insulin receptor substrate 1 modulates insulin receptor signaling. *J Biol Chem*, 269, 6051–6057.

Tauber, A.I. (1992). The birth of immunology III: The fate of the phagocytosis theory. *Cellular Immunol*, 139, 505–530.

Taunton, J., Hassig, C.A., Schreiber, S.L. (1996). A mammalian histone deacetylase related to the yeast transcriptional regulator Rpd3p. *Science*, 272, 408–411.

Taylor, G., Lehrer, M.S., Jensen, P.J., Sun, T.T., Lavker, R.M. (2000). Involvement of follicular stem cells in forming not only the follicle but also the epidermis. *Cell*, 102, 451–461.

Tena-Sempere, M. (2005). Exploring the role of ghrelin as novel regulator of gonadal function. *Growth Horm IGF Res*, 15, 83–88.

Tenuta, M., Carlomagno, F., Cangiano, B., Kanakis, G., Pozza, C., Sbardella, E., Isidori, A.M., Krausz, C., Gianfrilli, D. (2021). Somatotropic-Testicular Axis: A crosstalk between GH/IGF-I and gonadal hormones during development, transition, and adult age. *Andrology*, 9, 168–184.

Thomas, S.H., Willey, K., Wisher, M.H. (1978). Evidence that anti-lipolytic and lipogenic effects of insulin are mediated in rat adipocytes by same receptor. *Diabetologia*, 276–276.

Thomas, S.H., Wisher, M.H., Brandenburg, D., Sönksen, P.H. (1979). Insulin action on adipocytes. Evidence that the anti-lipolytic and lipogenic effects of insulin are mediated by the same receptor. *Biochem J*, 184, 355–360.

Thorens, B. (2012). Sensing of glucose in the brain. In *Appetite Control*, Joost, H.-G. (ed.). Springer, Berlin, Heidelberg.

Thorens, B., Guillam, M.-T., Beermann, F., Burcelin, R., Jaquet, M. (2000). Transgenic reexpression of GLUT1 or GLUT2 in pancreatic β cells rescues GLUT2–null mice from early death and restores normal glucose-stimulated insulin secretion. *J Biol Chem*, 275, 23751–23758.

Tigerstedt, R. and Bergman, P. (1898). Niere und Kreislauf 1. *Skandinavisches Archiv für Physiologie*, 8, 223–271.

Till, J.E. and McCulloh, E.A. (1961). A direct measurement of the radiation sensitivity of normal mouse bone marrow cells. *Radiat Res*, 14, 213–222.

Todaro, G., De Larco, J., Marquardt, H., Bryant, M., Sherwin, S., Sliski, A. (1979). Polypeptide growth factors produced by tumor cells and virus-transformed cells: A possible growth advantage for the producer cells. *Horm Cell Cult*, 6, 113–127.

Toft, D. and Gorski, J. (1966). A receptor molecule for estrogens: Isolation from the rat uterus and preliminary characterization. *Proc Natl Acad Sci USA*, 55, 1574.

Toker, A. and Marmiroli, S. (2014). Signaling specificity in the Akt pathway in biology and disease. *Adv Biol Regul*, 55, 28–38.

Tracey, K.J. (2002). The inflammatory reflex. *Nature*, 420, 853–859.

Tracey, K.J. (2007). Physiology and immunology of the cholinergic antiinflammatory pathway. *J Clin Invest*, 117, 289–296.

Tremaroli, V. and Bäckhed, F. (2012). Functional interactions between the gut microbiota and host metabolism. *Nature*, 489, 242–249.

Tsai, C.L., Pai, M.C., Ukropec, J., Ukropcová, B. (2019). Distinctive effects of aerobic and resistance exercise modes on neurocognitive and biochemical changes in individuals with mild cognitive impairment. *Curr Alzheimer Res*, 16, 316–332.

Tsuchiya, A. and Lu, W.Y. (2019). Liver stem cells: Plasticity of the liver epithelium. *World J Gastroenterol*, 25, 1037–1049.

Tubbs, A. and Nussenzweig, A. (2017). Endogenous DNA damage as a source of genomic instability in cancer. *Cell*, 168, 644–656.

Turner, N.C., Reis-Filho, J.S., Russell, A.M., Springall, R.J., Ryder, K., Steele, D., Savage, K., Gillett, C.E., Schmitt, F.C., Ashworth, A. et al. (2007). BRCA1 dysfunction in sporadic basal-like breast cancer. *Oncogene*, 26, 2126–2132.

Turpin-Nolan, S.M. and Brüning, J.C. (2020). The role of ceramides in metabolic disorders: When size and localization matters. *Nat Rev Endocrinol*, 16, 224–233.

Turton, M.D., O'Shea, D., Gunn, I., Beak, S.A., Edwards, C.M., Meeran, K., Choi, S.J., Taylor, G.M., Heath, M.M., Lambert, P.D. et al. (1996). A role for glucagon-like peptide-1 in the central regulation of feeding. *Nature*, 379, 69–72.

Tycko, B. (1997). DNA methylation in genomic imprinting. *Mutat Res*, 386, 131–140.

Tzatsos, A. and Kandror, K.V. (2006). Nutrients suppress phosphatidylinositol 3–kinase/Akt signaling via raptor-dependent mTOR-mediated insulin receptor substrate 1 phosphorylation. *Mol Cell Biol*, 26, 63–76.

Uciechowski, P. and Dempke, W.C.M. (2020). Interleukin-6: A masterplayer in the cytokine network. *Oncology*, 98, 131–137.

Ulloa, L., Doody, J., Massagué, J. (1999). Inhibition of transforming growth factor-beta/SMAD signalling by the interferon-gamma/STAT pathway. *Nature*, 397, 710–713.

Ullrich, A., Coussens, L., Hayflick, J.S., Dull, T.J., Gray, A., Tam, A.W., Lee, J., Yarden, Y., Libermann, T.A., Schlessinger, J. et al. (1984). Human epidermal growth factor receptor cDNA sequence and aberrant expression of the amplified gene in A431 epidermoid carcinoma cells. *Nature*, 309, 418–425.

Umbarger, H.E. (1956). Evidence for a negative-feedback mechanism in the biosynthesis of isoleucine. *Science*, 123, 848.

Umpleby, A.M. and Sönksen, P.H. (1987). Measurement of the turnover of substrates of carbohydrate and protein metabolism using radioactive isotopes. *Baillière's Clin Endocrinol Metab*, 1, 773–796.

Unger, R.H. (1971). Glucagon and the insulin: Glucagon ratio in diabetes and other catabolic illnesses. *Diabetes*, 20, 834–838.

Urakawa, I., Yamazaki, Y., Shimada, T., Iijima, K., Hasegawa, H., Okawa, K., Fujita, T., Fukumoto, S., Yamashita, T. (2006). Klotho converts canonical FGF receptor into a specific receptor for FGF23. *Nature*, 444, 770–774.

Valenstein, E.S. (2007). *The War of the Soups and the Sparks: The Discovery of Neurotransmitters and the Dispute over How Nerves Communicate*. Columbia University Press, New York.

Vallabhapurapu, S. and Karin, M. (2009). Regulation and function of NF-kappaB transcription factors in the immune system. *Annu Rev Immunol*, 27, 693–733.

Van Lint, C., Emiliani, S., Verdin, E. (1996). The expression of a small fraction of cellular genes is changed in response to histone hyperacetylation. *Gene Expression J Liver Res*, 5, 245–253.

Van Putten, L.M. (1958). The life span of red cells in the rat and the mouse as determined by labeling with DFP32 in vivo. *Blood*, 13, 789–794.

Van Schaftingen, E., Vandercammen, A., Detheux, M., Davies, D. (1992). The regulatory protein of liver glucokinase. *Adv Enzyme Regul*, 32, 133–148.

Vaupel, P., Schmidberger, H., Mayer, A. (2019). The Warburg effect: Essential part of metabolic reprogramming and central contributor to cancer progression. *Int J Radiat Biol*, 95, 912–919.

Veiga-Fernandes, H., Coles, M.C., Foster, K.E., Patel, A., Williams, A., Natarajan, D., Barlow, A., Pachnis, V., Kioussis, D. (2007). Tyrosine kinase receptor RET is a key regulator of Peyer's patch organogenesis. *Nature*, 446, 547–551.

Veldhuis, J.D., Iranmanesh, A., Erickson, D., Roelfsema, F., Bowers, C.Y. (2012). Lifetime regulation of growth hormone (GH) secretion. In *Handbook of Neuroendocrinology*, Fink, G., Pfaff, D.W., Levine, J.E. (eds). Academic Press, Cambridge.

Vesely, M.D., Kershaw, M.H., Schreiber, R.D., Smyth, M.J. (2011). Natural innate and adaptive immunity to cancer. *Annu Rev Immunol*, 29, 235–271.

Virtanen, K.A., Lidell, M.E., Orava, J., Heglind, M., Westergren, R., Niemi, T., Taittonen, M., Laine, J., Savisto, N.-J., Enerbäck, S. (2009). Functional brown adipose tissue in healthy adults. *N Eng J Med*, 360, 1518–1525.

Visse, R. and Nagase, H. (2003). Matrix metalloproteinases and tissue inhibitors of metalloproteinases: Structure, function, and biochemistry. *Circ Res*, 92, 827–839.

Vivanco, I. and Sawyers, C.L. (2002). The phosphatidylinositol 3–kinase AKT pathway in human cancer. *Nat Rev Cancer*, 2, 489–501.

Vizi, E.S. (1974). Interaction between adrenergic and cholinergic systems: Presynaptic inhibitory effect of noradrenaline on acetylcholine release. *J Neural Transm*, Suppl 11, 61–78.

Vizi, E.S., Fekete, A., Karoly, R., Mike, A. (2010). Non-synaptic receptors and transporters involved in brain functions and targets of drug treatment. *Br J Pharmacol*, 160, 785–809.

Von Zglinicki, T. (2000). Role of oxidative stress in telomere length regulation and replicative senescence. *Ann NY Acad Sci*, 908, 99–110.

Vousden, K.H. and Lu, X. (2002). Live or let die: The cell's response to p53. *Nat Rev Cancer*, 2, 594–604.

Vulpian, A. (1882). *Leçons sur l'action physiologique des substances toxiques et médicamenteuses*. Octave Doin, Paris.

Waddington, C.H. (1940). *Organisers and Genes*. The University Press, Cambridge.

Waddington, C.H. (2016). *An Introduction to Modern Genetics*. Routledge, London.

Wadhwa, R., Song, S., Lee, J.S., Yao, Y., Wei, Q., Ajani, J.A. (2013). Gastric cancer-molecular and clinical dimensions. *Nat Rev Clin Oncol*, 10, 643–655.

Wai, T. and Langer, T. (2016). Mitochondrial dynamics and metabolic regulation. *Trends Endocrinol Metab*, 27, 105–117.

Walter, P., Green, S., Greene, G., Krust, A., Bornert, J.M., Jeltsch, J.M., Staub, A., Jensen, E., Scrace, G., Waterfield, M. et al. (1985). Cloning of the human estrogen receptor cDNA. *Proc Natl Acad Sci USA*, 82, 7889–7893.

Walther, T.C. and Farese Jr., R.V. (2012). Lipid droplets and cellular lipid metabolism. *Annu Rev Biochem*, 81, 687–714.

Wang, G.L., Jiang, B.H., Rue, E.A., Semenza, G.L. (1995). Hypoxia-inducible factor 1 is a basic-helix-loop-helix-PAS heterodimer regulated by cellular O2 tension. *Proc Natl Acad Sci USA*, 92, 5510–5514.

Wang, Q., Zhang, Y., Yang, C., Xiong, H., Lin, Y., Yao, J., Li, H., Xie, L., Zhao, W., Yao, Y. et al. (2010). Acetylation of metabolic enzymes coordinates carbon source utilization and metabolic flux. *Science*, 327, 1004–1007.

Wang, J., Chen, X., Li, P., Su, L., Yu, B., Cai, Q., Li, J., Yu, Y., Liu, B., Zhu, Z. (2013). CRKL promotes cell proliferation in gastric cancer and is negatively regulated by miR-126. *Chem Biol Interact*, 206, 230–238.

Wang, C., Lutes, L.K., Barnoud, C., Scheiermann, C. (2022). The circadian immune system. *Sci Immunol*, 7, eabm2465.

Wankhade, U.D., Zhong, Y., Kang, P., Alfaro, M., Chintapalli, S.V., Thakali, K.M., Shankar, K. (2017). Enhanced offspring predisposition to steatohepatitis with maternal high-fat diet is associated with epigenetic and microbiome alterations. *PLoS One*, 12, e0175675.

Warburg, O. (1956). On respiratory impairment in cancer cells. *Science*, 124, 269–270.

Warburg, O., Posener, K., Negelein, E. (1924). Über den stoffwechsel der carcinomzelle. *Naturwissenschaften*, 12, 1131–1137.

Ward, P.S. and Thompson, C.B. (2012). Metabolic reprogramming: A cancer hallmark even Warburg did not anticipate. *Cancer Cell*, 21, 297–308.

Wasserman, D.H. (1995). Regulation of glucose fluxes during exercise in the postabsorptive state. *Annu Rev Physiol*, 57, 191–218.

Watanabe, M., Fukuda, A., Nabekura, J. (2014). The role of GABA in the regulation of GnRH neurons. *Front Neurosci*, 8, 387.

Waterland, R.A. and Jirtle, R.L. (2003). Transposable elements: Targets for early nutritional effects on epigenetic gene regulation. *Mol Cell Biol*, 23, 5293–5300.

Waterland, R.A., Travisano, M., Tahiliani, K.G., Rached, M.T., Mirza, S. (2008). Methyl donor supplementation prevents transgenerational amplification of obesity. *Int J Obes (Lond)*, 32, 1373–1379.

Way, E.L. (1985). Review of reviews. *Annu Rev Pharmacol Toxicol*, 25, 769–775.

Weaver, I.C., Cervoni, N., Champagne, F.A., D'alessio, A.C., Sharma, S., Seckl, J.R., Dymov, S., Szyf, M., Meaney, M.J. (2004). Epigenetic programming by maternal behavior. *Nat Neurosci*, 7, 847–854.

Webb, A.E. and Brunet, A. (2014). FOXO transcription factors: Key regulators of cellular quality control. *Trends Biochem Sci*, 39, 159–169.

Wei, J.W., Huang, K., Yang, C., Kang, C.S. (2017). Non-coding RNAs as regulators in epigenetics (Review). *Oncol Rep*, 37, 3–9.

Weigert, C., Hoene, M., Plomgaard, P. (2019). Hepatokines – A novel group of exercise factors. *Pflugers Arch*, 471, 383–396.

Weikum, E.R., Liu, X., Ortlund, E.A. (2018). The nuclear receptor superfamily: A structural perspective. *Protein Sci*, 27, 1876–1892.

Wein, M.N. (2018). Parathyroid hormone signaling in osteocytes. *JBMR Plus*, 2, 22–30.

Weinberg, R.A. (1994). Oncogenes and tumor suppressor genes. *CA Cancer J Clin*, 44, 160–170.

Weir, E.K., López-Barneo, J., Buckler, K.J., Archer, S.L. (2005). Acute oxygen-sensing mechanisms. *N Engl J Med*, 353, 2042–2055.

Wellen, K.E., Hatzivassiliou, G., Sachdeva, U.M., Bui, T.V., Cross, J.R., Thompson, C.B. (2009). ATP-citrate lyase links cellular metabolism to histone acetylation. *Science*, 324, 1076–1080.

Welte, Y., Adjaye, J., Lehrach, H.R., Regenbrecht, C.R. (2010). Cancer stem cells in solid tumors: Elusive or illusive? *Cell Commun Signal*, 8, 6.

West, D.C., Rees, C.G., Duchesne, L., Patey, S.J., Terry, C.J., Turnbull, J.E., Delehedde, M., Heegaard, C.W., Allain, F., Vanpouille, C. et al. (2005). Interactions of multiple heparin binding growth factors with neuropilin-1 and potentiation of the activity of fibroblast growth factor-2. *J Biol Chem*, 280, 13457–13464.

Whipps, J.M., Lewis, K., Cooke, R. (1988). Mycoparasitism and plant disease control. *Fungi in biological control systems*, 161–187.

Wiener, N. (2019). *Cybernetics or Control and Communication in the Animal and the Machine*. MIT Press, Cambridge.

Wilkinson, A.C., Igarashi, K.J., Nakauchi, H. (2020). Haematopoietic stem cell self-renewal in vivo and ex vivo. *Nat Rev Genet*, 21, 541–554.

Wilmut, I., Schnieke, A.E., McWhir, J., Kind, A.J., Campbell, K.H. (1997). Viable offspring derived from fetal and adult mammalian cells. *Nature*, 385, 810–813.

Wilson, J.E. (2003). Isozymes of mammalian hexokinase: Structure, subcellular localization and metabolic function. *J Exp Biol*, 206, 2049–2057.

Winner, B., Cooper-Kuhn, C.M., Aigner, R., Winkler, J., Kuhn, H.G. (2002). Long-term survival and cell death of newly generated neurons in the adult rat olfactory bulb. *Eur J Neurosci*, 16, 1681–1689.

Wise, D.R., Deberardinis, R.J., Mancuso, A., Sayed, N., Zhang, X.Y., Pfeiffer, H.K., Nissim, I., Daikhin, E., Yudkoff, M., McMahon, S.B. et al. (2008). Myc regulates a transcriptional program that stimulates mitochondrial glutaminolysis and leads to glutamine addiction. *Proc Natl Acad Sci USA*, 105, 18782–18787.

Wolff, J. (1892). *The Law of Bone Transformation*. Hirschwald, Berlin.

Wolff, G. (1987). Body weight and cancer. *Am J Clin Nutr*, 45, 168–180.

Wolff, G., Roberts, D., Galbraith, D. (1986). Prenatal determination of obesity, tumor susceptibility, and coat color pattern in viable yellow (A vy/a) mice: The yellow mouse syndrome. *J Heredity*, 77, 151–158.

Wolff, G.L., Kodell, R.L., Moore, S.R., Cooney, C.A. (1998). Maternal epigenetics and methyl supplements affect agouti gene expression in Avy/a mice. *FASEB J*, 12, 949–957.

Wolffe, A.P. and Matzke, M.A. (1999). Epigenetics: Regulation through repression. *Science*, 286, 481–486.

Woods, S.C., Lotter, E.C., McKay, L.D., Porte Jr., D. (1979). Chronic intracerebroventricular infusion of insulin reduces food intake and body weight of baboons. *Nature*, 282, 503–505.

Worthington, J.J., Klementowicz, J.E., Travis, M.A. (2011). TGFβ: A sleeping giant awoken by integrins. *Trends Biochem Sci*, 36, 47–54.

Wrenshall, G.A. (1955). Working basis for the tracer measurement of transfer rates of a metabolic factor in biological systems containing compartments whose contents do not intermix rapidly. *Can J Biochem Physiol*, 33, 909–925.

Wu, Q. and Palmiter, R.D. (2011). GABAergic signaling by AgRP neurons prevents anorexia via a melanocortin-independent mechanism. *Eur J Pharmacol*, 660, 21–27.

Wu, X., Bayle, J.H., Olson, D., Levine, A.J. (1993). The p53–mdm-2 autoregulatory feedback loop. *Genes Dev*, 7, 1126–1132.

Xie, W., Stribley, J.A., Chatonnet, A., Wilder, P.J., Rizzino, A., McComb, R.D., Taylor, P., Hinrichs, S.H., Lockridge, O. (2000). Postnatal developmental delay and supersensitivity to organophosphate in gene-targeted mice lacking acetylcholinesterase. *J Pharmacol Exp Therapeut*, 293, 896–902.

Xu, J., Ji, J., Yan, X.H. (2012). Cross-talk between AMPK and mTOR in regulating energy balance. *Crit Rev Food Sci Nutr*, 52, 373–381.

Xu, J., Mathur, J., Vessières, E., Hammack, S., Nonomura, K., Favre, J., Grimaud, L., Petrus, M., Francisco, A., Li, J. et al. (2018). GPR68 senses flow and is essential for vascular physiology. *Cell*, 173, 762–775.e16.

Yalow, R.S. and Berson, S.A. (1959). Assay of plasma insulin in human subjects by immunological methods. *Nature*, 184(Suppl 21), 1648–1649.

Yamamoto, K.R. (1985). Steroid receptor regulated transcription of specific genes and gene networks. *Annu Rev Genet*, 19, 209–252.

Yang, B., Guo, M., Herman, J.G., Clark, D.P. (2003). Aberrant promoter methylation profiles of tumor suppressor genes in hepatocellular carcinoma. *Am J Pathol*, 163, 1101–1107.

Yaribeygi, H., Farrokhi, F.R., Butler, A.E., Sahebkar, A. (2019). Insulin resistance: Review of the underlying molecular mechanisms. *J Cell Physiol*, 234, 8152–8161.

Yeo, G.S.H. (2014). The role of the FTO (fat mass and obesity related) locus in regulating body size and composition. *Mol Cell Endocrinol*, 397, 34–41.

Yeung, S.J., Pan, J., Lee, M.H. (2008). Roles of p53, MYC and HIF-1 in regulating glycolysis – The seventh hallmark of cancer. *Cell Mol Life Sci*, 65, 3981–3999.

Yip, S.H., Romanò, N., Gustafson, P., Hodson, D.J., Williams, E.J., Kokay, I.C., Martin, A.O., Mollard, P., Grattan, D.R., Bunn, S.J. (2019). Elevated prolactin during pregnancy drives a phenotypic switch in mouse hypothalamic dopaminergic neurons. *Cell Rep*, 26, 1787–1799.e5.

Yoo, S.-H., Yamazaki, S., Lowrey, P.L., Shimomura, K., Ko, C.H., Buhr, E.D., Siepka, S.M., Hong, H.-K., Oh, W.J., Yoo, O.J. (2004). PERIOD2:: LUCIFERASE real-time reporting of circadian dynamics reveals persistent circadian oscillations in mouse peripheral tissues. *Proc Natl Acad Sci*, 101, 5339–5346.

Yoshino, J., Mills, K.F., Yoon, M.J., Imai, S. (2011). Nicotinamide mononucleotide, a key NAD(+) intermediate, treats the pathophysiology of diet- and age-induced diabetes in mice. *Cell Metab*, 14, 528–536.

Yuan, H.X., Xiong, Y., Guan, K.L. (2013). Nutrient sensing, metabolism, and cell growth control. *Mol Cell*, 49, 379–387.

Zhang, W. and Bi, S. (2015). Hypothalamic regulation of brown adipose tissue thermogenesis and energy homeostasis. *Front Endocrinol (Lausanne)*, 6, 136.

Zhang, Y., Proenca, R., Maffei, M., Barone, M., Leopold, L., Friedman, J.M. (1994). Positional cloning of the mouse obese gene and its human homologue. *Nature*, 372, 425–432.

Zhang, Y., Hu, X., Tian, R., Wei, W., Hu, W., Chen, X., Han, W., Chen, H., Gong, Y. (2006). Angiopoietin-related growth factor (AGF) supports adhesion, spreading, and migration of keratinocytes, fibroblasts, and endothelial cells through interaction with RGD-binding integrins. *Biochem Biophys Res Commun*, 347, 100–108.

Zhao, F., Mancuso, A., Bui, T.V., Tong, X., Gruber, J.J., Swider, C.R., Sanchez, P.V., Lum, J.J., Sayed, N., Melo, J.V. et al. (2010). Imatinib resistance associated with BCR-ABL upregulation is dependent on HIF-1alpha-induced metabolic reprograming. *Oncogene*, 29, 2962–2972.

Zhu, T., Gobeil, F., Vazquez-Tello, A., Leduc, M., Rihakova, L., Bossolasco, M., Bkaily, G., Peri, K., Varma, D.R., Orvoine, R. et al. (2006). Intracrine signaling through lipid mediators and their cognate nuclear G-protein-coupled receptors: A paradigm based on PGE2, PAF, and LPA1 receptors. *Can J Physiol Pharmacol*, 84, 377–391.

Zierold, C., Nehring, J.A., Deluca, H.F. (2007). Nuclear receptor 4A2 and C/EBPbeta regulate the parathyroid hormone-mediated transcriptional regulation of the 25–hydroxyvitamin D3–1alpha-hydroxylase. *Arch Biochem Biophys*, 460, 233–239.

Zimmermann, L.-M.A., Correns, A., Furlan, A.G., Spanou, C.E., Sengle, G. (2021). Controlling BMP growth factor bioavailability: The extracellular matrix as multi skilled platform. *Cell Signal*, 85, 110071.

Zsarnovszky, A., Horvath, T.L., Garcia-Segura, L.M., Horvath, B., Naftolin, F. (2001). Oestrogen-induced changes in the synaptology of the monkey (*Cercopithecus aethiops*) arcuate nucleus during gonadotropin feedback. *J Neuroendocrinol*, 13, 22–28.

Zuckerman, V., Wolyniec, K., Sionov, R.V., Haupt, S., Haupt, Y. (2009). Tumour suppression by p53: The importance of apoptosis and cellular senescence. *J Pathol*, 219, 3–15.

Index

A, B

C, D

E, F

G, H

I, J

K, L

M, N

O, P

R, S

T, V

W, X

Other titles from

in

Biology and Biomedical Engineering

2023

VERNA Emeline
Asymptomatic Osseous Variations of the Postcranial Human Skeleton (Comparative Anatomy and Posture of Animal and Human Set – Volume 5)

WALCH Jean-Paul, BLAISE Solange
Phyllotaxis Models: A Tool for Evolutionary Biologists

2022

DAMBRICOURT MALASSÉ Anne
Embryogeny and Phylogeny of the Human Posture 2: A New Glance at the Future of our Species (Comparative Anatomy and Posture of Animal and Human Set – Volume 4)

FOSSÉ Philippe
Structures and Functions of Retroviral RNAs: The Multiple Facets of the Retroviral Genome (Nucleic Acids Set – Volume 1)

2021

DAMBRICOURT MALASSÉ Anne
Embryogeny and Phylogeny of the Human Posture 1: A New Glance at the Future of our Species (Comparative Anatomy and Posture of Animal and Human Set – Volume 3)

GRANDCOLAS Philippe, MAUREL Marie-Christine
Systematics and the Exploration of Life

HADJOUIS Djillali
The Skull of Quadruped and Bipedal Vertebrates: Variations, Abnormalities and Joint Pathologies (Comparative Anatomy and Posture of Animal and Human Set – Volume 2)

HULLÉ Maurice, VERNON Philippe
The Terrestrial Macroinvertebrates of the Sub-Antarctic Îles Kerguelen and Île de la Possession

2020

CAZEAU Cyrille
Foot Surgery Viewed Through the Prism of Comparative Anatomy: From Normal to Useful
(Comparative Anatomy and Posture of Animal and Human Set – Volume 1)

DUJON Bernard, PELLETIER Georges
Trajectories of Genetics

2019

BRAND Gérard
Discovering Odors

BUIS Roger
Biology and Mathematics: History and Challenges

2016

CLERC Maureen, BOUGRAIN Laurent, LOTTE Fabien
Brain-Computer Interfaces 1: Foundations and Methods
Brain-Computer Interfaces 2: Technology and Applications

FURGER Christophe
Live Cell Assays: From Research to Health and Regulatory Applications

2015

CLARYSSE Patrick, FRIBOULET Denis
Multi-modality Cardiac Imaging: Processing and Analysis

2014

CHEZE Laurence
Kinematic Analysis of Human Movement

CLARYSSE Patrick, FRIBOULET Denis
Multi-modality Cardiac Imaging: Processing and Analysis

DAO Tien Tuan, HO BA THO Marie-Christine
Biomechanics of the Musculoskeletal: Modeling of Data Uncertainty and Knowledge

FANET Hervé
Medical Imaging Based on Magnetic Fields and Ultrasounds

FARINAS DEL CERRO Luis, INOUE Katsumi
Logical Modeling of Biological Systems

MIGONNEY Véronique
Biomaterials

TEBBANI Sihem, LOPES Filipa, FILALI Rayen, DUMUR Didier, PAREAU Dominique
CO_2 Biofixation by Microalgae: Modeling, Estimation and Control

Printed and bound by CPI Group (UK) Ltd, Croydon, CR0 4YY

18/09/2023

08117277-0004